Current Progress in Biorobotics, Mechatronics and Automation

Current Progress in Biorobotics, Mechatronics and Automation

Edited by **Alfred Silva**

Willford Press

New York

Published by Willford Press,
118-35 Queens Blvd., Suite 400,
Forest Hills, NY 11375, USA
www.willfordpress.com

Current Progress in Biorobotics, Mechatronics and Automation
Edited by Alfred Silva

International Standard Book Number: 978-1-68285-327-6 (Hardback)

Printed in China.

Contents

Preface

The purpose of the book is to provide a glimpse into the dynamics and to present opinions and studies of some of the scientists engaged in the development of new ideas in the field from very different standpoints. This book will prove useful to students and researchers owing to its high content quality.

Biorobotics is a recent field which has undergone rapid advancement in a very short span of time. This book contains some path-breaking studies in the fields of robotics, mechatronics and automation. The various studies that are constantly contributing towards advancing technologies and evolution of this field are examined in detail. Some of the topics covered in this extensive book are intelligent robotics, biomimetics, modeling and control of robotic systems, fuzzy logic and artificial intelligence in robotics, multimodality human-machine interactions, etc. Scientists and students actively engaged in this field will find this book full of crucial and unexplored concepts. A number of latest researches have been included to keep the readers up-to-date with the global concepts in biorobotics, mechatronics and automation.

At the end, I would like to appreciate all the efforts made by the authors in completing their chapters professionally. I express my deepest gratitude to all of them for contributing to this book by sharing their valuable works. A special thanks to my family and friends for their constant support in this journey.

Editor

1

Sensor-Based Trajectory Generation for Advanced Driver Assistance System

Christopher James Shackleton, Rahul Kala and Kevin Warwick *

School of Systems Engineering, University of Reading, Whiteknights, Reading, Berkshire, RG6 6AY, UK; E-Mails: c.j.shackleton@pgr.reading.ac.uk (C.J.S); rkala001@gmail.com (R.K.)

* Author to whom correspondence should be addressed; E-Mail: k.warwick@reading.ac.uk

---

Abstract: This paper investigates the trajectory generation problem for an advanced driver assistance system that could sense the driving state of the vehicle, so that a collision free trajectory can be generated safely. Specifically, the problem of trajectory generation is solved for the safety assessment of the driving state and to manipulate the vehicle in order to avoid any possible collisions. The vehicle senses the environment so as to obtain information about other vehicles and static obstacles ahead. Vehicles may share the perception of the environment via an inter-vehicle communication system. The planning algorithm is based on a visibility graph. A lateral repulsive potential is applied to adaptively maintain a trade-off between the trajectory length and vehicle clearance, which is the greatest problem associated with visibility graphs. As opposed to adaptive roadmap approaches, the algorithm exploits the structured nature of the environment for construction of the roadmap. Furthermore, the mostly organized nature of traffic systems is exploited to obtain orientation invariance, which is another limitation of both visibility graphs and adaptive roadmaps. Simulation results show that the algorithm can successfully solve the problem for a variety of commonly found scenarios.

Keywords: advanced driver assistance systems; trajectory generation; intelligent vehicles; path planning; visibility graphs

---

1. Introduction

Advanced Driver Assistance Systems (ADASs) [1] are seen to be the bridge between the current, driver oriented automotive design and future autonomous vehicle design. ADASs come in a variety of formats, from pedestrian detection [2], to lane keeping assistance/lane departure warning [3]. These systems are in place to reduce the amount that human drivers have to do in order to control a vehicle; this is a particularly necessary task, as 34,826 road casualties occurred in 2009 within the European Union [4]. The same reference shows that personal cars and taxis account for the largest percentage of these, in total 48.88%, with pedestrians in second, marking pedestrian detection and avoidance systems as a necessity for future vehicles. However, in order for these systems to function, sensors must be employed for the collection of information, which can then be used as inputs to trajectory planning algorithms. The interconnection of these advanced driver assistance systems is what is likely to lead to the first commercially available autonomous vehicle.

ADAS make an interesting example of human centric computing, wherein technology is used to assist humans [5]. The aim of ADAS is to use intelligent devices to aid in the decision making of human drivers. The chief motivation is to make vehicles secure and avoid any possible collision, even if the human makes errors. An important aspect of these systems is hence to enable technology to work hand-in-hand with a human operator, wherein any information should be presented in a manner acceptable to the human, while any decisions or actions of the assistance system should be in consensus with the human driver's preferences. Lack of consensus or lack of trust between the human and the assistance system can be a severe threat.

ADAS may be information-based or manipulation-based. Information based systems use intelligent sensors and data processing to provide information to the user useful for his/her driving. Manipulation based systems, in turn, use sensor and vision information to actually control the car in scenarios that seem to be dangerous. The human, on the other hand, may still be required for general driving, depending upon his/her preferences. Manipulation-based systems are harder to design, as they have additional considerations of when, what and how much assistance is to be given—they are, however, safer. In real life, a safe state may become dangerous in a split second, owing to a poor driving decision or a large reaction time to changes in the environment. Information-based systems warn the driver, allowing them to take the necessary safety measures. Considering high operating speeds, the time spent by a human in perceiving the warning sign, interpreting it and deciding on the preventative action may be too large [6]. Assistance systems can, however, take precautionary actions well in time, as well as prepare to reduce the effects on a driver should a crash occur [7]. It should be remembered, however, that false positives in sensing provide additional concerns [8].

1.1. Sensing

The first important task in an ADAS is sensing, wherein the vehicle perceives the other vehicles and static obstacles around. The sensing results in a local map of the environment, which is used for further processing. Sensors, such as radar, LIDaR, ultrasonics and cameras are used within the automotive industry to provide information to a vehicle's control systems about its surroundings. It is these sensors that are used for the ADAS systems. In this case, many, multiple sensors are used at once for the same task in order to verify information [9] or to make measurements where the primary sensing modality fails [10]. This is known as sensor fusion. Benefits and problems with the interconnection of systems are highlighted in Darms and Winners' work [11], where some applications of sensors are also highlighted.

Many sensors operating at boundaries can prevent systems from working correctly in particular environments, and therefore, this restricts the use of that particular sensor as a solution to a problem. This is evident in the case of Automatic Cruise Control (ACC) in which it is possible to implement camera-based systems; however, most commercially available ACC units operate using radar. This is due to the fact that radar is unaffected by lighting conditions and weather, whilst still having sufficient range. The sensor also satisfies the requirements for following a vehicle at speed. Cameras, however, may be ideal for multi-object tracking ACC.

Typically, sensors within an automotive application are required to be low cost, as well as reliable. The most common sensor found on entry-level vehicles is the ultrasonic parking aid, of which many variants exist. The range for this particular sensor is relatively small, at approximately 4 m, which may be ideal for close following applications; however, it may be necessary to measure larger distances, in which case a camera (up to 40 m range), LIDaR or radar (both 150 m range) may be appropriate.

A modern perspective is to use inter-vehicle communication [12,13] between intelligent vehicles. When operating in a grid of mixed traffic consisting of both intelligent vehicles and non-intelligent vehicles, the intelligent vehicles can transmit information about other vehicles or obstacles. This enables limited sensing capability vehicles to obtain information about the traffic ahead and vehicles to "see" beyond their vision range. On top of this, collaborative data checking can refine vehicle sensing errors.

1.2. Trajectory Generation and Assistance

Given a map, the aim of trajectory generation is to construct a short, safe and smooth trajectory. Safety not only accounts for the fact that no collision should occur, but also makes the vehicle maintain the correct safety distance. The safety distance covers for any sensing and actuation errors that may appear. Further, this is in consensus with human driving, wherein drivers prefer to maintain wide gaps between themselves and vehicles all around. Important considerations in the choice of trajectory planning algorithms are completeness, optimality and computation time. Reactive planning techniques (e.g., [14]) assess the immediate scenario and compute the immediate move. Such techniques may well have small computation times; however, they are almost always neither complete nor optimal. Hence, deliberative techniques (e.g., [15]) are preferred, which, at the expense of computation time, are better in completeness and optimality.

The environment may be structured or non-structured. When planning in a structured environment, it is assumed that the complete environment is known, with the different obstacles depicted as polygons or circles with known sizes. This is naturally true in the case of traffic scenarios, with other vehicles being mostly rectangular, whose geometry can be sensed. Search techniques then fit a lot of applications, as they assure both optimality and completeness. A structured environment can be easily converted into a graph (or similar structure) with a limited number of nodes for fast planning, although search-based planning in an unstructured environment would be too computationally expensive. Typical approaches include Voronoi maps [16], velocity obstacles [17] and visibility graphs [18,19].

Unlike mainstream mobile robotics, the aim is not to make the vehicle physically move by a computed trajectory, since the human may have a different plan in mind. Instead, the aim is to assess the vehicle's safety state, depending upon which it is decided whether the assistance system should intervene in the human driving to correct his/her trajectory and, if so, by what magnitude. Should evasive action need to be taken, this could be conveyed to the driver by means of a force feedback steering wheel, as used in [20,21]; interestingly, acceleration reduction can also be conveyed to the driver in a similar way [22].

1.3. Proposed Solution and Main Contributions

This paper deals with the design of an assistance system that ensures user safety and takes preventative steps for the same. The algorithm assumes that a map produced by sensors and in cooperation with the other vehicles is already available. The map is used for generation of the trajectory, which is the chief part of the problem tackled in this paper. The constructed trajectory is then used for deciding the control action to be applied based on the assessed risk. Due to the current constraints, the sensing and manipulation aspects of the assistance system cannot be physically implemented and tested; they are for motivation only.

The developed solution is a hybrid of visibility graphs [18,19] and adaptive roadmaps [23–26]. The visibility graphs are well-suited for structured environments for which they perform fast and effective planning. The general idea is to place graph nodes across the obstacle points. The nodes are then assessed for connectivity to produce a graph, which is used for planning. The biggest problem with this approach is the assumption of a structured environment, as well as the ability to control the trade-off between the trajectory length and clearance. Clearance denotes the (more than minimal) safety distance available to the vehicle. The assumption of a structured environment is not a bad assumption to make in a traffic scenario. However it is important to intelligently place the vehicles depending upon how much space is available.

Adaptive roadmap based approaches [23–26] can easily model the potential functions to trade-off between the trajectory length and clearance. Being widely used for mobile robotics, these generally sample out random points from the map, which are later checked for connectivity, and the resultant graph produced is called as a roadmap. The paper uses the potential-based modelling of these approaches applied to a graph based on the visibility graphs. As a result, lesser and more strategically placed nodes are produced.

Orientation is a major factor, which decides the feasibility of a node (and the associated clearance or length of a path) in such a graph or roadmap-based approach. The factor is even more useful in a road scenario in which vehicles are tightly packed on roads instead of having wide open spaces, as in many mobile robotics cases. Hence, it is not possible to focus on diagonal or maximum length for node placement. The paper handles this problem using the mostly organized nature of a general traffic landscape, as compared to that in mobile robotics. For the same reason, the application of the potential for alteration of a visibility graph node is restricted to the lateral direction of the road. Section 3.3 elaborates this point.

The key contributions of the approach are: (i) using a hybrid of potential fields and visibility graphs for trajectory planning, (ii) using heuristics to solve the problem of rotational dependence associated with such techniques in environments with narrow spaces, (iii) interpreting all traffic behaviours and casting them into a visibility graph framework, rather than only using nodes around obstacles and (iv) interpreting the human driver's driving intentions (through the heading direction) for the problem of trajectory planning, thereby making the system computing trajectory close to the human desired trajectory.

The remainder of this paper is organized as follows: in Section 2, some of the related works are presented. Section 3 describes the problem and goes forward with the modelling of the complete algorithm. Experimental results are given in Section 4, and some concluding remarks are made in Section 5.

2. Related Works

In a recent work, Anderson *et al.* [27] studied a similar problem. The authors used Delaunay triangles to compute all possible homotopies in a given map. The authors also used Dijkstra's algorithm for computing the trajectory. The greatest limitation of the approach, however, is that the central points were used for trajectory generation. This means that for scenarios having wide segments between obstacles, the vehicle would drive at the centre, over-compromising its distance to clearance. The proposed algorithm uses the potential function to model the trade-off. Furthermore, the authors did not model the behaviour of a vehicle following another vehicle (as other vehicles were treated as static obstacles), whereas this is considered in the proposed approach.

Our prior work focused on the use of a Rapidly-exploring Random Tress (RRT) Connect [28,29] algorithm for the task of navigation of multiple autonomous vehicles. The vehicles were assumed to be connected via an inter-vehicle communication system, allowing all vehicles to be planned in a prioritized manner. The search was biased towards the areas around the current lateral position of the vehicles. In a related work [30], the problem was solved using RRT. The RRT was sampled using the vehicle's control model, which ensured that the trajectory generated was safely navigable. The proposed approach is, however, modelled as an ADAS instead of as an autonomous vehicle. RRT and similar approaches can be computationally expensive and, hence, are good models for autonomous driving, where planning frequency is not large. The proposed approach meanwhile assumes the structured nature of the environment for faster planning.

A related problem is decision making in intelligent vehicles. Schubert *et al.* [31] sensed the vehicles ahead, behind and the distance from the lane markings for decision making regarding lane change. The authors used Bayesian networks for the task. In another approach, Hegeman *et al.* [32] computed the feasibility of overtaking based on which a human could initiate an overtaking manoeuvre. For the task of construction of the overtaking trajectory, Naranjo *et al.* [33] developed a fuzzy rule-based system. The system was divided into stages of change to the overtaking lane: complete an overtake and return to the original lane. All these systems perform well when the road is marked with lanes and the entire traffic strictly operates in lanes. In reality, some segments of traffic on some roads may get unorganized, where the vehicles partly slip between lanes. Further, the problem of obstacle avoidance cannot be perfectly solved by lane changes. Hence, generalized planners (like the one proposed) that do not necessarily assume lanes are considered to be better.

Significant work has been done in the domain of mobile robotics for the task of trajectory planning. Gayle *et al.* [34] used a social potential field to differentiate between types of agents in a multi-agent framework. Using this, along with the general potential field, the authors carried out the planning of agents, which moved under the guidance of an adaptive roadmap [35]. A general graph search cannot be employed for the problem, due to computational constraints. Kala *et al.* [36] proposed a multi-layer graph search, which made the algorithm iterative and computationally fast. The authors initially carried the graph search on lower resolution maps, and based on the results, the resolution of promising areas was increased. Similar work in the domain of multiple autonomous vehicles can be found in Kala and Warwick [37], which consisted of four layers of hierarchy.

In another approach, the hierarchical D* algorithm was presented by Cagigas and Abascal [38]. The D* algorithm is better suited to a dynamic environment, and its hierarchical nature makes it computationally less intensive. Even though the modifications result in making these approaches computationally less demanding, they cannot be used in such real time systems. Further, it is not possible to hierarchically construct the trajectory of a vehicle, which is a concept suited for open space-like environments. Decisions about overtaking and lane changes are only possible knowing the actual available separations between vehicles and the obstacles. It is not possible to construct a coarser map and make such decisions, as employed in [38].

3. Algorithm

This section talks about the complete design of the assistance system. First, the problem statement is defined, and later, the different segments of the algorithm are discussed.

3.1. Problem Definition

Consider that a vehicle is travelling at a speed v and is currently located at position s with an orientation of Φ. Let the vehicle be a rectangle of length L and width W. A road segment ahead of the vehicle with a length of Ω is considered. It is assumed that the vision algorithms can sense the road ahead and differentiate it from forbidden zones, the zone for vehicles travelling in the opposite direction, pavements, ditches *etc.* The first task associated with the algorithm is to sense the obstacles and other vehicles around. Consider that the vehicle is fitted with appropriate sensors to sense these or that the vehicles are intelligent and can sense each other (and the obstacles) and share the information.

Hence, let R be the set of vehicles or obstacles, each with a position p_i and orientation θ_i. Since, for a forward travelling vehicle, a collision is only possible with vehicles ahead with smaller speeds, only these are considered. Cases, such as verging, make collisions with vehicles to the side possible. However, such collisions are handled by measuring and tracking side distances and are broadly not dealt with by trajectory-based warning systems. For simplicity, all other vehicles and obstacles are assumed to be rectangles of length l_i and width w_i. Only vehicles and obstacles within the road segment are considered. The only vehicles to be considered are those that the vehicle being controlled looks like it will overtake in the future. These vehicles generally have a lower speed than the vehicle being controlled. The human driver may control his/her speed, so as to clearly indicate the intentions of overtaking or following the vehicle ahead [39]. The algorithm is, therefore, largely active only in the case of overtaking.

Given such a map, the first problem is to construct a trajectory, τ. Since the subsequent motion of the other vehicles cannot be ascertained, they are treated as static obstacles. Hence, subsequent text will use the term vehicles and obstacles interchangeably. The trajectory planning is instantaneous, and hence, as these vehicles move, the trajectory adapts itself. In general, the attempt is to compute a trajectory, which is feasible $(\tau(t) \in \xi^{free} \forall t)$, is short in length (*minimize* $\|\tau\|$), has a high average clearance (*maximize* $\|C(\tau)\|$) and has a high smoothness (or low curvature) at the steepest turn (*minimize* $max(\kappa(\tau))$). Here, ξ^{free} denotes the obstacle free configuration space, which considers all the other obstacles as static, $\|.\|$ denotes the Euclidian norm, C denotes the clearance and κ denotes the curvature.

The other problem is to consider a control action. It is assumed that the user applies a control action of u at the current state. The trajectory, τ, is assessed to compute the safety of the current state. Let the desired input to trace the constructed trajectory be u_d. The algorithm, hence, needs to modify the control input to produce a control input, u', used for the navigation of the vehicle, such that the user barely feels the difference, while the control used for navigation is still safe. This means when the vehicle is in a very safe state, the user input, u, is used for navigation. However, in a very collision-prone state, effectively, the vehicle drives itself until a safe state is reached. The general framework is given by Figure 1.

Figure 1. General architecture of the assistance system.

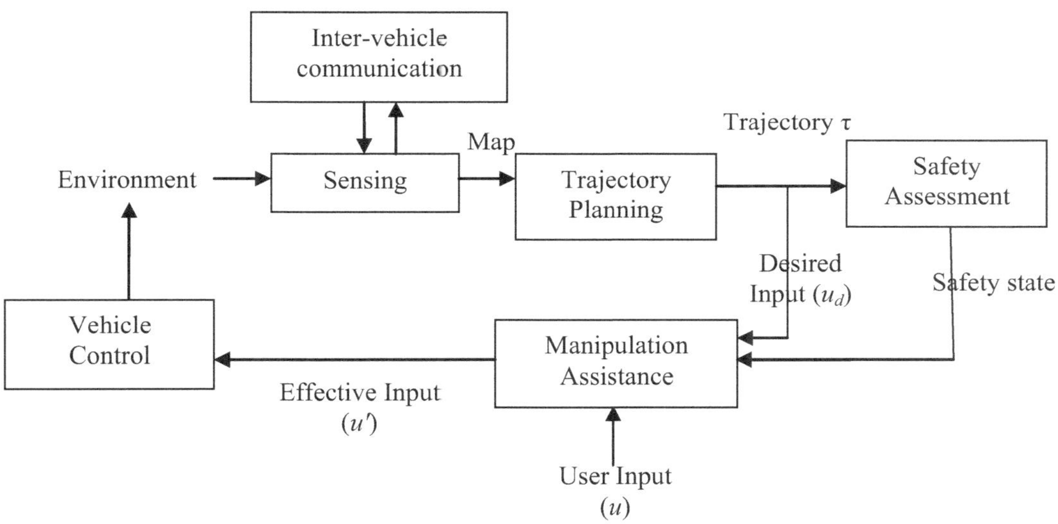

3.2. Initializing Visibility Graph

The visibility graph $G(V, E)$ needs to be constructed based upon the sensed obstacles. This sub-section deals with the computation of the node set, V. The first type is obstacle nodes. Using these nodes, the vehicle can avoid an obstacle. Since, in a structured environment, the optimal (length only) trajectory of a point vehicle goes through the obstacle corners, the initial position of these nodes is taken to be just outside the obstacle corners. Let an obstacle be positioned at p_i with orientation θ_i, such that its four corners are at C_i^1, C_i^2, C_i^3 and C_i^4. The obstacle nodes are placed just outside the obstacle, given by Equation (1), where δ_i^j is a small vector pointing radially outwards from the corner, C_i^j. Here, i covers vehicles, while j covers the corners of the ith vehicle:

$$O = \bigcup_{i,j} C_i^j + \delta_i^j \tag{1}$$

The second type of node is the vehicle following nodes. It may not be possible for a vehicle to avoid all the other vehicles before the end of the road, and hence, it may have to slow down and follow another vehicle. The purpose of the graph is to admit all the possible plans of the vehicle. While obstacle nodes admit the overtaking and obstacle avoidance plans, the vehicle following nodes are supposed to admit the plans, where the vehicle decides to follow some other vehicle. These nodes are taken at a distance of q behind every vehicle in R. Here, q is the safety distance, which allows the vehicle to actually slow down and follow.

Consider a vehicle located at p_i with orientation θ_i. The lateral position (Y-axis, along the width of the road) is the same as that of p_i. Let C_i^j be the corner of the vehicle at p_i, which has the least longitudinal occupancy (most behind longitudinally, along the X axis, the length of the road). For the node to be admissible, it is necessary that it lies longitudinally ahead of the vehicle's current longitudinal occupancy and, subsequently, further by a distance, so as to allow a turn (currently equal to the vehicles length). The longitudinal position of the node is taken at a distance q behind C_i^j. These nodes may hence be given by Equation (2). Throughout the paper, for a point $P(x,y)$, $P[X]$ refers to the X axis component (x) and $P[Y]$ refers to the Y axis component (y).

$$F = \bigcup_i \left\{ \left(C_i^j[X] - q, p_i[Y] \right), j = \arg\min(C_i^j[X]) \right\} \tag{2}$$

3.3. Applying Lateral Potentials

Since the vehicle is not point-sized, it is evident that the obstacle nodes (as initialized) cannot be used for navigation and need to be moved in proportion to the vehicle size. The movement should first cater to the feasibility considerations, such that a vehicle placed at the node does not collide with the obstacle. Subsequently, if additional distance is available, the node should be moved, so as to maintain a trade-off between path length and clearance. Excessive movement would make the paths too long, while small movements would result in small clearances. Obstacle nodes are placed very close to the obstacles and, hence, placement of the vehicle at the obstacle node implies zero clearance. As these nodes are moved away, the clearance increases at the cost of path length. Each node is affected by a repulsive potential from all the obstacles and the road boundaries. Such a motion of the nodes is carried out iteratively for a few iterations. In the small regions around the obstacles, the potential is large, and hence, the node is pushed back strongly until it reaches a point far enough, when potential almost dies off. If sufficient distance is not available, the node would lie in the middle of the obstacles. This is explained in Figure 2(a).

The other major issue is that the vehicle is rectangular, and its feasibility at a position (and hence the clearance) depends upon the orientation of the vehicle (Figure 2(b)). A popular approach [18] is to maintain a minimal distance equal to half the diagonal, which ensures that any orientation would lead to feasibility. Road scenarios are tightly packed, and hence, such extra space cannot always be kept. However, we exploit here the generally organized nature of a traffic landscape, where vehicles are generally driving along the road, unlike mobile robotics, where robots can be heading just about anywhere.

Consider, for example, a close overtake/obstacle avoidance. A vehicle would slide in from its current position to a position laterally just next to the vehicle/obstacle being avoided (Figure 2(c)). Hence, in the closest case, wherein no extra distance is available, the separation between the vehicle's central position and the obstacle boundary would be half the vehicles width (say $W/2$). In other words, potentials can be applied in order to keep a distance of $W/2$ from obstacles to ensure feasibility.

Figure 2. Application of lateral Potentials. (**a**) Length and clearance trade-off. (**b**) Problem of rotation. (**c**) Heuristic of keeping minimum separation as half the width. (**d**) Sources of potential at a point o_i.

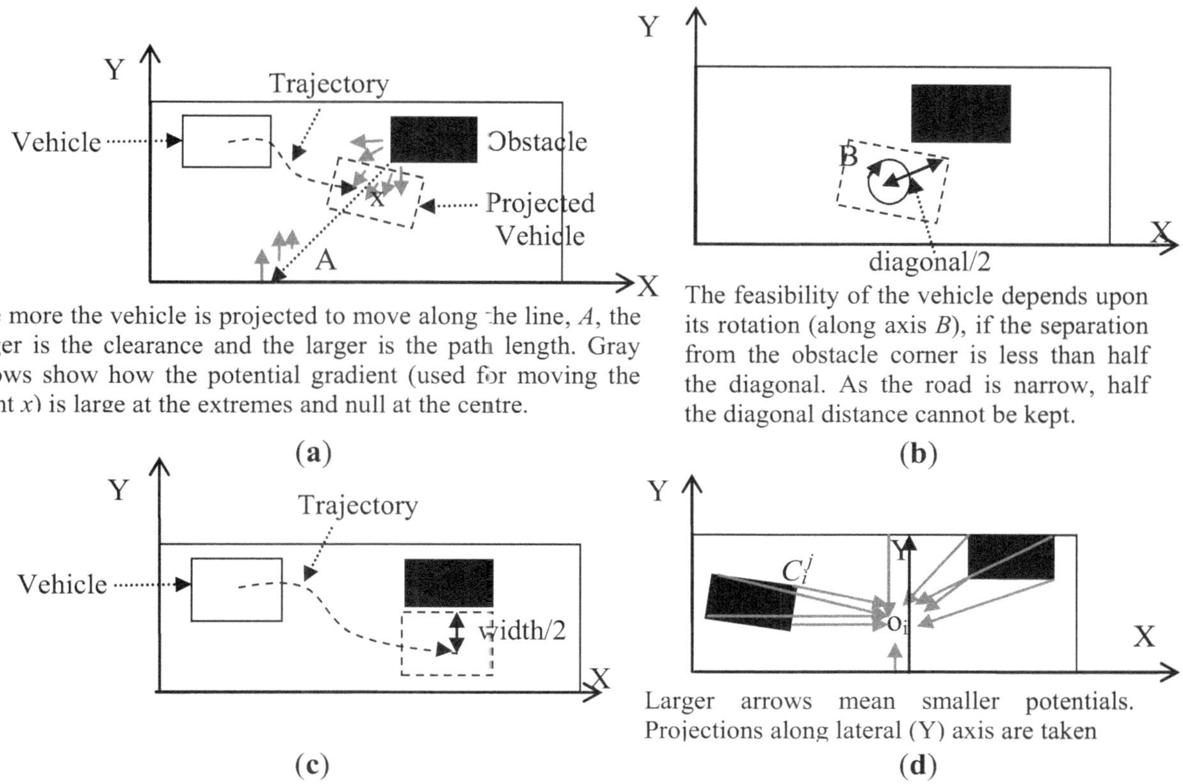

The more the vehicle is projected to move along the line, A, the larger is the clearance and the larger is the path length. Gray arrows show how the potential gradient (used for moving the point x) is large at the extremes and null at the centre.

(**a**)

The feasibility of the vehicle depends upon its rotation (along axis B), if the separation from the obstacle corner is less than half the diagonal. As the road is narrow, half the diagonal distance cannot be kept.

(**b**)

(**c**)

Larger arrows mean smaller potentials. Projections along lateral (Y) axis are taken

(**d**)

This heuristic, however, only holds when the obstacle avoidance point is located laterally next to the obstacle, unlike the diagonal version of a visibility graph approach or a potential direction of adaptive roadmaps (where multiple points are deployed per obstacle for framing an avoidance strategy). Hence, the potentials used for motion of the nodes are applied only in the lateral direction.

Consider an obstacle node located at a position o_i. It is repelled by all the obstacles and the road boundaries. For computational constraints, the obstacles are assumed to be represented by only the corner points C_i^j, each of which repels the node by a magnitude inversely proportional to the square of the distance. The road boundaries also act as obstacles and repel the node. The repulsion is, however, proportional to the shortest distance between the vehicle and the road boundaries. The resultant potential is given by Equation (3).

$$Z_i = \sum_{C_i^j} \begin{pmatrix} \left(\frac{1}{\max\left(\left\|C_i^j - o_i\right\| - W/2, 1\right)}\right)^2 \hat{u}(C_i^j - o_i).\hat{Y} + \left(\frac{1}{\max\left(o_i[Y] - W/2, 1\right)}\right)^2 \\ - \left(\frac{1}{\max\left(M - o_i[Y] - W/2, 1\right)}\right)^2 \end{pmatrix} \quad (3)$$

The first term in Equation (3) denotes the potential due to obstacles, while the second and third terms denote the potentials from the left and right boundaries. One is kept as a minimum distance to avoid excessively large numbers as distances approach zero. $\hat{u}(C_i^j - o_i)$ is the unit vector in the direction C_i^j to o_i, and the projection of the resultant potential in the Y axis is considered. M is the road width. The sources of potential are explained in Figure 2(d).

At each iteration, obstacle nodes are moved as per the immediate potential, given by Equation (4).

$$o_i'[Y] = \max\left(o_i[Y] + \alpha Z_i, \beta\right) \tag{4}$$

Here, α scales the potential to the immediate movement of the node, while β restricts the maximum amount by which the node may be moved.

Additional nodes are added. The first is the source node (s), which is the current position of the vehicle. This node has a single edge to a direction maintenance node, which ensures the initial trajectory is generated in the current heading direction of the vehicle (Φ). This node is taken at a distance of L from the current position s of the vehicle (or $s + L\hat{u}(\Phi)$). The last category is destination nodes (D), which are used to navigate the vehicle from obstacle avoidance points to the end of the road segment, so as to complete the trajectory within the segment, if feasible. A vehicle in the absence of any obstacle aims to maintain its lateral position on the road. This set of nodes is hence given by Equation (5), where Ω is the length of the road and o_i' is the obstacle node after the application of the lateral potential:

$$D = \bigcup_i (\Omega, o_i'[Y]) \tag{5}$$

The vertex set V of the graph is hence given by Equation (6):

$$V = \{s, s + L\hat{u}(\Phi)\} \cup O' \cup F \cup D \tag{6}$$

3.4. Graph Search

The source node has a single edge, which is to the directional maintenance node. The rest of $(|V| - 1)^2$ possible edges between all vertices are checked for feasibility. A configuration space, ξ^{free}, is constructed, treating all the obstacles as static. The path between node V_i and node V_j traversed by the vehicle in the direction V_i to V_j is checked for feasibility in this configuration space. If the path is feasible, an edge is added.

A uniform cost search algorithm is applied over the graph to compute the best path. The trajectory cost function is taken as the trajectory length; however, a penalty is applied for small clearances. The lateral potential measured at the node is taken as the indicator of the clearance loss. This encourages the algorithm to find smaller and clearer paths. Since lateral potentials are already applied, all nodes, which could have reasonable clearances, obtain positions to allow these clearances. This means that the graph search practically works only on length, avoiding any node that could not obtain a reasonable clearance. Minimizing the length automatically results in maximizing smoothness. Further infeasible nodes (if any) have a very high penalty and are hence not used in the optimal trajectory.

Every node is associated with three types of cost. These are path length from the source (L), total clearance from the source (C) and total cost ($Cost$). In an expansion of a node V_j from node V_i, the costs are updated by Equations (7–9):

$$L(V_j) = L(V_i) + ||V_j - V_i|| \tag{7}$$

$$C(V_j) = C(V_i) + Z(V_j) \tag{8}$$

$$Cost(V_j) = L(V_j) + \rho C(V_j) \tag{9}$$

Here, $Z(V_j)$ is the potential measured at the point V_j and ρ is the penalty constant. Reasonably far from the obstacle, the potential is nearly zero and, hence, so is the penalization.

The search may not always end in a destination node, as it may not be possible to reach the end of the road segment, and instead, a vehicle may end up by following another vehicle. In such cases, the most distant node is chosen, and ties are broken on the basis of the total cost. This results in a path (τ') from the source to goal.

The path returned by the graph search (τ') needs to be additionally smoothed at the joints of the nodes; this is done by using spline curves. A coarser level trajectory is sampled and passed as control points for the construction of the spline curve. The resultant curve is taken to be the trajectory (τ) of the vehicle.

3.5. Trajectory Control

The trajectory obtained is assessed for a vehicle's safety state. An unsafe state requires a greater manoeuvre, and hence, the trajectory is not very smooth. The minimum curvature along the trajectory is measured. In a discrete trajectory, at any general point at a distance of t on the vehicle's trajectory (say $\tau(t)$) the curvature ($\kappa(\tau(t))$) is given by Equation 10, where d is a small number:

$$\kappa(\tau(t)) = ||\tau(t+d) + \tau(t-d) - 2\tau(t)|| \tag{10}$$

Lesser curvatures give a safer state. This factor is normalized, so as to lie between zero and one.

Let the minimum curvature recorded on the trajectory be $\kappa(\tau)$. Consider at any instance the user gives an input, u, to the system. Based on the computed trajectory, let the desired input of the system, as per the computed trajectory, be u_d, the magnitude of which depends upon the kinematic modelling and control system. The resultant input (u') given to the vehicle is then found from Equation (11):

$$u' = \begin{cases} u & \kappa(\tau) \le \kappa_{\min} \\ \kappa(\tau)u_d + (1-\kappa(\tau))u & \kappa_{\min} < \kappa(\tau) < \kappa_{\max} \\ u_d & \kappa(\tau) \ge \kappa_{\max} \end{cases} \tag{11}$$

Here, κ_{min} is the minimum threshold below, which the system has considered safe enough and the user is allowed to drive. κ_{max} is the maximum threshold above which the system is considered unsafe and the human is disallowed to drive. In intermediate states, the resultant input is given as a weighted average of the desired and user inputs, which means that in this interval, the resultant input is gradually taken over from the user, as he/she drives with less of a safety margin.

4. Results

The algorithm was tested through simulations. The simulation tool took as input the sensed obstacles. Each of these had its own size and orientation with respect to the road. The obstacles were placed nearly in lanes, so as to make the scenario more realistic. However, the difference in sizes and orientations necessitated a non-lane-based trajectory planning. The initial position and orientation of the vehicle was also fed into the tool. The simulation tool assessed the scenario and computed the trajectory, which was displayed.

We first discuss here a step-by-step solution to the simplest task, which is that of obstacle avoidance. The vehicle had an obstacle in front of it, which it had to overcome. First, the obstacle nodes and the vehicle-following nodes were placed as shown in Figure 3(a). These nodes were acted upon by the lateral potentials and, hence, were moved, as shown in Figure 3(b). Figure 3(b) also shows the source node, direction maintenance node and the destination nodes. The edges were connected by feasibility analysis. The optimal path is shown separately. The smoothened trajectory is shown in Figure 3(d).

Figure 3. Results for obstacle avoidance. (**a**) Initial nodes. (**b**) Nodes after application of lateral potential. (**c**) Resultant trajectory.

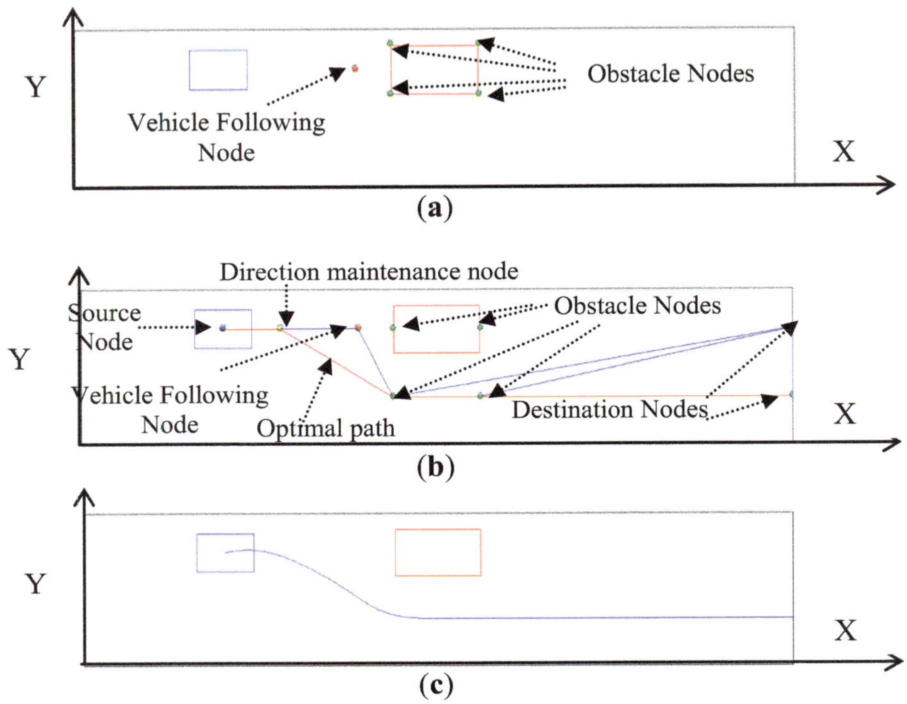

Figure 4. Experimental Results.

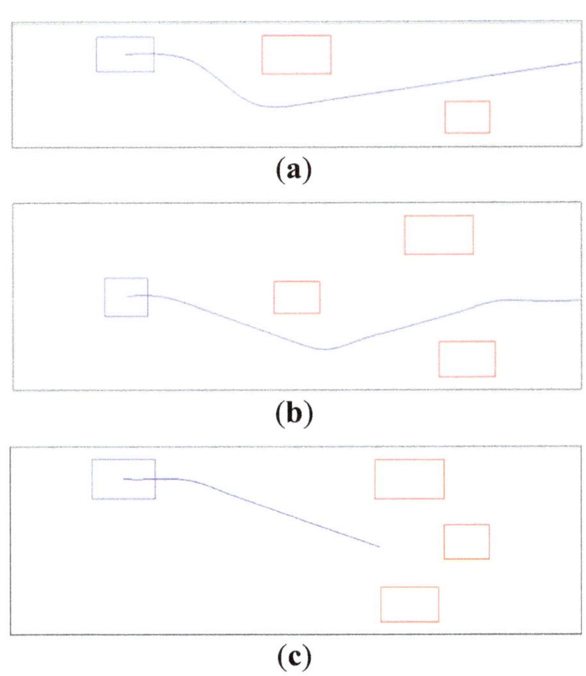

The second scenario consisted of two obstacles (or vehicles). The first obstacle was larger than the second. The algorithm decided to pass the first obstacle on its right-hand side and the second obstacle on its left-hand side, which is (arguably) the best strategy to take. The resultant trajectory is shown in Figure 4(a).

To test the scalability of the approach, another obstacle was added into the scenario. The resultant scenario consisted of an obstacle just ahead of the vehicle to avoid, which meant that it must turn to the left or right. However, other obstacles were positioned on either side. The algorithm decided to make the vehicle turn right, as it would later find a smaller obstacle, which was easier to avoid. Taking a left turn initially could have made the subsequent traversal risky. The resultant trajectory for the scenario is shown in Figure 4(b).

In the last scenario, the road was completely blocked by vehicles. Hence, the vehicle under control needed to decide which vehicle in front to follow. The choice was such that the vehicle reached the most distant point, and hence, the central position was chosen. This trajectory is shown in Figure 4(c).

An attempt was made to gradually take control of the vehicle from the human driver, for which curvature was used as an indicator. The best way to study this effect is the obstacle-avoidance scenario. Experiments were performed over a set of points, which varied in their distance from the obstacle. As the vehicle went near to the obstacle, there was an increase in the curvature, which indicated a higher risk. Hence, if the human driver did not see the obstacle and kept driving, the algorithm would gradually intervene. The trajectories for different positions are shown in Figure 5(a), while the corresponding change in the curvature values is shown in Figure 5(b).

Figure 5. Effect of change in heading distance to the obstacle. (**a**) Trajectories (**b**) Curvature.

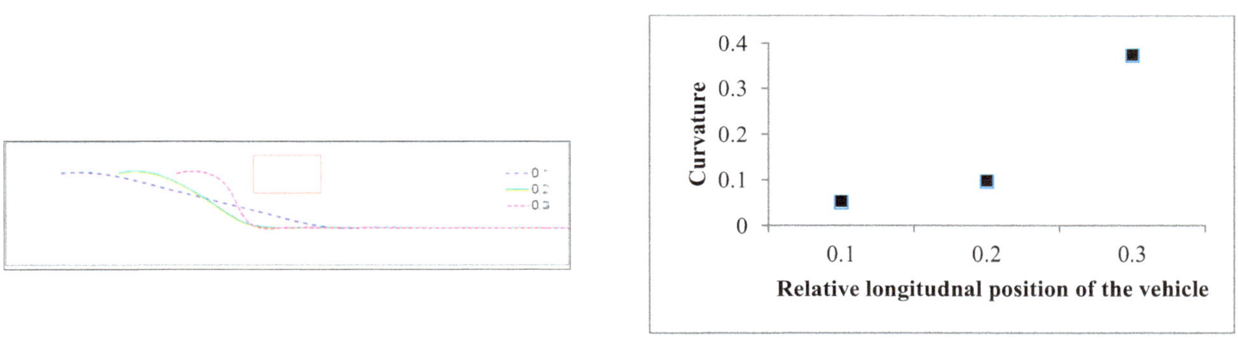

(**a**) (**b**)

Figure 6. Effect of change in orientation to the obstacle. (**a**) Trajectories (**b**) Curvature.

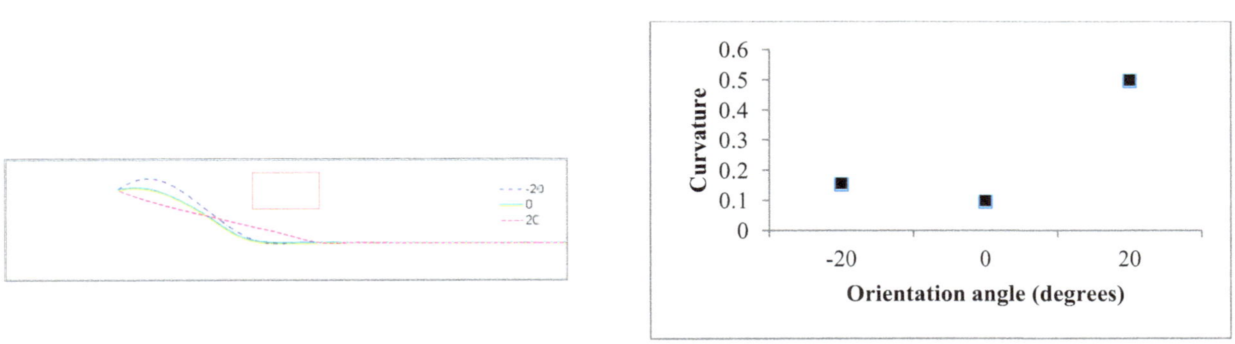

(**a**) (**b**)

The risk is not always due to the distance of the vehicle from the obstacle; it can also be due to the heading direction. This factor was also tested. The same position of the vehicle was tested for safety for various values of the heading direction. The corresponding trajectories are shown in Figure 6(a), and the curvatures are plotted in Figure 6(b). An anti-clockwise turn is obviously risky, since the vehicle has to turn by a greater magnitude. This is confirmed by a significant rise in curvature. The figure also shows that a clockwise turn is also associated with a slightly higher risk, as it makes the turn a little more difficult.

5. Conclusions

In this paper, it was assumed that the environment can be sensed using a variety of sensors, in cooperation with the other vehicles using an inter-vehicle communication system. The sensed environment was used by a trajectory planner. The planning algorithm employed a hybrid of visibility graphs and an adaptive roadmap. A number of additional nodes were added, as per the problem requirements.

The motion of the nodes of the graph was restricted to the lateral direction only, which gives rotational invariance to the algorithm. Uniform cost search was used on the resultant map, and the path was smoothened using spline curves. The problem was solved on a variety of typical scenarios. In each scenario, the task was to construct a trajectory to successfully avoid all the obstacles, failing which it was preferred to follow a vehicle instead. As the vehicle approached an obstacle, a rise in curvature was observed. This can be used to smoothly change the control from a human driver to the assistance system. A similar observation was made on the rotation of the vehicle.

The motivation was to design and implement the complete assistance system. Currently, the biggest limitation of the work is that the system cannot be simulation-based on human inputs. Hence, the physical manipulation is restricted to motivation only. The simulation needs to be extended to a virtual driving system, over which a human can be made to control a vehicle in assistance with the designed system. In terms of trajectory generation, better trajectory cost functions need to be considered, which match perfectly with the human preferences. Based on the human generated inputs, it may be necessary to assess the intent of the human, rather than just assuming the human takes the best decisions based on the current pose. Ultimately, testing on a physical vehicle is necessary to validate performance.

It should also be noted that, due to the simplified modelling of the vehicle, the trajectory planning algorithm in its current form would not be suitable for application in a real traffic scenario. This is due to the fact that considerations, such as side slip (swerving) are not taken into account; however, this may be improved by use of a more sophisticated vehicle model. Traffic rules govern the decision whether a lane change is possible or not, which in the current system is decided purely by the human driver. The system cannot alter the decision, and hence, the interpretation of the traffic rules for safety consideration is entirely up to the human driver. Currently, the system makes every such change safe by modification of steering or travel speeds. Disallowing lane changes or overtaking depending upon the traffic rules directly by the algorithm may well be taken into account in future versions of the algorithm.

One important question, which has not been dealt with here, is how much a human driver would be willing to allow a computer-based system to take over vehicle control. Clearly, this is a much bigger problem than can be tackled in a paper of this type, which is concerned primarily with the technical aspects of bringing this possibility about. In such a situation, many different social pressures and requirements come into effect and, as with all computer-based control systems, a vitally important aspect is ultimate confidence in performance delivery on the part of the computer system. Hence, there is a need for realistic simulation runs and subsequent practical scenario trials in order to prove the validity and safety of the computer system.

Acknowledgments

The authors wish to thank the Commonwealth Scholarship Commission in the United Kingdom and the British Council for their support of the second named author through the Commonwealth Scholarship and Fellowship Program—2010—UK award number INCS-2010-161.

References

1. Hummel, T.; Kühn, M.; Bende, J.; Lang, A. *Advanced Driver Assistance Systems: An Investigation of Their Potential Safety Benefits Based on an Analysis of Insurance Claims In Germany*; Research report FS 03; German Insurance Association (GDV): Berlin, Germany, 2011.
2. Gerónimo, D.; López, A.M.; Sappa, A.D.; Graf, T. Survey of pedestrian detection for advanced driver assistance systems. *IEEE Trans. Pattern Anal. Mach. Intell.* **2010**, *32*, 1239–1258.
3. Sarshar, M. A Novel System for Advanced Driver Assistance Systems. In Proceedings of the 4th Annual IEEE Systems Conference, San Diego, CA, USA, April 2010; pp. 529–534.
4. Trivedi, M.M.; Cheng, S.Y. Holistic sensing and active displays for intelligent driver support systems. *Computer* **2007**, *40*, 60–68.
5. Trivedi, M.M.; Gandhi, T.; McCall, J. Looking-in and looking-out of a vehicle: Computer-vision-based enhanced vehicle safety. *IEEE Trans. Intell. Transport. Syst.* **2007**, *8*, 108–120.
6. Cellario, M. Human-centered intelligent vehicles: Toward multimodal interface integration. *IEEE Intell. Syst.* **2001**, *16*, 78–81.
7. Hannan, M.A.; Hussain, A.; Samad, S.A. Sensing systems and algorithms for airbag deployment decision. *Sensors* **2011**, *11*, 888–890.
8. Parasuraman, R.; Hancock, P.A.; Olofinboba, O. Alarm effectiveness in driver-centred collision-warning systems. *Ergonomics* **1997**, *40*, 390–399.
9. Stein, G.P.; Mano, O.; Shashua, A. Vision-Based ACC with a Single Camera: Bounds on Range and Range Rate Accuracy. In Proceedings of the IEEE Intelligent Vehicles Symposium, Columbus, OH, USA, 9–11 June 2003; pp. 120–125.
10. Schlegl, T.; Bretterklieber, T.; Neumayer, M.; Zangl, H. Combined capacitive and ultrasonic distance measurement for automotive applications. *Sensors* **2011**, *11*, 2636–2642.
11. Dams, M.; Winner, H. A Modular System Architecture for Sensor Data Processing of ADAS Applications. In Proceedings of the IEEE Intelligent Vehicles Symposium, Las Vegas, NV, USA, 6–8 June 2005; pp. 729–734.
12. Tsugawa, S. Inter-Vehicle Communications and Their Applications to Intelligent Vehicles: An Overview. In Proceedings of the IEEE Intelligent Vehicle Symposium, Versailles, France, 17–21 June 2002; pp. 564–569.

13. Reichardt, D.; Miglietta, M.; Moretti, L.; Morsink, P.; Schulz, W. CarTALK 2000: Safe and Comfortable Driving Based upon Inter-Vehicle-Communication. In Proceedings of the IEEE Intelligent Vehicle Symposium, Versailles, France, 17–21 June 2002; pp. 545–550.
14. Khatib, O. Real-Time Obstacle Avoidance for Manipulators and Mobile Robots. In Proceedings of the 1985 IEEE International Conference on Robotics and Automation, St. Louis, MO, USA, March 1985; pp. 500–505.
15. Stentz, A. Optimal and Efficient Path Planning for Partially-Known Environments. In Proceedings of the 1994 IEEE International Conference on Robotics and Automation, San Diego, CA, USA, 8–13 May 1994; pp. 3310–3317.
16. Choset, H.; Burdick, J. Sensor Based Planning. I. The Generalized Voronoi Graph. In Proceedings of the 1995 IEEE International Conference on Robotics and Automation, Nagoya, Japan, 21–27 May 1995; pp. 1649–1655.
17. Fiorini, P.; Shiller, Z. Motion Planning in Dynamic Environments Using Velocity Obstacles. *Int. J. Roboti. Res.* **1998**, *17*, 760–772.
18. Wesley, M.A.; Lozano-Pérez, T. An algorithm for planning collision-free paths among polyhedral obstacles. *Comm. ACM* **1979**, *22*, 560–570.
19. Oommen, B.; Iyengar, S.; Rao, N.; Kashyap, R. Robot navigation in unknown terrains using learned visibility graphs. Part I: The disjoint convex obstacle case. *IEEE J. Robot. Autom.* **1987**, *3*, 672–681.
20. Keller, C.G.; Dang, T.; Fritz, H.; Joos, A.; Rabe, C.; Gavrila, D.M. Active pedestrian safety by automatic braking and evasive steering. *IEEE Trans. Intell. Transport. Syst.* **2011**, *12*, 1292–1304.
21. Jensen, M.J.; Tolbert, A.M.; Wagner, J.R.; Member, S.; Switzer, F.S.; Finn, J.W. A customizable automotive steering system with a Haptic feedback control strategy for obstacle avoidance notification. *IEEE Trans. Veh. Tech.* **2011**, *60*, 4208–4216.
22. Mulder, M.; Abbink, D.A.; van Paassen, M.M.; Mulder, M. Design of a Haptic gas pedal for active car-following support. *IEEE Trans. Intell. Transport. Syst.* **2011**, *12*, 268–279.
23. Kavraki, L.E.; Kolountzakis, M.N.; Latombe, J.C. Analysis of probabilistic roadmaps for path planning. *IEEE Trans. Robot. Autom.* **1998**, *14*, 166–171.
24. Kavraki, L.E.; Svestka, P.; Latombe, J.C.; Overmars, M.H. Probabilistic roadmaps for path planning in highdimensional configuration spaces. *IEEE Trans. Robot. Autom.* **1996**, *12*, 566–580.
25. Gayle, R.; Sud, A.; Lin, M.C.; Manocha, D. Reactive Deformation Roadmaps: Motion Planning of Multiple Robots in Dynamic Environments. In Proceedings of the 2007 IEEE/RSJ International Conference on Intelligent Robots and Systems, San Diego, CA, USA, 29 October–2 November 2007; pp. 3777–3783.
26. Quinlan, S.; Khatib, O. Elastic Bands: Connecting Path Planning and Control. In Proceedings of the 1993 IEEE International Conference on Robotics and Automation, Atlanta, GA, USA, 2–6 May 1993; pp. 802–807.
27. Anderson, S.J.; Karumanchi, S.B.; Iagnemma, K. Constraint-Based Planning and Control for Safe, Semi-Autonomous Operation of Vehicles. In Proceedings of the 2012 IEEE Intelligent Vehicles Symposium, Madrid, Spain, 3–7 June 2012; pp. 383–388.
28. Kala, R.; Warwick, K. Multi-vehicle planning using RRT-connect. *Paladyn J. Behav. Robot.* **2012**, *2*, 134–144.

29. Kala, R.; Warwick, K. Planning of Multiple Autonomous Vehicles Using RRT. In Proceedings of the 10th IEEE International Conference on Cybernetic Intelligent Systems, London, UK, September 2011; pp. 20–25.
30. Kuwata, Y.; Karaman, S.; Teo, J.; Frazzoli, E.; How, J.P.; Fiore, G. Real-time motion planning with applications to autonomous urban driving. *IEEE Trans. Contr. Syst. Tech.* **2009**, *17*, 1105–1118.
31. Schubert, R.; Schulze, K.; Wanielik, G. Situation assessment for automatic lane-change maneuvers. *IEEE Trans. Intell. Transport. Syst.* **2010**, *11*, 607–616.
32. Hegeman, G.; Tapani, A.; Hoogendoorn, S. Overtaking assistant assessment using traffic simulation. *Transport. Res. C* **2009**, *17*, 617–630.
33. Naranjo, J.E.; González, C.; García, R.; de Pedro, T. Lane-change fuzzy control in autonomous vehicles for the overtaking maneuver. *IEEE Trans. Intell. Transport. Syst.* **2008**, *9*, 438–450.
34. Gayle, R.; Moss, W.; Lin, M.C.; Manocha, D. Multi-Robot Coordination Using Generalized Social Potential Fields. In Proceedings of the 2009 IEEE International Conference on Robotics and Automation, Kobe, Japan, 12–17 May 2009; pp.106–113.
35. Gayle, R.; Manocha, D. Navigating Virtual Agents in Online Virtual Worlds. In Proceedings of the 13th International Symposium on 3D Web Technology, Los Angeles, CA, USA, April 2008; ACM: New York, NY, USA; pp. 53–56.
36. Kala, R.; Shukla, A.; Tiwari, R. Robotic path planning in static environment using hierarchical multi-neuron heuristic search and probability based fitness. *Neurocomputing* **2011**, *74*, 2314–2335.
37. Kala, R.; Warwick, K. Multi-level planning for semi-autonomous vehicles in traffic scenarios based on separation maximization. *J. Intell. Robotic Syst.* **2013**, doi: 10.1007/s10846-013-9817-7.
38. Cagigas, D.; Abascal, J. A hierarchical extension of the D* algorithm. *J. Intell. Robotic Syst.* **2005**, *42*, 393–413.
39. Kala, R.; Warwick, K. Motion planning of autonomous vehicles in a non-autonomous vehicle environment without speed lanes. *Eng. Appl. Artif. Intell.* **2013**, doi: 10.1016/j.engappai.2013.02.001.

2

An Improved Reinforcement Learning System Using Affective Factors

Takashi Kuremoto [1,*], Tetsuya Tsurusaki [1], Kunikazu Kobayashi [2], Shingo Mabu [1] and Masanao Obayashi [1]

[1] Graduate School of Science and Engineering, Yamaguchi University, Tokiwadai 2-16-1, Ube, Yamaguchi 755-8611, Japan; E-Mails: staff-nn-len@ml.cc.yamaguchi-u.ac.jp (T.T.); mabu@yamaguchi-u.ac.jp (S.M.); m.obayas@yamaguchi-u.ac.jp (M.O.)

[2] School of Information Science & Technology, Aichi Prefectural University, Ibaragabasama 152203, Ngakute, Aichi 480-1198, Japan; E-Mail: kobayashi@ist.aichi-pu.ac.jp

* Author to whom correspondence should be addressed; E-Mail: wu@yamaguchi-u.ac.jp

---

Abstract: As a powerful and intelligent machine learning method, reinforcement learning (RL) has been widely used in many fields such as game theory, adaptive control, multi-agent system, nonlinear forecasting, and so on. The main contribution of this technique is its exploration and exploitation approaches to find the optimal solution or semi-optimal solution of goal-directed problems. However, when RL is applied to multi-agent systems (MASs), problems such as "curse of dimension", "perceptual aliasing problem", and uncertainty of the environment constitute high hurdles to RL. Meanwhile, although RL is inspired by behavioral psychology and reward/punishment from the environment is used, higher mental factors such as affects, emotions, and motivations are rarely adopted in the learning procedure of RL. In this paper, to challenge agents learning in MASs, we propose a computational motivation function, which adopts two principle affective factors "Arousal" and "Pleasure" of Russell's circumplex model of affects, to improve the learning performance of a conventional RL algorithm named Q-learning (QL). Compared with the conventional QL, computer simulations of pursuit problems with static and dynamic preys were carried out, and the results showed that the proposed method results in agents having a faster and more stable learning performance.

Keywords: multi-agent system (MAS); computational motivation function; circumplex model of affect; pursuit problem; reinforcement learning (RL)

1. Introduction

"Reinforcement" is first used by Pavlov in his famous conditioned reflex theory of the 1920s. It explains that the stimulus from the external environment can be categorized as rewards or punishments and they change the nature of the brain and the behavior of animals. The concept of reinforcement has been introduced in artificial intelligence (AI) from the 1950s and, as a bio-inspired machine learning method, reinforcement learning (RL) has been developed rapidly since the 1980s [1]. As analyzed by Doya, RL may take place in the basal ganglia of the brain: even the parameters used in RL may involve neuromodulators such as dopamine, serotonin, noradrenaline and acetylcholine [2]. In recent years, RL has been widely used in game theory [1], autonomous robotics [3,4], intelligent control [5,6], nonlinear forecasting [7–10], multi-agent systems (MASs) [11–18], and so on.

In [3], Asada *et al.* proposed a vision-based RL method to acquire cooperative behaviors of mobile robots in dynamically changing real worlds. In [4], Kollar and Roy combined RL with extended Kalman filter (EKF) to realize trajectory optimization of autonomous mobile robots in an unknown environment. Jouffe proposed a fuzzy inference system with RL to solve nonlinear control problems in [5], meanwhile Obayashi *et al.* realized robust RL for control systems adopting slide mode control concept in [6]. In our previous works, several kinds of neuro-fuzzy network types RL systems have been proposed as the time series predictors [7–10], and the internal models of agents to solve the goal-directed exploration problems in unknown environments [11–14]. Kobayashi *et al.* adopted attention degree into an RL named Q learning (QL) for multi-agent system (MAS) [15–17] acquiring adaptive cooperative behaviors of agents and confirmed the effectiveness of their improved RL by simulations of the pursuit problem (hunting game) [18].

The principle of RL can be summarized as using the adaptive state-action pairs to realize the optimal state transition process where optimal solution means that the maximum rewards are obtained by the minimum costs. The rewards from the environment are changed to be the values of states, actions, or state-action pairs in RL. Almost all well-known RL methods [1,19–22] use the value functions to change the states of the learner to find the optimized state transitions. However, when the learner (agent) observes the state of environment partially or the state of the environment is uncertain, it is difficult to select adaptive actions (behaviors). For example, in a multi-agent system (MAS), *i.e.*, multiple agents exploring an unknown environment, neighborhood information is dynamic and the action decision needs to be given dynamically, the autonomy of agents makes the state of the environment uncertain and not completely observable [11–18]. When RL is applied to MASs, problems such as "curse of dimension" (the explosion of state-action space), "perceptual aliasing problem" (such as the state transition in partially observable Markov decision process (POMDP)), and uncertainty of the environment come to be high hurdles.

The action selection policy in RL plays a role of motivation of the learner. The learning process of RL is to find the optimal policy to decide the valuable actions during the transition of states.

The behavior decision process by RL is clear and logical, based on the reward/punishment prediction. Meanwhile, to decide an action/behavior, high order animals, especially human beings may not only use the thinking brain, *i.e.*, logical judgment, but also the emotional brain, *i.e.*, instinctive response. Recently, neuroscience suggests that there are two paths for producing emotion: a low road is from the amygdala to the brain and body, and it is twice as fast as another high road which carries the same emotional information from the thalamus to the neocortex [23,24]. So it is possible for our brains to register the emotional meaning of a stimulus before that stimulus has been fully processed by the perceptual system, and the initial precognitive, perceptual, emotional processing of the low road, fundamentally, is highly adaptive because it allows people to respond quickly to important events before complex and time-consuming processing has taken place [23].

So, in contrast with the value based RL, behavior acquisition for autonomous mobile robots have also been approached by computational emotion models recently. For example, the Ide and Nozawa groups proposed a series of emotion-to-behavior rules for the goal exploration in unknown environment of plural mobile robots [25,26]. They used a simplified circumplex model of emotion given by Larson and Diener [27] which comes from a circumplex model of affect by Russell [28]. The affect model of Russell [28] is concluded by the statistical categorization of psychological (self-evaluate) experiments, and it suggests eight main affects named "Arousal", "Excitement", "Pleasure", "Contentment", "Sleepiness", "Depression", "Misery", and "Distress" with a relationship map of circulus. These affect factors are abstracted in a two-basic dimension space with "Pleasant-Unpleasant" (valence dimension) and "High Activation-Low Activation" (arousal dimension) axes by Larson and Diener [27]. Using these emotional factors, Ide and Nozawa introduced a series of rules to drive robots to pull or push each other to realize the avoidance of obstacles and cooperatively find a goal in an unknown environment [25,26]. To overcome problems such as dead-lock and multiple goal exploration when the complex environment was applied, Kuremoto *et al.* improved the emotion-to-behavior rules by adding another psychological factor: "curiosity" in [29,30]. However, all these emotional behavior rules for mobile robots only drive robots to move towards the static goals. Moreover, learning function, which is an important characteristic of intelligent agent, is not equipped with the model. So, agents can explore the goal(s), but cannot find the optimal route to the goal(s).

In this paper, we propose adopting affect factors into conventional RL to improve the learning performance of RL in MAS. We suggest that the fundamental affect factors "Arousal" and "Pleasure" are multiplied to produce an emotion function, and the emotion function is linearly combined with Q function which is the state-action value function in Q-learning (QL), a well-known RL method, to compose a motivation function. The motivation function is adopted into the stochastic policy function of RL instead of Q function in QL. Agents select available behaviors not only according to the states they observed from the environment, but also referring to their internal affective responses. The emotion function is designed by calculating the distance from agent to the goal, and the distance between the agent and other agents is perceived in the field of view. So the cooperative behaviors may be generated during the goal exploration of plural agents.

The rest of this paper is structured as follows. In Section 2, Russell's affect model is introduced simply at first, then, a computational emotion function is described in detail. Combining the emotion function with a well-known reinforcement learning method "Q-learning" (QL), a motivation function is constructed and it is used as the action selection policy as the improved RL of this paper.

In Section 3, to confirm the effectiveness of the proposed method, we applied it to two kinds of pursuit problems: a pursuit problem simulation with a static prey and a simulation with a dynamic prey, and the results of simulations are reported. Discussions and conclusions are stated in Sections 4 and 5, respectively.

2. An Improved Q Learning

2.1. Russell's Affect Model

Though there are various description and models of human being's emotion given by psychologists and behaviorists, circumplex models are famous and popularly studied [31–33]. Wundt proposed a 3-D emotion model with dimensions of "pleasure-displeasure", "arousal-calmness" and "tension-relaxation" 100 years ago [33]. Psychologists recently distinguish "emotion" and "affect" as "emotions", which are "valenced reactions to events, agents, or objects, with their particular nature being determined by the way in which the eliciting situation is construed" and "affect means evaluative reactions to situations as good or bad" [34]. In this study, we adopt basic dimensions of affects of circumplex model to invoke the emotional responses for the decision of autonomous agents.

Russell's circumplex model of affect [28] (See Figure 1) is simple and clear. It is given by a statistic method (principal-components analysis) to categorize human being's affects using 343 subjects' self-reports judgments. In the circumplex model, eight kinds of affective states including pleasure (0°), excitement (45°), arousal (90°), distress (135°), displeasure (180°), depression (225°), sleepiness (270°) and relaxation (315°) are located in a circular order. The interrelationships of these states are also represented by the arrangement, e.g., high values of arousal and pleasure result in a high value of excitement.

Figure 1. A circumplex model of affect proposed by Russell [27,28].

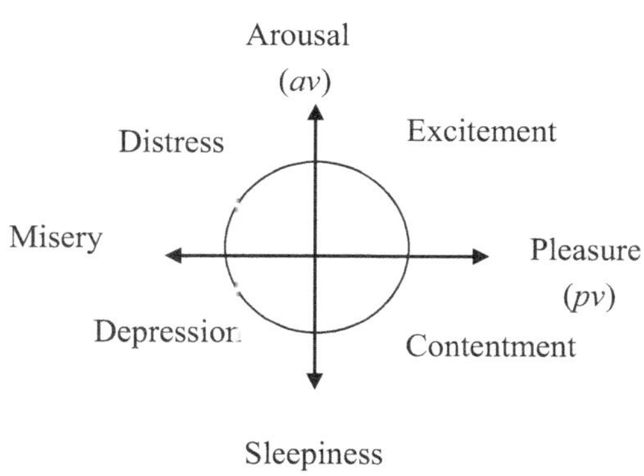

2.2. Emotion Function

In contrast with using the value function to decide an action/behavior of agents in conventional reinforcement learning methods [1–22], we propose using affective factors to invoke emotional response and reaction during the state transition process of agents in multi-agent systems (MASs). The reason for this proposal is according to the principle of interacting brain systems: the emotional brain and the thinking brain make decisions cooperatively [23,24]. Here, we use "Arousal" and

"Pleasure" of Russell's basic affects to construct "eliciting situation-oriented emotion" of agents. Moreover, the goal-directed problem that needs to be solved by RL is limited to pursuit problems or unknown environment explorations using local observable information.

Now, suppose an eliciting situation for an agent i, $i \in \{1,2,\ldots,N\}$ is e, $e \in E \equiv \{1,2,\ldots,|E|\}$, and the corresponding emotion is $Emo_i(e)$. When the agent perceives objects with high rewards (e.g., a goal) in the situation e, then $Emo_i(e) = toPv_i(e)$ (see Equation (1)), where $toPv_i(e)$ (see Equation (2)) is a mulptiplication of *Arousal* $av_i(e)$ (see Equation (4)) and *Pleasure* $Pv_i(e)$ (see Equation (5)). Meanwhile, if the agent finds other agents in the situation e, the emotion of the agent is totally effected by the emotion of other (perceived) agents $Emo_i(a) = getPv_i(a)$ (see Equation (1)). Consequently, the emotion function of the agent i in the state a is given by Equations (1–5).

$$Emo_i(e) = \begin{cases} toPv_i(e) & \text{if } 0 < d_{goal}(e) \le Depth \\ getPv_i(e) & \text{otherwise} \end{cases} \tag{1}$$

$$toPv_i(e) = av_i(e) \cdot pv_i(e) \quad \text{if } 0 < d_{goal}(e) \le Depth \tag{2}$$

$$getPv_i(e) = av_j(e) \cdot pv_j(e) \quad \text{if } 0 < d_{agent}(e) \le Depth \tag{3}$$

$$av_i(e) \leftarrow \begin{cases} av_i(e) + \Delta av & \text{if } e_i = e_i' \\ av_i(e) - \Delta av & \text{otherwise} \end{cases} \tag{4}$$

$$pv_i(e) = \begin{cases} Pv \cdot \exp\left(-d_{goal}^2(e)/\sigma^2\right) & \text{if } 0 < d_{goal}(e) \le Depth \\ -Pv & \text{otherwise} \end{cases} \tag{5}$$

where $d_{goal}(e)$ is Eclidean distance from the agent to a goal in enviroment exploration problem or a prey in pursuit problem if the goal is perceivable, and $d_{agent}(e)$ denotes Eclidean distance between agent i and agent j if the agent j is perceivable in situlation e. *Depth* is the threshold value of perceivable depth of space for sensory stimuli of visual, auditory, or olfactory signals, Δav, Pv, σ are positive parameters, e_i and e_i' are the current and next eliciting situations respectively.

So the formulas designed above can be incorporated into the following inferences rules:

Rule A: *Pleasure* $pv_i(e)$ concerns with the distance between the agent i, $i \in \{1,2,\ldots,N\}$ and the goal (Equation (5)) if the situation e means that the goal is perceived .

Rule B: *Arousal* $av_i(e)$ increases when the eliciting situation is continued (Equation (4)).

Rule C: Emotion state $Emo_i(e)$ (Equations (1–3)) of agent i in the eliciting situation e is expressed by the multiplication of *Pleasure* $pv_i(e)$ or $pv_j(e)$ (Equation (5)) and *Arousal* $av_i(e)$ or $av_j(e)$ (Equation (4)).

Rule D: Emotion state $Emo_i(e)$ is constructed and changed by stimuli from objects and events: perceiving the goal or other agents dynamically (Equations (1–5)).

2.3. A Motivation Function

When the Q learning algorithm (QL) [1,21] of the convetional reinforcement learning (RL) is used to find the optimal solution of a Markov decision process (MDP) which describes the stochastic state transition of agents, a state-action value function $Q_i(s,a)$ is calculated as follows:

$$Q_i(s,a) \leftarrow Q_i(s,a) + \alpha[r + \gamma \max_{a'} Q_i(s',a') - Q_i(s,a)] \tag{6}$$

where $s \in R^{N\times|E|}$ is a state oberved by the agent i, $a \in R^{|E|}$ is an action selected by the agent according to the value of $Q_i(s,a)$ (higher $Q_i(s,a)$ serves higer probability of action a at the state a), s', a' are the next state and selectable action, and r is a scalar of reward from the environment, $0 < \alpha < 1$ is a learning rate, and $0 < \gamma < 1$ a discount constant.

MDP, defined as the transition of states that only depends on the current state and the selected action [1], showed good convergence for the problems in MDP. However, in many real-world cases, Markov assumption is not satisfied. For example, when a learner observes its local environment or the state indirectly, different actions need to be selected for the optimal state transition. This case is called partially observable Markov decision process (POMDP) [1]. In MASs, an agent even needs to decide its optimal action according to the decision of other agents, so the problem of MASs belongs to non-Markov nature. To improve QL for POMDP and non-Markovian problems, approaches to change the environment to be MDP-like have been proposed, such as using belief state which uses the information of transition history [1,21], Monte-Carlo policy evaluation [35], and the different rewards of agents [36]. Furthermore, additional information such as the relationship between agents is also adopted into state-action value function in [18].

Here, agents learning of a MAS, we propose that it not only is decided by the value of observed state of environment, *i.e.*, Equation (6), but also situation-oriented emotion of agent given by Equations (1–5). A motivation function of agent i to express state-emotion-action value is proposed as follows.

$$M_i(s, Emo_i(e), a) = L \cdot Q_i(s,a) + (1.0 - L) \cdot Emo_i(e) \tag{7}$$

where $0 \leq L \leq 1.0$ is a constant to adjust the balance of emotional response $Emoi(e)$ and knowledge exploitation $Q_i(s,a)$. Specifically, the action a, $a \in A \equiv \{1,2,\ldots,|A|\}$ denotes the action of agent i in each available situation e, $|A| = |E|$ here.

2.4. Policy to Select an Action

As for the traditional QL, we use the soft-max seletion policy to decide an action according to the Bolzmann (Gibbs) distribution [1].

$$p_i(a \mid s, Emo_i(e)) = \frac{\exp\{M_i(s, Emo_i(e), a)/T\}}{\sum_a \exp\{M_i(s, Emo_i(e), a)/T\}} \tag{8}$$

where $p_i(a|s,Emo(e))$ is the probability of selecting action a at state s and emotion $Emo(e)$ of an eliciting situation e. T is a positive constant called "temperature" in the Blzmann distribution function. Higher T encourages the agent to select an available action more randomly. Here, we suggest using $T \leftarrow T^{Episode}$, $0 < T < 1$, $Episode = 1,2,\ldots Episode$ to reduce T according to the increase of learning iteration ($Episode$) to obtain stable learning convergence.

2.5. Learning Algorithm

The improved QL algorithm then can be given as follows:

Step 1 Initialize $Q_i(s,a) = 0$; $av_i = 0$ for each agent i, $i \in \{1,2,\ldots,N\}$ and all obserable states $s \in S$ and actions.

Step 2 Repeat following in each episode.

(1) Return to the initial state.

(2) Repeat the following until the final state.

i. Observe the state $s_i \in S$ of the environment, judge the situation $e_i \in E$ if the goal or other agents appear in perceivable area, calculate emotion $Emo_i(e)$ and motivation $M_i(s, Emo_i(e), a)$ at the state s_i and situation e_i of each selectable actions, select an action $a_i \in A$ according to the stochastic policy (Equation (8)).

ii. Execute the action a_i; observe a reward r and next state s_i'.

iii. Renew $Q_i(s,a)$ according to Equation (6).

(3) $s_i \leftarrow s_i$'

Step 3 Stop if episodes are repeated enough.

3. Computer Simulation

3.1. Definition of Pursuit Problem

To confirm the effectiveness of the proposed method, computer simulations of pursuit problems were executed. In Figure 2, a case of 2-hunter-1-static-prey problem with its initial state of the exploration environment is shown. The size of the environment is 17 × 17 grids. Prey (ridstays at the position (9, 9) at all times, while two hunters (○) find and capture the prey from (2, 2) and (14, 14) respectively. The perceivable area of a hunter is a square field around the hunter as shown in Figure 2, and the depth limitation was set to be 4 grids in 4 directions. So, the global coordinate information was not used but local observable information, the environment belongs to PDMDP.

The pursuit problem, or hunting game, can be designed in various conditions [18], and for convenience, it is defined as a static or dynamic goal (prey) that needs to be captured by all plural agents (hunters) simultaneously as one of the final states, as shown in Figure 3.

A hunter, which is an autonomous agent, observes its local environment as a state in five dimensions: Wall or Passage in four directions (up, down, left, and right) and Prey information in one dimension. For the first four dimensions, there are three discrete values: near, far, and unperceivable. Prey dimension has five values: up, down, left, right, and unperceivable. So the number of states s_i observed by a hunter i is $4 \times 3 \times 5 = 60$. The number of action is designed as $|A| = 4$ (up, down, left, right), and the number of state-action value function values $Q_i(s,a)$ (Q table called in QL) is $60 \times 4 = 240$. Additionally, the emotional response $Emo_i(e)$ to the eliciting situations, *i.e.*, perception of prey or other agents, are also in four directions, as for the action: up, down, left, and right $|E| = 4$. So the number of motivation function $M_i(s, Emo_i(e), a)$ is $240 \times 4 = 960$.

Figure 2. An environment for 2-hunter-1-static-prey problem.

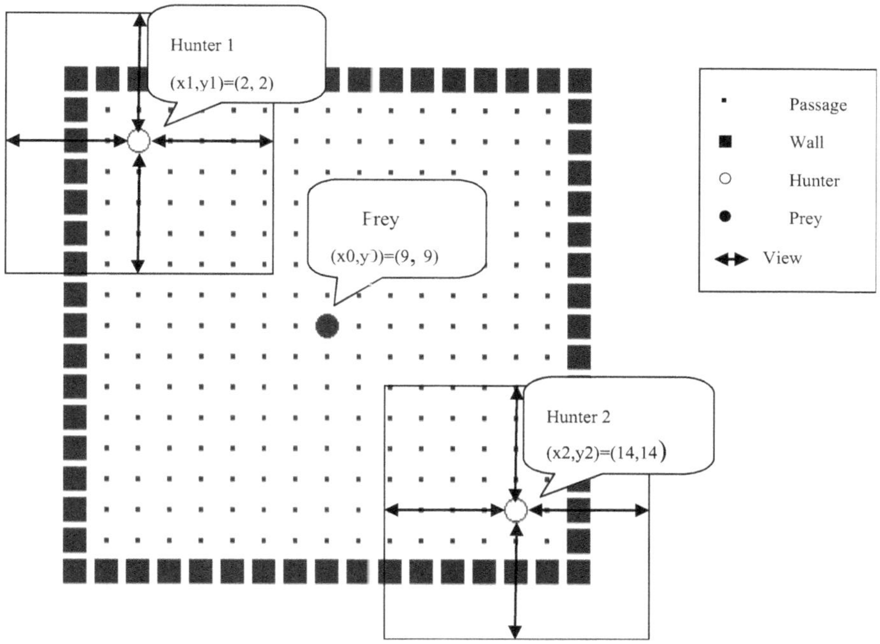

Figure 3. Available final states of a 2-hunter-1-static-prey problem. Prey (le final states of a 2-hunter-1-static-pr.

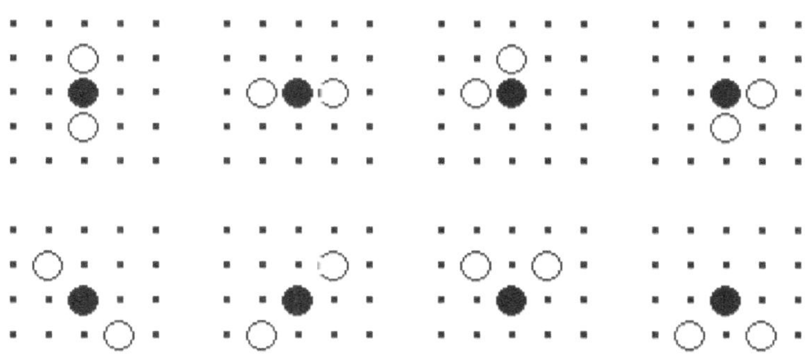

The learning performance of the Q learning method and the proposed method were compared for the pursuit problem in simulation. A part of the setting of simulations and parameters are listed in Tables 1 and 2. Special parameters used in the proposed method are listed in Table 3.

Table 1. Common conditions in the simulations.

Description	Symbol	Value
Episode limitation	*Episode*	200 times
Exploration limitation in one episode	step	10,000 steps
Size of the environment	$X \times Y$	17 × 17 grids
Threshold (depth) of perceive field	*Depth*	4 grids
Number of hunter	i	2, 3, 4
Number of action/situation	a/e	4

Table 2. Parameters used in the simulation of 2-hunter-1-static-prey problem.

Parameter	Symbol	Q learning	Proposed method
Learning rate	α	0.9	0.9
Damping constant	γ	0.9	0.9
Temperature (initial value)	T	0.99	0.994
Reward of prey captured by 2 hunters	r_1	10.0	10.0
Reward of prey captured by 1 hunter	r_2	1.0	1.0
Reward of one step movement	r_3	−0.1	−0.1
Reward of wall crash	r_4	−1.0	−1.0

Table 3. Parameters of the proposed method used in the simulations.

Parameter	Symbol	Value
Coefficient of *Emo*	L	0.5
Coefficient of *Pleasure*	Pv	200.0
Initial value of *Arousal*	Av	1.0
Modification of *Arousal*	Δav	0.01
Constant of Gaussian function	σ	8.0

3.2. Results of Simulation with a Static Prey

Both QL and the proposed method achieved the final states when learning process converged in the simulation. Figure 4 shows the different tracks of two hunters given by conventional QL (Figure 4(a)) and the proposed method (Figure 4(b)) where "S" denotes the start position of hunters, "○" denotes the start position of hunters and "●" is the position of the static prey. The convergence of the proposed method was faster and better than Q learning, as shown as in Figure 5 where "Q" indicates conventional QL results, and "M" is by the proposed motivation function used method.

Figure 4. Comparison of different tracks of two hunters captured a static prey. (**a**) Results of Q learning (QL); (**b**) Results of the proposed method.

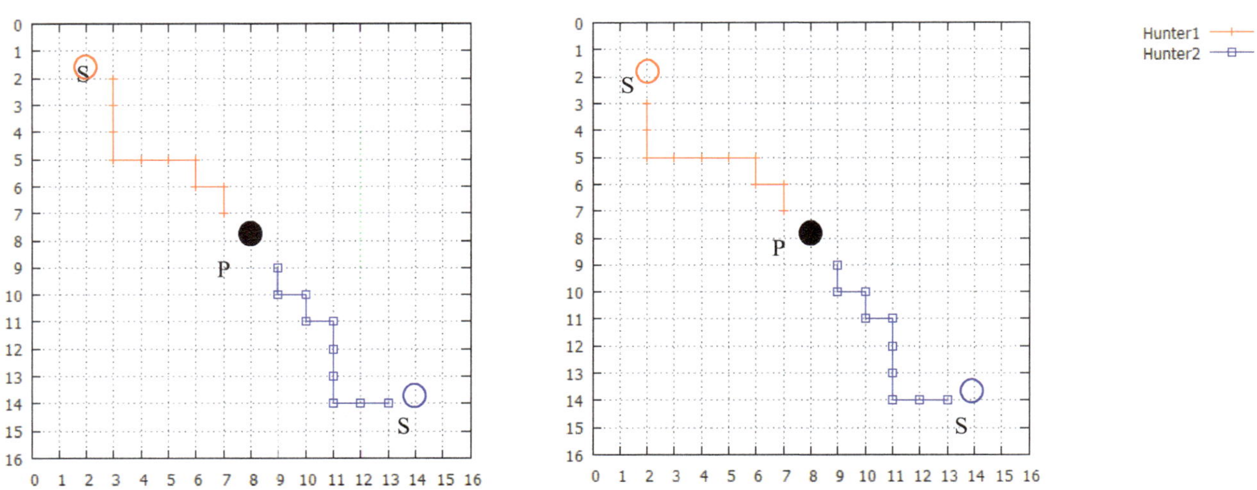

Figure 5. Comparison of the change of exploration costs (the number of steps from start to final states) during learning process in 2-hunter-1-static-prey problem simulation (average of 10 simulations).

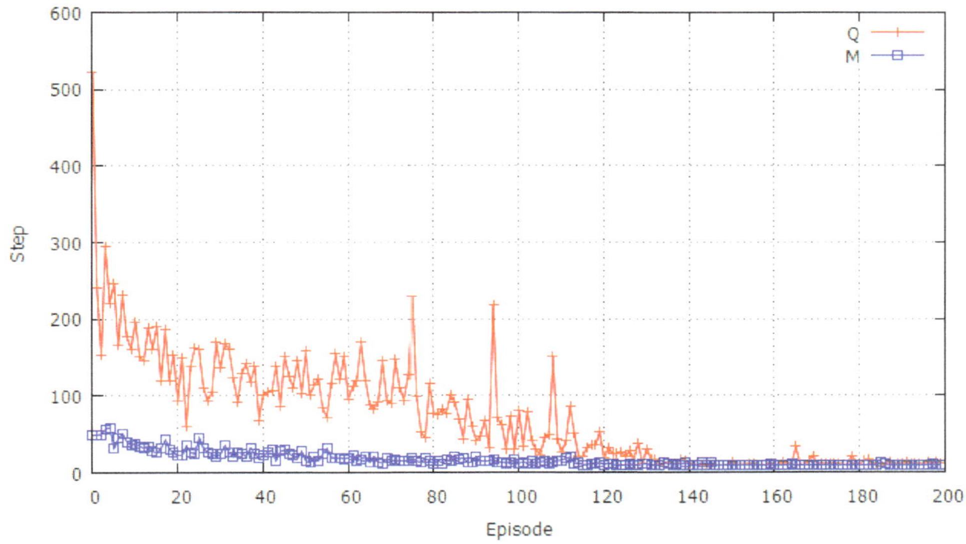

The average steps of one episode (from start to final states) during total learning process (200 episodes) given by 10 times simulations of the conventional QL and the proposed method were 72.9 and 17.1 as it shown in the first row of Table 4. Similar simulations using three and four hunters to find/capture a static prey were also performed and the learning performance of different methods was compared in Figure 6 and Table 4. The proposed method was confirmed to be superior in terms of learning performance, *i.e.*, it involves shorter steps to capture the prey in all simulations.

Table 4. Average exploration steps during learning in static prey problem simulations (200 episodes, 10 trials).

Number of hunter	Q learning	Proposed method
2	72.9	17.1
3	41.1	15.0
4	50.7	18.6

Figure 6. Comparison of average exploration steps during learning in static prey problem simulations (200 episodes, 10 trials).

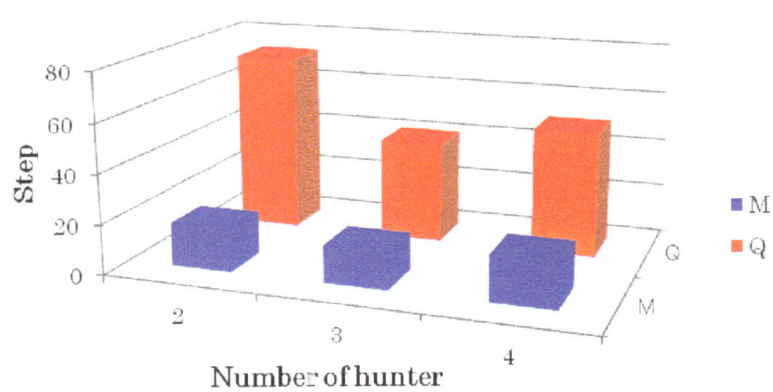

3.3. Results of Simulation with a Dynamic Prey

When a prey is not static, *i.e.*, moves during the exploration process, the learning of hunters' adaptive actions become more difficult. We designed a simulation of dynamic prey problem with a periodic movement process of the prey. The simulation was performed at the same start state, as shown in Figure 3. After the start state, the prey moved one step to the left before they arrived at the left wall. Then, it returned to the right direction until the right wall, and repeated the left-right movement periodically.

Table 5 gives the parameters used in a simulation of 2-hunters-1-dynamic-prey problem, and in Figure 7, the difference between the learning processes of different methods was shown, where "Q" means the result of conventional QL, and "M" depicts the change of exploration cost (steps) during learning (episode) using the proposed method.

Table 5. Parameters used in the simulation of 2-hunter-1-dynamic-prey problem.

Parameter	Symbol	QL	Proposed method
Learning rate	α	0.9	0.9
Damping constant	γ	0.9	0.9
Temperature (initial value)	T	0.99	0.99
Reward of prey captured by 2 hunters	r_1	10.0	10.0
Reward of prey captured by 1 hunter	r_2	1.0	1.0
Reward of one step movement	r_3	−0.1	−0.1
Reward of wall crash	r_4	−1.0	−1.0
Coefficient of *Emo*	L	-	0.5
Coefficient of *Pleasure*	Pv	-	10.0
Initial value of *Arousal*	Av	-	1.0
Modification of *Arousal*	Δav	-	0.1
Constant of Gaussian function	σ	-	8.0

Figure 7. Comparison of the change of exploration costs (the number of steps from start to final states) during learning process in 2-hunter-1-dynamic-prey problem simulation (average of 10 simulations).

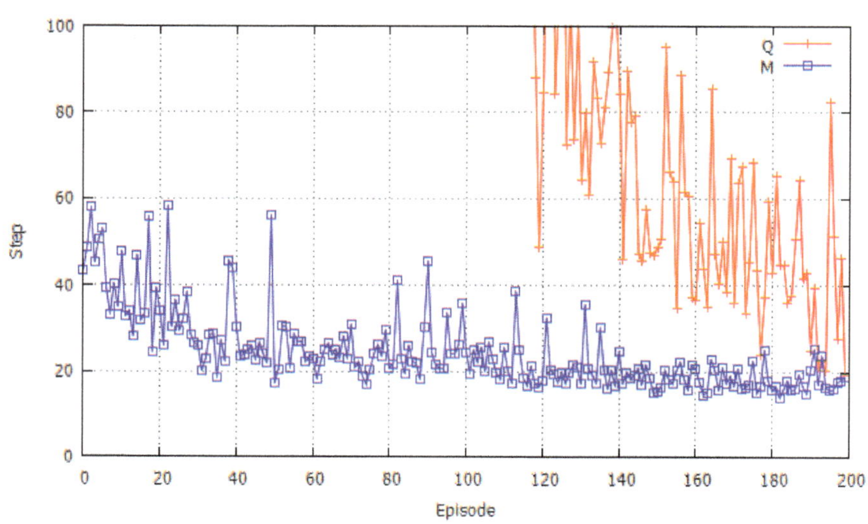

In Figure 8, the trajectories of two hunters that pursued the dynamic prey were shown. In contrast to Figure 6, the trajectories of two hunters in Figure 8 were shown where they pursued the prey who moved to left or right directions (the broken lines). The trajectories were the solutions after 200 training (episodes) by different methods: Figure 8(a) shows conventional QL and Figure 8(b) the proposed method. The length of each trajectory was the same: 13 steps (the theoretical optimal solution is 11 steps).

Figure 8. Comparison of different tracks of two hunters capturing a dynamic prey. (**a**) Results of Q learning (QL); (**b**) Results of the proposed method.

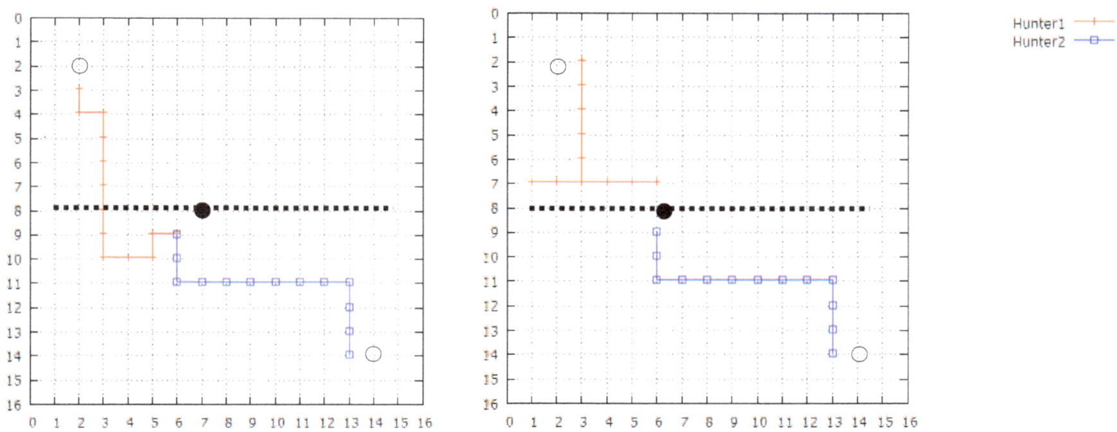

Additionally, all average exploration steps during learning of different numbers of hunters for the dynamic prey problem are summarized in Table 6. As a result, the proposed method showed higher efficiency than conventional QL, and the effectiveness was more obvious than the case of static prey problems (Also see Figure 9).

Table 6. Average exploration steps during learning in dynamic prey problem simulations (200 episodes, 10 trials).

Number of hunter	Q learning (QL)	Proposed model (DEM)
2	202.7	24.4
3	65.1	13.3
4	96.0	15.3

Figure 9. Comparison of average exploration steps during learning in dynamic prey problem simulations (200 episodes, 10 trials).

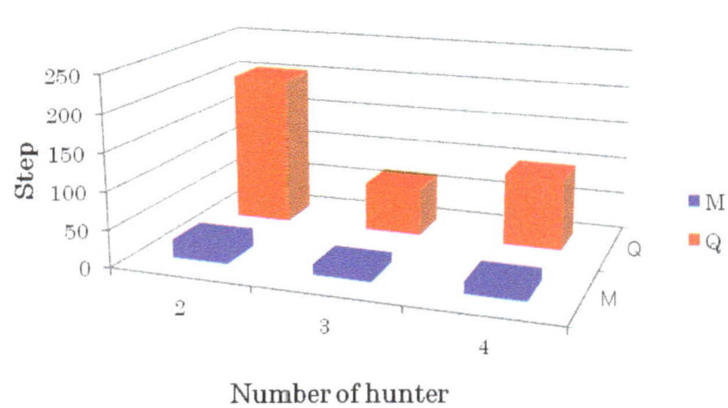

4. Discussions

As the results of pursuit problem simulations, the proposed learning method which combined QL and affect factors enhanced the learning efficiency compared to the conventional QL. Parameters used in all methods were optimal values from experiments. For example, different values of balance coefficient L in Equation (7) may yield different learning performance. $L = 0.5$ was used either in static prey simulation as shown in Table 3, or in dynamic prey simulation as shown in Table 5 because the optimal value yielded the shortest path for each simulation. In Figure 10, a case of a simulation with three hunters and one dynamical prey was shown. The value of L increased from 0 to 1.0 by a difference of 0.1 in the horizontal axis, and the lowest value of the vertical axis shows the average length (steps) of the 10 simulations of the capture process to be 13.11 steps at $L = 0.8$, so $L = 0.8$ was used as the optimal parameter value in this case.

Figure 10. Comparison of the change of exploration costs (the number of steps from start to final states) by different Q-M balance coefficient L in Equation (7) during learning process in 3-hunter-1-dynamic-prey problem simulation (average of 10 simulations).

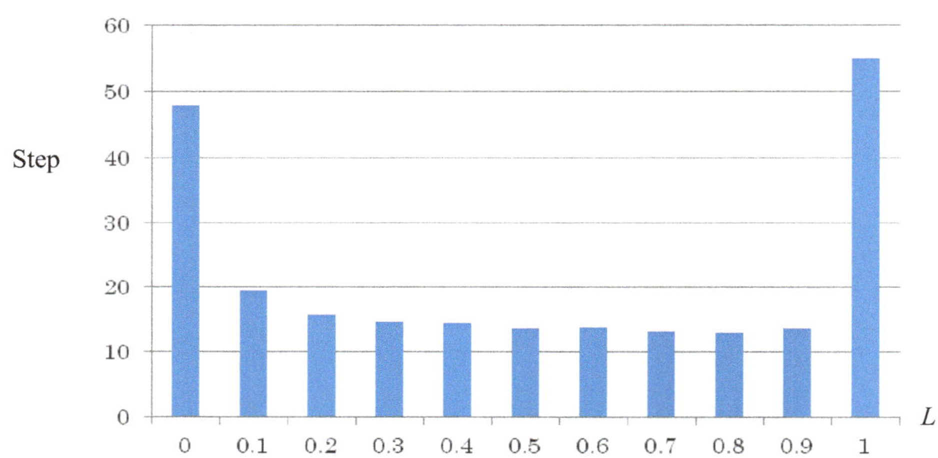

It is interesting to investigate how *Arousal av* and *Pleasure pv* values changed during the exploration and learning process in the simulation of pursuit problem. In Figure 11, the change of these affect factors in the 2-hunter-1-dynamic-prey simulation is depicted. In Figure 11(a), Arousal values of Hunters 1 and 2 changed together in the first three-step period, and then separated. This suggests that the state of the local environment observed by Hunter 1 changed more dramatically than the situation of Hunter 2, so *Arousal av* of Hunter 1 dropped according to the exploration steps. This is the result of Equation (4), the definition of *Arousal av*. In contrast, in Figure 11(b), *Pleasure pv* of Hunter 1 rose steeply to high values from step 2, corresponding to the dramatic change of the observed state: it might find and move to the prey straight ahead. From the 9th step, the *Pleasure* value of Hunter 2 also rose to high values for finding the prey or perceiving the high *Pleasure* value of Hunter 1.

From this analysis of the change of affect factors, it can be judged that the proposed method worked efficiently in the reinforcment learning process and it results in the improvement of learning performance of the MAS.

Figure 11. The change of *Arousal av* and *Pleasure pv* values of two hunters during the exploration in dynamic prey problem simulation. (**a**) The change of *Arousal av*; (**b**) The change of *Pleasure pv*.

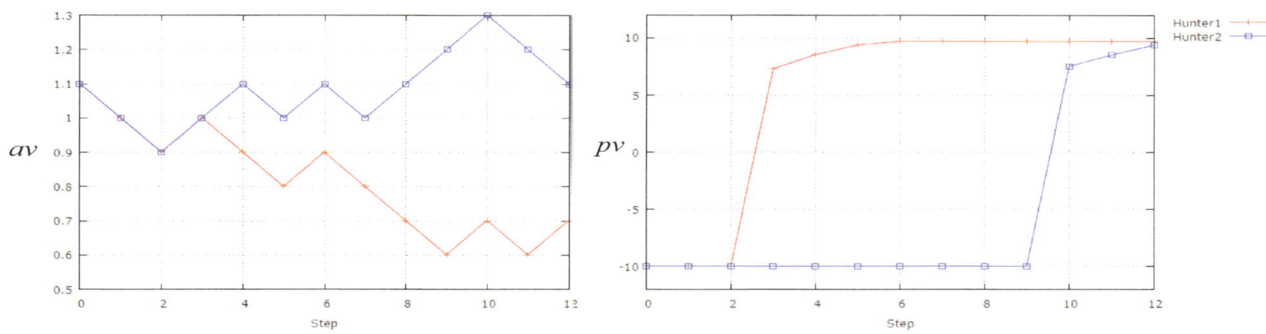

5. Conclusions and Future Works

To improve the learning performance of reinforcement learning for multi-agent systems (MASs), a novel Q-learning (QL) was proposed in this paper. The main idea of the improved QL is the adoption of affect factors of agents which constructed "situation-oriented emotion" and a motivation function which is the combination of conventional state-action value function and the emotion function.

Compared with the conventional QL, the effectiveness of the proposed method was confirmed by simulation results of pursuit problems with static and dynamic preys in the sense of learning costs and convergence properties.

The fundamental standpoint of this study to use affective and emotional factors in the learning process is in agreement with Greenberg [24]: "Emotion moves us and reason guides us". Conventional QL may only pay attention to "reasoning" from "the thinking brain" [24], or basal ganglia [2]. Meanwhile, in the proposed method we also considered the role of "the emotion brain" [23], amygdala or limbic system [24].

Therefore, as expected, future works will identify functions such as the neuro-fuzzy networks [9–14], and more emotion functions such as fear, anger, *etc.* [31,32,37,38], and other behavior psychological views [39] may be added to the proposed method. All these function modules may contribute to a higher performance of autonomous agents and it is interesting to apply these agents to develop intelligent robots.

Acknowledgments

A part of this study was supported by Foundation for the Fusion of Science and Technology (FOST) 2012-2014 and Grant-in-Aid for Scientific Research (No. 23500181) from JSPS, Japan.

Conflict of Interest

The authors declare no conflict of interest.

References

1. Sutton, R.S.; Barto, A.G. *Reinforcement Learning: An Introduction*; The MIT Press: Cambridge, MA, USA, 1998.

2. Doya, K. Metalearning and neuromodulation. *Neural Netw.* **2002**, *15*, 495–506.
3. Asada, M.; Uchibe, E.; Hosoda, K. Cooperative behavior acquisition for mobile robots in dynamically changing real worlds via vision-based reinforcement learning and development. *Artif. Intell.* **1999**, *110*, 275–292.
4. Kollar, T.; Roy, N. Trajectory optimization using reinforcement learning for map exploration. *Int. J. Robot. Res.* **2008**, *27*, 175–196.
5. Jouffe, L. Fuzzy inference system learning by reinforcement learning. *IEEE Trans. Syst. Man Cybern. B* **1998**, *28*, 338–355.
6. Obayashi, M.; Nakahara, N.; Kuremoto, T.; Kobayashi, K. A robust reinforcement learning using concept of slide mode control. *Artif. Life Robot.* **2009**, *13*, 526–530.
7. Kuremoto, T.; Obayashi, M.; Yamamoto, A.; Kobayashi, K. Predicting Chaotic Time Series by Reinforcement Learning. In Proceedings of the 2nd International Conference on Computational Intelligence, Robotics, and Autonomous Systems, Singapore, 15–18 December 2003.
8. Kuremoto, T.; Obayashi, M.; Kobayashi, K. Nonlinear prediction by reinforcement learning. *Lect. Note. Comput. Sci.* **2005**, *3644*, 1085–1094.
9. Kuremoto, T.; Obayashi, M.; Kobayashi, K. Forecasting Time Series by SOFNN with Reinforcement Learning. In Proceedings of the 27th Annual International Symposium on Forecasting, Neural Forecasting Competition (NN3), New York, NY, USA, 24–27 June 2007.
10. Kuremoto, T.; Obayashi, M.; Kobayashi, K. Neural forecasting systems. In *Reinforcement Learning, Theory and Applications*; Weber, C., Elshaw, M., Mayer, N.M., Eds.; InTech: Vienna, Austria, 2008; pp. 1–20.
11. Kuremoto, T.; Obayashi, M.; Kobayashi, K.; Adachi, H.; Yoneda, K. A Reinforcement Learning System for Swarm Behaviors. In Proceedings of IEEE World Congress Computational Intelligence (WCCI/IJCNN 2008), Hong Kong, 1–6 June 2008; pp. 3710–3715.
12. Kuremoto, T.; Obayashi, M.; Kobayashi, K. Swarm behavior acquisition by a neuro-fuzzy system and reinforcement learning algorithm. *Int. J. Intell. Comput. Cybern.* **2009**, *2*, 724–744.
13. Kuremoto, T.; Obayashi, M.; Kobayashi, K.; Adachi, H.; Yoneda, K. A neuro-fuzzy learning system for adaptive swarm behaviors dealing with continuous state space. *Lect. Notes Comput. Sci.* **2008**, *5227*, 675–683.
14. Kuremoto, T.; Obayashi, M.; Kobayashi, K. An improved internal model for swarm formation and adaptive swarm behavior acquisition. *J. Circuit. Syst. Comput.* **2009**, *18*, 1517–1531.
15. Sycara, K.P. Multi-agent systems. *Artif. Intell. Mag.* **1998**, *19*, 79–92.
16. Mataric, J. Reinforcement learning in multi-robot domain. *Auton. Robot.* **1997**, *4*, 77–93.
17. Makar, R.; Mahadevan, S. Hierarchical multi agent reinforcement learning. *Adv. Neural Inf. Process. Syst.* **2000**, *12*, 345–352.
18. Kobayashi, K.; Kurano, T.; Kuremoto, T.; Obayashi, M. Cooperative behavior acquisition using attention degree. *Lect. Notes Comput. Sci.* **2012**, *7665*, 537–544.
19. Barto, A.G.; Sutton, R.S.; Anderson, C.W. Neuron-like adaptive elements that can solve difficult learning control problems. *IEEE Trans. Syst. Man. Cybern.* **1983**, *13*, 834–846.
20. Sutton, R.S. Learning to predict by the method of temporal difference. *Mach. Learn.* **1988**, *3*, 9–44.
21. Watkins, C.; Dayan, P. Technical note: Q-learning. *Mach. Learn.* **1992**, *8*, 55–68.

22. Konda, V.R.; Tsitsiklis, J.N. Actor-critic algorithms. *Adv. Neural Inf. Process.* **2000**, *12*, 1008–1014.
23. LeDoux, J.E. *The Emotional Brain: The Mysterious Underpinnings of Emotional Life*; Siman & Schuster: New York, NY, USA, 1996.
24. Greenberg, L. Emotion and cognition in psychotherapy: The transforming power of affect. *Can. Psychol.* **2008**, *49*, 49–59.
25. Sato, S.; Nozawa, A.; Ide, H. Characteristics of behavior of robots with emotion model. *IEEJ Trans. Electron. Inf. Syst.* **2004**, *124*, 1390–1395.
26. Kusano, T.; Nozawa, A.; Ide, H. Emergent of burden sharing of robots with emotion model (in Japanese). *IEEJ Trans. Electron. Inf. Syst.* **2005**, *125*, 1037–1042.
27. Larsen, R.J.; Diener, E. Promises and problems with the circumplex model of emotion. In *Review of Personality and Social Psychology*; Clark, M.S., Ed.; Sage: Newbury Park, CA, USA, 1992; Volume 13, pp. 25–59.
28. Russell, J.A. A circumplex model of affect. *J. Personal. Soc. Psychol.* **1980**, *39*, 1161–1178.
29. Kuremoto, T.; Obayashi, M.; Kobayashi, K.; Feng, L.-B. Autonomic behaviors of swarm robots driven by emotion and curiosity. *Lect. Notes Comput. Sci.* **2010**, *6630*, 541–547.
30. Kuremoto, T.; Obayashi, M.; Kobayashi, K.; Feng, L.-B. An improved internal model of autonomous robot by a psychological approach. *Cogn. Comput.* **2011**, *3*, 501–509.
31. Russell, J.A.; Feldman Barrett, L. Core affect, prototypical emotional episodes, and other things called emotion: Dissecting the elephant. *J. Personal. Soc. Psychol.* **1999**, *76*, 805–819.
32. Russell, J.A. Core affect and the psychological construction of emotion. *Psychol. Rev.* **2003**, *110*, 145–172.
33. Wundn, W. *Outlines of Psychology*; Wilhem Englemann: Leipzig, Germany, 1897.
34. Ortony, A.; Clore, G.; Collins, A. *The Cognitive Structure of Emotions*; Cambridge University Press: Cambridge, UK, 1988.
35. Jaakkola, T.; Singh, S.P.; Jordan, M.I. Reinforcement learning algorithm for partially observable Markov decision problems. *Adv. Neural Inf. Process. Syst.* **1994**, *7*, 345–352.
36. Agogino, A.K.; Tumer, K. Quicker Q-Learning in Multi-Agent Systems. Available online: http://archive.org/details/nasa_techdoc_20050182925 (accessed on 30 May 2013).
37. Augustine, A.A.; Hemenover, S.H.; Larsen, R.J.; Shulman, T.E. Composition and consistency of the desired affective state: The role of personality and motivation. *Motiv. Emot.* **2010**, *34*, 133–143.
38. Watanabe, S.; Obayashi, M.; Kuremoto, T.; Kobayashi, K. A New Decision-Making System of an Agent Based on Emotional Models in Multi-Agent System. In Proceedings of the 18th International Symposium on Artificial Life and Robotics, Daejeon, Korea, 30 January–1 February 2013; pp. 452–455.
39. Aleksander, I. Designing conscious systems. *Cogn. Comput.* **2009**, *1*, 22–28.

3

Recent Trends in Lower-Limb Robotic Rehabilitation Orthosis: Control Scheme and Strategy for Pneumatic Muscle Actuated Gait Trainers

Mohd Azuwan Mat Dzahir [1,2,*] and Shin-ichiroh Yamamoto [1]

[1] Shibaura Institute of Technology, Department of Bio-Science Engineering, 307 Fukasaku, Minuma-ku, Saitama City, Saitama 337-8570, Japan; E-Mail: yamashin@se.shibaura-it.ac.jp

[2] Universiti Teknologi Malaysia, Faculty of Mechanical Engineering, UTM Skudai, Johor Bahru 81310, Malaysia

* Author to whom correspondence should be addressed; E-Mail: nb11503@shibaura-it.ac.jp or azuwan@fkm.utm.my

---

Abstract: It is a general assumption that pneumatic muscle-type actuators will play an important role in the development of an assistive rehabilitation robotics system. In the last decade, the development of a pneumatic muscle actuated lower-limb leg orthosis has been rather slow compared to other types of actuated leg orthoses that use AC motors, DC motors, pneumatic cylinders, linear actuators, series elastic actuators (SEA) and brushless servomotors. However, recent years have shown that the interest in this field has grown exponentially, mainly due to the demand for a more compliant and interactive human-robotics system. This paper presents a survey of existing lower-limb leg orthoses for rehabilitation, which implement pneumatic muscle-type actuators, such as McKibben artificial muscles, rubbertuators, air muscles, pneumatic artificial muscles (PAM) or pneumatic muscle actuators (PMA). It reviews all the currently existing lower-limb rehabilitation orthosis systems in terms of comparison and evaluation of the design, as well as the control scheme and strategy, with the aim of clarifying the current and on-going research in the lower-limb robotic rehabilitation field.

Keywords: pneumatic muscle-type actuators; co-contraction strategy of antagonistic actuators

---

1. Introduction

The outcomes of rehabilitation therapy that implements body weight support treadmill training for incomplete spinal cord injuries (SCIs) and stroke patients have been reported in several previous studies since the 1990s. SCI involves damage to any component of the nerves or spinal cord located at the end of the spinal canal, which is either complete or incomplete. However, it often causes permanent changes in strength, sensation and other body functions below the site of the injury. The symptoms vary widely, beginning with pain to paralysis and, then, to incontinence. The paralysis may be identified as a weakness, which might occur with abnormal tone (e.g., spasticity or rigidity). During the stance phase, leg instability (*i.e.*, hyperextension or knee buckling) may result in unsafe walking, pain and inefficient energy. Moreover, inadequate limb clearance, impaired balance, sensory deficits and pain during the swing phase may contribute to falls, loss of balance and increased nervousness associated with walking. Furthermore, the loss of motor control prevents a patient from performing a precise movement in coordination with the timing and intensity of the muscle action.

Previously, a patient's paralyzed legs were physically operated by two therapists in manual training. In accordance with treadmill training therapy, based on the rules of spinal locomotion, research carried out by Wernig *et al.* for the incomplete paralysation of paraplegic and tetraplegic patients confirmed that the training was able to improve most of the patients' walking capability [1,2]. The patients involved in this training were provided with motor-driven treadmill training therapy, along with a body weight support (BWS) and assisted limb movements by therapists, for daily upright walking training. Based on the rehabilitation sessions, nearly 80% of patients with incomplete spinal cord injuries (a total of 33 individuals) were capable of walking independently after the treadmill training, with partial body weight support. However, this training procedure was physically difficult for therapists to execute for long durations of time. Recently, robot-assisted therapy devices have become increasingly used in SCI rehabilitation therapy. This assistive robot either compensates for the functionalities that a patient does not have or tries to recover the impaired functionalities. Even though it may not be able to fully compensate for impairments, or even provide a cure, it should be able to enhance or extend certain impaired functions, consequentially increasing the quality of life, encouraging independent living, as well as supporting the need for social interactions and communication. Depending on the degree and location of the injury, the actual rehabilitation or treatment can vary widely. In many cases, substantial rehabilitation and physical therapy are required for spinal cord injuries, particularly if the patient's injuries interfere with the activities of daily life.

Since SCI patients frequently have difficulties with daily functional movements and activities, it is possible to decrease their loss of function through rehabilitation therapy during the critical stage. This rehabilitation therapy engages carefully designed repetitive exercises, which are either passive or active. In a passive exercise, the therapist or a robot will actively assist the patient with moving the affected lower-limb repetitively, as prescribed. In an active exercise, the patients themselves will put effort into moving their legs, with no physical assistance. With the contribution of therapists, assistive robotic technology has had a significant ability to provide novel means for motivating, monitoring and coaching. In addition, many lower-limb leg orthoses for rehabilitation have been developed to assist in human locomotion training; they can be used for a long time and for varying degrees of spasticity or paresis [1–14]. According to Dietz *et al.*, who performed lower-limb-assisted gait training using a developed orthosis system with BWS and treadmill training on patients with incomplete SCI, advocated that the afferent participation from the lower limb and hip joint movements are essential for the activation of the central pattern generator for locomotion rehabilitation training in SCI patients [3,4].

Consequently, the interest in this field has grown exponentially in recent years, mainly due to the demand for a much more compliant and interactive human-robotics system. Therefore, this work will appraise all of the current existing lower-limb rehabilitation orthoses, based on compliant actuator systems, in terms of their evaluation, design, control scheme and strategy. They will then be compared to each another, with the intent of clarifying current and on-going research in the lower-limb robotics rehabilitation field.

2. Existing Lower-Limb Orthoses for Gait Rehabilitations and Evaluations

Numerous assistive orthosis systems for gait rehabilitation have been developed that delve into several types of lower-limb rehabilitation, such as treadmill gait trainers, over-ground gait trainers, stationary gait and ankle trainers, foot-plate-based gait trainers and active foot orthoses for the neurologically impaired (including stroke and spinal cord injury (SCI) patients) [5–8]. These systems implement very unique mechanical structures, designs, actuators, methods, control schemes and rehabilitation strategies, as well as various procedures to ensure the reliability and robustness of the systems when compared to others. The rapid development of rehabilitation robotics over the last decade is working toward fully restoring or improving the mobility of affected limb functions and helping patients achieve a better quality of life.

2.1. Motorized Lower-Limb Orthosis Systems for Rehabilitation

The driven gait orthosis (DGO), also known as LOKOMAT (Hocoma AG, Volketswill Switzerland), is currently available on the market and has been extensively researched in many rehabilitation centers as one of the best examples for a gait orthosis that can be used for lower-limb disabilities [9–11]. This orthosis system is shown in Figure 1a. It consists of three main parts: body weight support, treadmill and powered leg orthosis. A direct current (DC) motor, with helical gears, was used for the actuation power of the system to precisely control the trajectory of the hip and knee joints. Considerable control algorithms have been implemented in this system to improve its performance, such as position, adaptability, impedance controllers, *etc.* To stimulate the locomotor function of the spinal cord and to activate leg muscles that have lost the capacity to actuate voluntary movement, it is important to provide adequate afferent input to the affected lower limb. It could be anticipated that the afferent input produced using automatic-based training is at least as efficient as that generated using manual training.

Figure 1b shows the treadmill gait trainer system, which incorporated an electromechanical gait device with the treadmill/gait training, known as the LokoHelp (LokoHelp Group, Germany). The LokoHelp used a different mechanical system compared to the LOKOMAT, which implemented a powered leg orthosis. The foot-powered orthosis, known as "Pedago", uses an electromechanical gait device that was designed to provide a gait motion during the training session [12]. The control device helps to move the patients' foot trajectory with a fixed step length of 400 mm, in which the gait cycle

(GC) speed can be varied from zero up to 5 km/h. Based on the research findings, it was proven that walking ability could be improved by incorporating task-oriented gait training with mechanical gait training devices or with treadmill training.

Figure 1. (**a**) LOKOMAT [10]; (**b**) LokoHelp (picture courtesy of LokoHelp group); and (**c**) ReoAmbulator (picture courtesy of Motorika Ltd.).

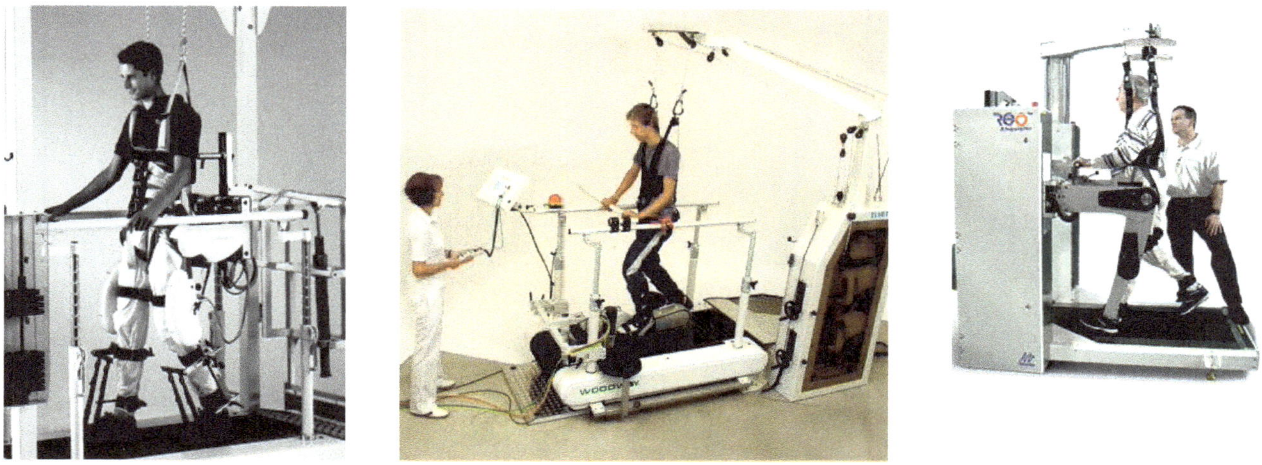

(**a**) (**b**) (**c**)

The ReoAmbulator robotic system (Motorika Ltd., Mount Laurel, NJ, USA), which is also known as "AutoAmbulator", is another example of existing treadmill gait trainers for lower-limb rehabilitation therapy, as shown in Figure 1c. This system has been used in research centers and medical hospitals for rehabilitation therapies and educational research studies [13,14]. This system also implements a powered leg orthosis, "robotic arms", which enables patients to contribute during the gait motion, but also provides the remaining force necessary for walking. The robotic arms are attached to the thigh and ankle of the patient's leg before a stepping pattern is performed using the implemented control scheme and strategy. In previous research on this system, it was concluded that robot-assisted gait training was able to provide improvements in balance and gait that are comparable to conventional/ manual physical rehabilitation therapies.

Apart from the available commercialized rehabilitation orthosis systems, the growth of the ReoAmbulator system has been rather immense with the development of different prototypes. The development of LOPES increased researchers' interest in developing a humanlike musculoskeletal assistive orthosis system. This gait rehabilitation orthosis employs the Bowden-cable driven series of elastic actuators (SEA), with the servomotors as the actuation system, to implement low weight (pure) force sources at both the posterior and anterior sides of the leg orthosis, as illustrated in Figure 2a [15,16]. It implemented impedance control (as opposed to admittance control), which is based on a combination of position sensing with force actuation to operate the lower-limb leg orthosis. The training effect of this orthosis was enhanced by emphasizing the implementation of an assist as needed (AAN) control algorithm. This enabled an increment of the active voluntary participation of the patients. Moreover, it is also possible to imply unhindered walking practice in the orthosis device, where the required forces/torques for imposing a gait pattern are determine based on the system's evaluation.

Figure 2. (**a**) LOPES [15]; (**b**) active leg exoskeleton (ALEX) [17]; and (**c**) NEUROBike [20].

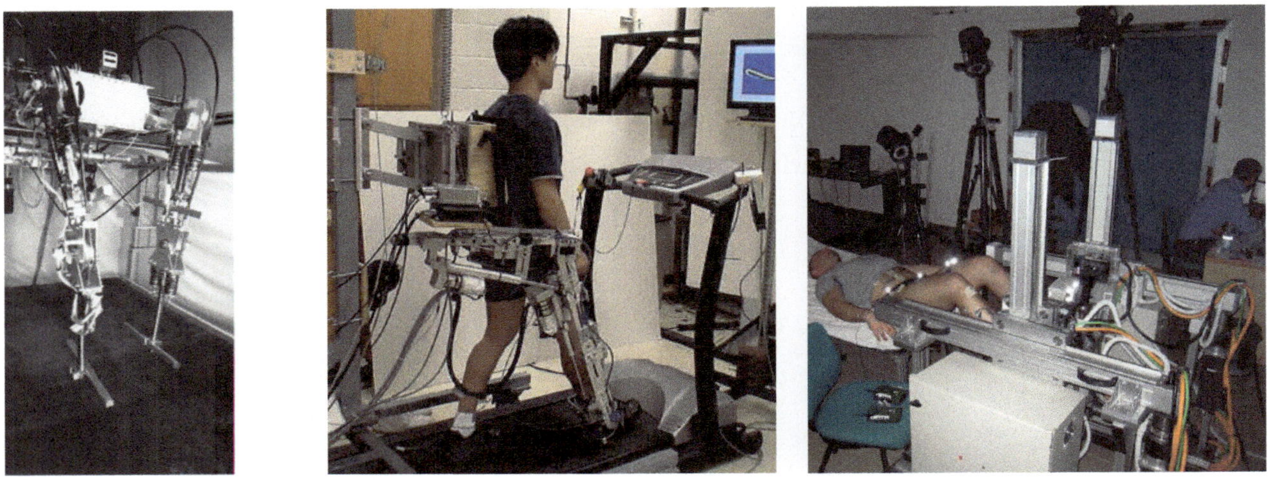

(**a**) (**b**) (**c**)

In the following years, a robot-assisted gait training (RAGT) was developed with an active leg exoskeleton (ALEX), also integrating the AAN rehabilitation strategy into the orthosis system. Compared to other existing robotic training methods, this strategy allows the patient to actively contribute during the retraining process of gait locomotion. This gait rehabilitation device is shown in Figure 2b. It implemented the use of linear actuators to actuate the hip joint thigh device and knee joint shank device of the leg orthosis [17,18]. It has been proven that an intensive gait retraining process has great potential to significantly provide benefits for the patients, including chronic stroke survivors. This can be achieved by effectively applying enough forces on the ankle of the subject through actuators placed at the hip and knee joints of the exoskeleton's leg orthosis, by means of a force-field controller.

Later, a stationary gait and ankle trainer system was developed to provide neural-rehabilitative treatments aimed at recovering walking abilities in post-stroke patients. This orthosis system employed the use of brushless servomotors and pulleys to actively control the angular excursions of the gait orthosis, known as the neural-rehabilitative platform for bedridden post-stroke patients (NEUROBike) [19]. The prototype of this system is shown in Figure 2c. The passive and active exercises were emphasized in this system by implementing the kinematic models of leg-joint angular excursions during both "sit-to-stand" and "walking" in the control algorithms. To summarize, providing a number of exercises at an early phase based on the intensity and the severity of the pathology is required by the programmed therapy. In addition, customized treatment adapted by this system may facilitate patients by increasing their flexibility in lower-limb control, which leads to significant improvements in motor control performance during locomotion.

In addition, a robotic gait rehabilitation (RGR) trainer prototype was also invented within the same year as the NEUROBike system, to assist treadmill gait retraining for patients with unusual gait patterns that were associated with exaggerated pelvis obliquity, illustrated in Figure 3a. This orthosis is composed of three subsystems: stationary frame, human-robot interface (HRI) and treadmill training. Servo-tube linear electromagnetic actuators were used to generate the power source for the exoskeleton [20]. Based on a hypothesis, the correction of a stiff-legged gait pattern entails addressing both the primary and secondary gait deviations to restore a physiological gait pattern. Therefore, an expanded impedance control strategy was used to generate the corrective moments, only when the leg was in swing motion, by switching the force field that affects the obliquity of the pelvis. It has been demonstrated that this system can be effective in guiding the pelvis to the frontal plane via force fields used for altering pelvic obliquity.

Figure 3. (**a**) Robotic gait rehabilitation (RGR) trainer [20]; and (**b**) LOKOIRAN [21].

(**a**) (**b**)

Recently, a new gait training robotic device (LOKOIRAN) was designed to be suitable for patients with various diagnoses, such as SCI, stroke, multiple sclerosis (MS) and sport injury cases, aging and people with balance and locomotion disorders. Figure 3b illustrates the system's prototype. This gait training device delves into several subsystems, consisting of body weight support, a leg exoskeleton, a driving system and a transmission system. It employs alternating current (AC) motors connected to a slide-crank mechanism via belts and pulleys to provide the energy for the system [21]. The implemented control system enables flexibility in motion and permits subjects to change the speed of the foot plates by engaging the speed control mode and the admittance control mode.

The evaluated motorized lower-limb gait rehabilitation orthosis systems mentioned are only a fraction of the currently existing rehabilitation orthoses. However, it could be summarized from these examples that their development has reached an advanced level; whereby, many of the lower-limb gait rehabilitation orthoses, based on electrical motors, have already been commercialized. With the speed of growth in their mechanical design, as well as, the implementation of advanced control schemes and strategies, the space available for enhancements might soon reach its peak.

2.2. Attributes of Pneumatic Muscle Actuators (PMA)

The implementation of pneumatic muscles enables pneumatic power to be transferred into mechanical power. This actuator will be shortened in the longitudinal direction and enlarged in the radial direction during the contraction stage, when it is being inflated; when being deflated, it will turn back to its original form. The pneumatic muscle is able to employ a tensile force to an attached load during the contraction stage. This force is unidirectional, whereby the original length of a certain designed diameter and the internal pressure will determine its value. Moreover, this actuator also inhibits nonlinear behaviors, such as hysteresis, compressibility and time variance. However, in exchange, this pneumatic muscle also has an inherently compliant attribute, which is suitable for a human-robotics system. This type of actuator is similar to the human muscle principle; a shorter muscle length produces a smaller contracting force and *vice versa*. Furthermore, it is comparable to electric actuators, due to the direct coupling to the load, the structural optimization and the power/weight ratio.

In addition to the abovementioned attributes, there exist two main weaknesses that limit the application of pneumatic muscles. The first weakness is the nonlinear behavior of pressure build-up, and the second weakness is the hysteresis effect, due to its geometric structure. These drawbacks cause complexity when scheming high-performance control systems. Therefore, this research is dedicated to solving these problems, using a simple paradigm and control strategy for handling the sudden increase in pressure and the hysteresis behavior of the PMA. Based on the proposed empirical-based static force mathematical model, which consist of a correction factor caused by the effect of the end caps, it showed an inconsistency of the high contracting ratios derived by the famous researcher, Tondu *et al.*, [22]. The extreme difficulty in constructing an accurate mathematical model was established by the fact that nearly all of the present models proposed were approximations. This model was later modified through various methods, used by other researches, to further improve the mathematical model [23–30].

2.3. Pneumatic Muscle Actuated Lower-Limb Rehabilitation Orthosis System

Compared to the motorized lower-limb rehabilitation orthosis systems (*i.e.*, DC motors, AC motors, linear actuators, SEA, servomotors, brushless motors and pneumatic cylinders), the growth of pneumatic muscle-actuated rehabilitation orthosis systems has been rather poor. This is also the description of the development of the control system for pneumatic muscles. However, numerous research studies in the last 10 years have tried to introduce these types of actuation systems into the lower-limb rehabilitation robotics field. This may indicate a significant shift of researchers' interests towards the implementation of a pneumatic muscle-actuated lower-limb rehabilitation orthosis.

A hip orthosis exoskeleton powered by pneumatic artificial muscle (PAM) was invented by Vimieiro *et al.*, at the Bioengineering Laboratory in 2004, as shown in Figure 4a [31,32]. This exoskeleton system was designed and modeled for patients with a motor deficit, a result of poliomyelitis. It consists of two main parts: the first is a polyethylene pelvic brace to provide the stability for the orthosis system, and the second is a polyethylene support for the thigh. This orthosis system implements position control using potentiometers for activating the control valves, either to pressurize the PAM or to return it to neutral status. Based on clinical tests, it was proven that this rehabilitation engineering was able to provide equipment and devices for aiding patients in recovering their movements or to improve their quality of life. A better gait pattern and an improvement of the left step transposition in the toe-off phase were reported by patients.

Later came the robotic gait trainer (RGT) for stroke rehabilitation, which is an ankle rehabilitation device powered by lightweight springs over muscle (SOM), proposed by Kartik *et al.* It was developed in 2006, as shown in Figure 4b [33]. The design is structurally based on a tripod mechanism with one fixed link. This orthosis device is able to provide the dorsiflexion and plantar-flexion, as well as the inversion and eversion when moving the foot about the ankle joint. It implements an angular position for the control system and uses two types of sensors (*i.e.*, potentiometer and pressure sensor). In this study, Kartik *et al.* suggested that the range and position of motion (ROM) is necessary for safe dorsiflexion/plantar-flexion and inversion/eversion movements. This was proven by the results from their analysis, which demonstrated that the tripod structure was able to generate a ROM that matches the safe anatomical range of the ankle joint during the gait cycle.

Figure 4. (**a**) Hip orthosis [32]; (**b**) robotic gait trainer (RGT) [33]; and (**c**) ankle-foot orthosis (AFO). SOM, springs over muscle [36].

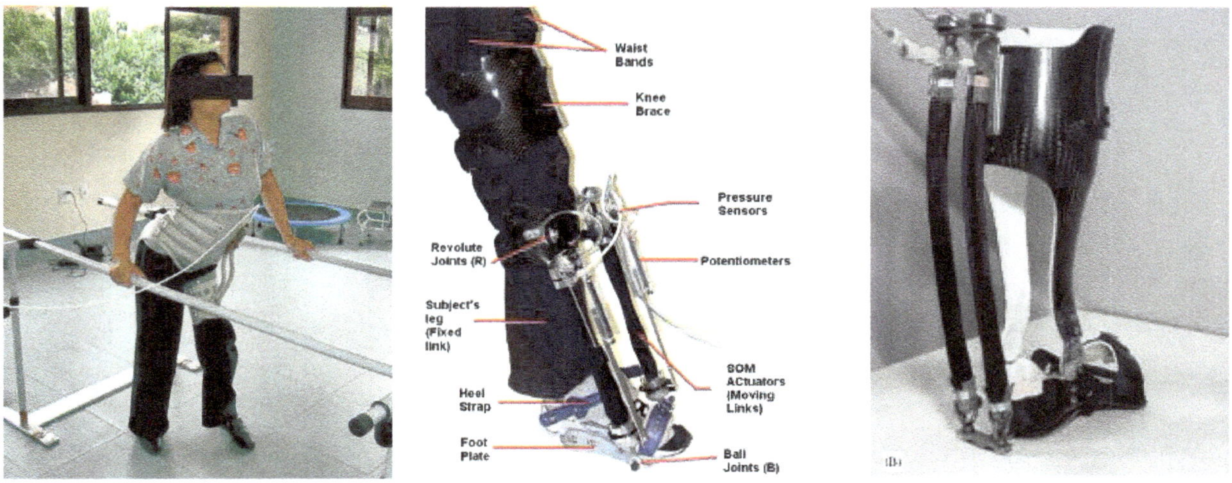

(**a**) (**b**) (**c**)

In contrast, the prototype of an ankle-foot orthosis (AFO) powered by artificial pneumatic muscle was also introduced by Ferris *et al.* in 2006. The prototype was of the human lower-limb that could comfortably provide dorsiflexion and plantar flexion torque during walking motion training, as illustrated in Figure 4 [34–36]. This orthosis is composed of a hinge joint, a carbon fiber shell and two pneumatic artificial muscles. The proportional myoelectric control, using a PC-based controller (real-time control), had been implemented in the control system. The performance of the novel controller enables naive wearers to promptly become accustomed to the orthosis, without the pneumatic muscle co-contraction. It is believed that this orthosis design will be useful in learning human walking biomechanics and in providing assistance of patients with neurological injuries during rehabilitation training.

Conversely, by focusing on the development of "human friendly" exoskeleton orthosis systems, Costa *et al.* in 2006 proposed a powered lower-limb orthosis, which can produce powerful, yet naturally safe, operations for paraplegic patients, as illustrated in Figure 5a [37]. This was realized by combining a highly compliant actuator (PMA) with an embedded intelligent control system (a three level PID joint torque control scheme) to manipulate the antagonistic actuators of the exoskeleton. It is difficult to provide a system with dependability and inherent safety, while utilizing a highly compliant actuator, using conventional designs alone. However, the design philosophy of this system may provide a significant insight into the development of rehabilitation orthosis systems and improve rehabilitative procedures for paraplegic patients.

Figure 5. (**a**) Powered lower-limb orthosis [37]; (**b**) Robotic Gait Trainer in Water (RGTW) [38]; and (**c**) powered ankle-foot exoskeleton [39].

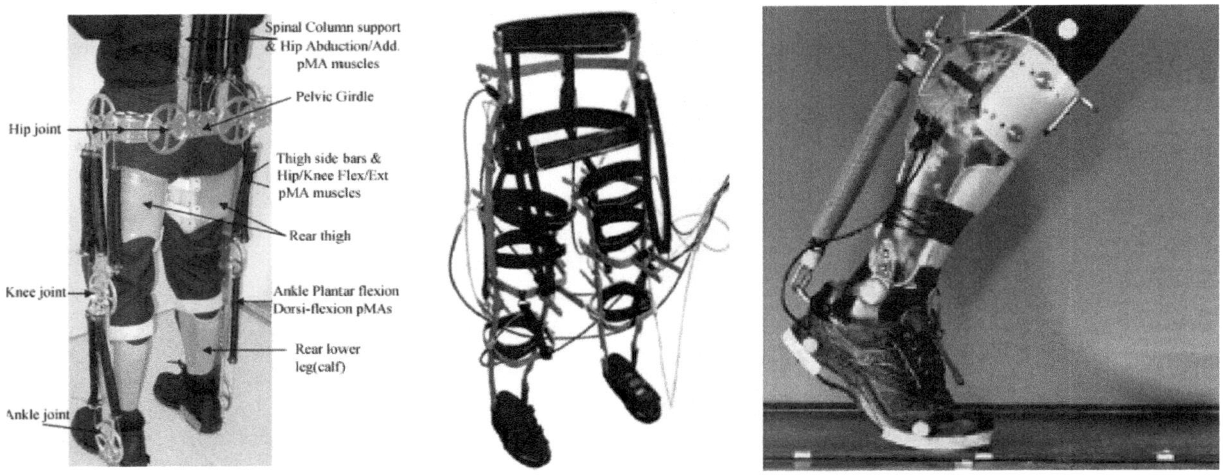

(**a**) (**b**) (**c**)

Contrastingly, Figure 5b shows the Robotic Gait Trainer in Water (RGTW). This system was designed for the development of an underwater gait training orthosis by Miyoshi *et al.* in 2008 [38]. The RGTW is a hip-knee-ankle-foot orthosis with pneumatic McKibben actuators as the actuation system. The basis of the angular motion for the control system was determined by a healthy subject walking under water. The aim for this study was to achieve repetitive physiological gait patterns to improve movement dysfunctions. By implementing this orthosis system device, not only the effect of hydrotherapy should be expected, but standard treadmill training is also included. This could also be sufficiently effective for patients undergoing hip-joint movement dysfunction treatments.

In 2009, Malcom *et al.* developed a powered ankle-foot exoskeleton, which investigated the role of the tibialis anterior (TA) in the walk-to-run condition, as shown in Figure 5c [39–42]. The pneumatic muscles are used to provide the dorsiflexion and plantar-flexion torques through the assisting orthosis for incomplete SCI patients during assist and resist conditions. This orthosis device implements an electromyography (EMG) control with a feed-forward algorithm; whereby, a set of rotary encoders and load cells are used to measure the treadmill belt speed, ankle angle and the dorsiflexion and plantar-flexion torques. Through a hypothesis developed from gait transitions and research evaluations, it was demonstrated that the powered exoskeleton had great potential in fundamental gait studies.

After the introduction of AFO by Ferris *et al.*, the development of this system was later continued by Sawicki *et al.*, a few years later. In 2009, the pneumatically powered knee-ankle-foot orthosis (KAFO) was proposed through the study of human motor adaptation, gait rehabilitation and locomotion energetics; as shown in Figure 6a [43]. Compared to the AFO control system, this system implements a physiologically-inspired controller that utilized the patient's muscle information, which is determined using electromyography to measure the timing and amount of the artificial muscle forces. Based on several research findings, it is believed that powered knee-ankle-foot orthoses are promising for basic science and clinical applications, since they have successfully assisted individuals with incomplete SCI during locomotor training, metabolic energy consumption and neural adaptation for neurologically intact human walkers.

Figure 6. (**a**) Knee-ankle-foot orthosis (KAFO) [43]; (**b**) continuous passive motion (CPM) [44]; and (**c**) power-assist lower-limb orthosis [45].

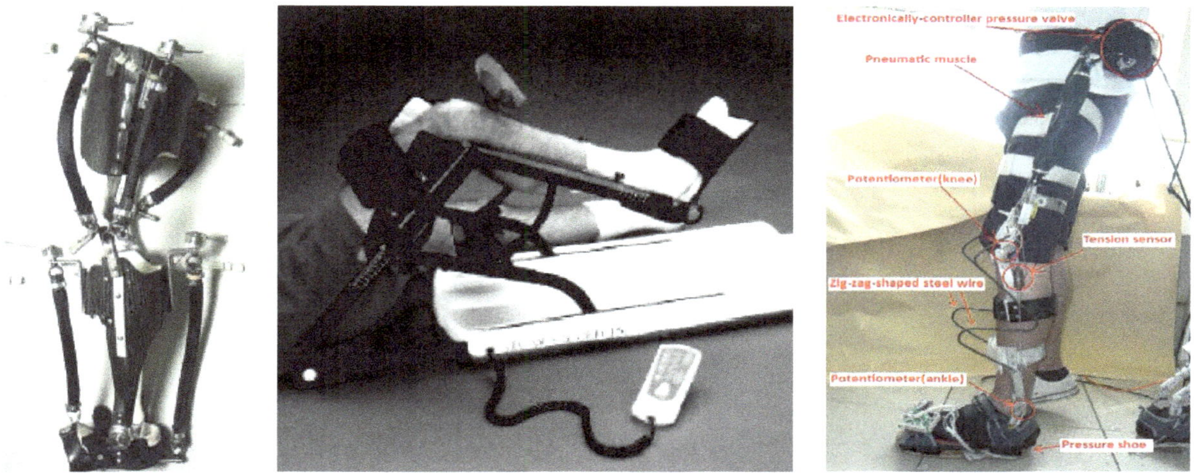

(**a**) (**b**) (**c**)

New high performance devices are required for applying continuous passive rehabilitation training for post-traumatic disabilities regarding the bearing joints of the inferior limbs; therefore, the introduction of a stationary gait and ankle trainer, known as continuous passive motion (CPM), was based on the rehabilitation system illustrated in Figure 6b [44]. This system was invented by Tudor *et al.* in 2009, using pneumatic muscles as the actuation system for providing a low-cost rehabilitation system. With the lower limb being immobilized during the rehabilitation (patient lying on a bed), it allows for the hip and knee joints to perform recovery exercises. When compared to the electro-mechanically-actuated rehabilitation system, which causes discomfort for the users, due to the introduction of shocks upon the reversion of the sensing of motion, this system utilizes a source of energy, namely air, which enables the shocks that occur to be completely absorbed.

Figure 6c shows a power-assist lower-limb orthosis, proposed by Yeh *et al.* in 2010, for assisting the elderly and individuals suffering from sport injuries with walking or climbing stairs using McKibben pneumatic muscles as the actuation system [45]. For achieving better tracking performance, an inverse control for the feed-forward compensation is constructed using the hysteresis model, which is then combined with loop transfer recovery (LTR) feedback control. In addition to ensuring smooth switching between different phases during operation, bump-less switching compensators are implemented in the combined control system. Based on the research findings, it was demonstrated that the orthosis was able to effectively accomplish the assistive function of human locomotion during walking and climbing stairs.

Moreover, the two-degrees of freedom active ankle-foot orthosis (AAFO) was designed and manufactured in 2011 by Carberry *et al.* for post stroke rehabilitation, exemplified in Figure 7a [46]. By implementing a novel actuator linkage using air muscles, a lightweight and discrete orthosis system was achieved. This design enables the entire actuation system to be placed behind the leg of the orthosis. A feedback control that utilizes a fuzzy logic gait phase detection system is implemented with the use of two types of sensory devices: the first is force sensitive resistors (FSRs), located under the insole of the shoe; the second is the rotary encoder for measuring the angular displacement of the ankle joint. However, it is unlikely that suitable methods of supplying air pressure to the device can be found, even though this system exhibits many desirable features. This system may well be beneficial to after-stroke patients, as it allows a more complete rehabilitation of the ankle joint.

Figure 7. (**a**) Active ankle-foot orthosis (AAFO) [46]; (**b**) bio-inspired active soft orthotic for ankle-foot pathologies [47]; and (**c**) active modular elastomer for soft wearable assistance robots [48].

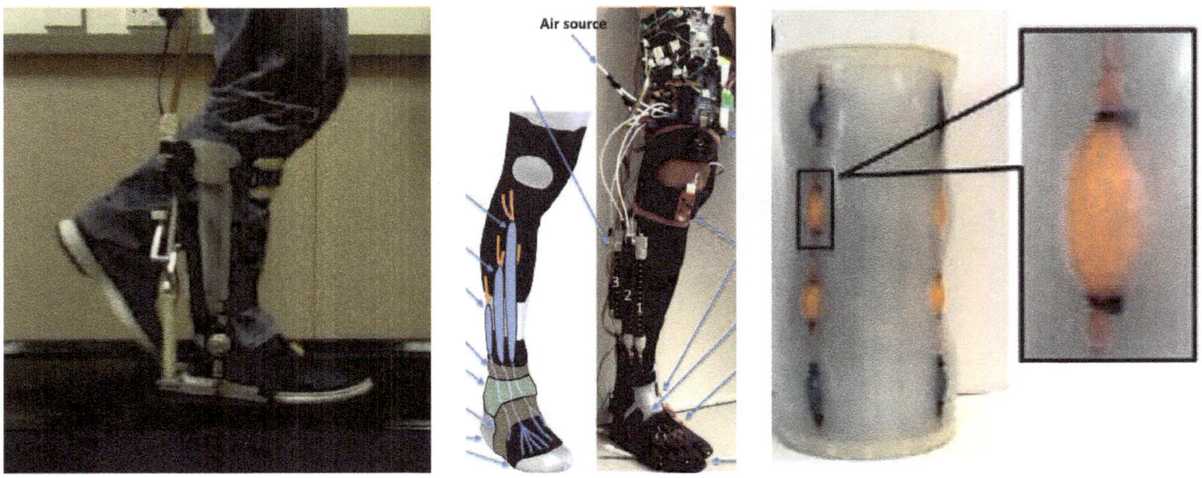

(**a**) (**b**) (**c**)

In 2011, a bio-inspired active soft orthotic device for ankle foot pathology was developed by Park *et al.* for treating gait pathologies associated with neuromuscular disorders, as shown in Figure 7b [47]. By utilizing the advantages of the pneumatic artificial muscle actuators, an inspired biological musculoskeletal system with a muscle-tendon-ligament structure was introduced as the design of this orthosis system. Three types of sensors are used for the control system: the first is a strain sensor for measuring ankle joint angle changes; the second is an internal measurement unit (IMU) to measure the orientations of the lower leg and the foot; and the third is a pressure sensor to identify the foot ground contacts and gait cycle events. The implemented feed-forward and feedback controllers were able to demonstrate a good repeatability of the ankle joint angle control. Based on the outcomes of the result, this research is believed to be capable of providing rich spaces for growth for rehabilitation techniques for ankle pathologies in the near future.

Furthermore, in 2012, Park *et al.* also developed another lower-limb rehabilitation orthosis, known as the active modular elastomer sleeve for soft wearable assistance robots, to support and monitor human joint motions, as illustrated in Figure 7c [48]. With a different system design proposal, this orthosis device implements a series of miniaturized pneumatically-powered McKibben-type actuators. These actuators are wrapped in between monolithic elastomer sheets, so as to exert tension. Through shape and rigidity control, simultaneous motion sensing and active force response are allowed by wrapping the material around the joint. The muscle contractions for the actuators are measured by placing the hyper-elastic strain sensor perpendicularly to the axial direction of each corresponding actuator. This strain sensor will detect the radial expansion of each actuator, which is then transformed to the contraction length of the muscle actuator. Based on the preliminary study of this device system, a few improvements should still be made within the design structure and control system.

Figure 8a presents a developed, inexpensive, pneumatically-powered assisted knee-ankle-foot orthosis (KAFO), using McKibben actuators, for providing assistance during gait training; proposed by Teng *et al.* in 2012 [49]. To determine the relationship between the inclination angles of each joint with pneumatic muscle displacement, the equation is expressed by using a trigonometry method; employed in the control system algorithm and strategy. However, this lower-limb orthosis system is still in the early development stage of design improvement; therefore, further evaluation on system performance has yet to be concluded.

Figure 8. (**a**) Inexpensive KAFO [49]; (**b**) orthosis for walking assistant [50]; and (**c**) a six degrees of freedom (DOF) robotic orthosis for rehabilitation [52].

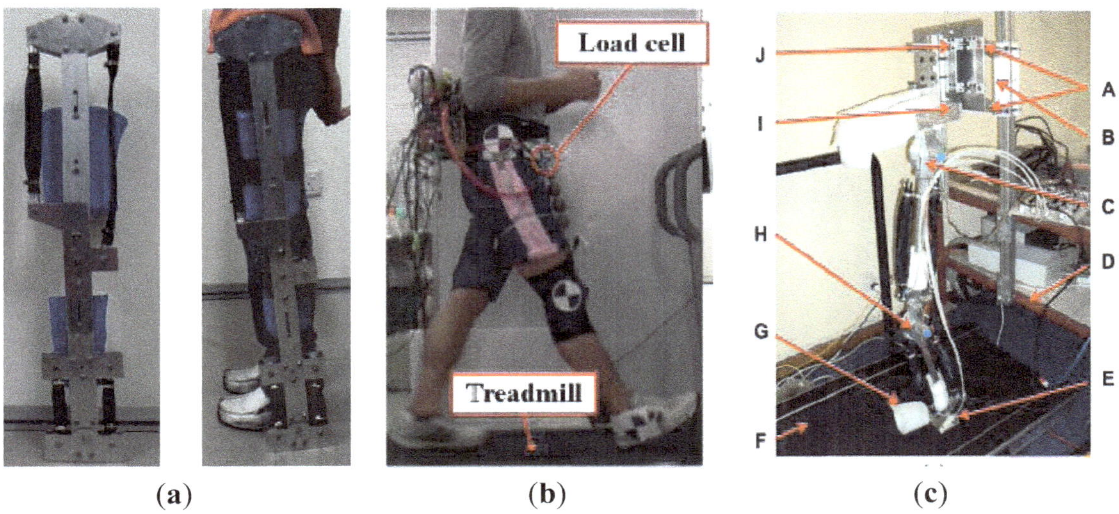

In 2013, Kawamura *et al.* initiated the development of an orthosis for walking assistance. It is designed using straight fiber pneumatic artificial muscles in assisting the forward swing of the leg and increasing the step length to further recuperate patients' walking abilities, as illustrated in Figure 8b [50]. The pressure control unit is implemented using the developed dual pneumatic control system (DPCS) by manipulating the pulse-width modulation (PWM) signal to control the valve. This orthosis system has yet to reach its completion and requires further improvements in its control scheme and strategy when handling the nonlinearity behavior of the actuator. The assistant force generated by the orthosis system is not adequate enough for driving the intended task.

Recently, in 2013, Hussain *et al.*, invented a six degree of freedom robotic orthosis for gait rehabilitation to encourage patients' voluntary contribution in the robotic gait training process, as shown in Figure 8c [51,52]. It implements four pneumatic muscle actuators, which are arranged as two pairs of antagonistic mono-articular muscles at the hip and knee joint angles. This system integrates the AAN gait training algorithm based on the adaptive impedance control, employing a boundary-layer-augmented sliding mode control (BASMC)-based position controller, to afford an interactive robotic gait training system. It was proven that the implementation of the adaptive impedance control scheme is able to provide gait motion training that is comparable to the one provided by physical therapists. Additionally, the result findings demonstrated that an increase/decrease in the human's voluntary participation during gait training will result in a decrease/increase of robotic assistance.

Table 1. Comparison of existing pneumatic muscle-actuated lower-limb rehabilitation orthosis systems.

Comparison of Existing Pneumatic Muscle Actuated Lower-Limb Rehabilitation Orthosis Systems						
Orthosis System	**Time Scale**	**Robotic System Types**	**Actuators**	**Antagonistic Actuators**	**Control System**	**References**
Hip orthosis exoskeleton	2004	Hip orthoses	McKibben pneumatic muscle	Mono-articular for hip joint (flexion)	Position control using the potentiometers for activating the control valves	[31,32]
Robotic gait trainer (RGT)	2006	Foot orthoses	Lightweight spring over muscle (SOM)	Mono-articular for ankle joint (dorsiflexion)	Angular position control system	[33]
Ankle-foot orthosis (AFO)	2006	Foot orthoses	McKibben pneumatic muscle	Mono-articular for ankle joint (dorsiflexion and plantar-flexion)	Proportional myoelectric control using a PC-based controller	[34–36]
Powered lower-limb orthosis	2006	Treadmill gait trainers	Pneumatic muscle actuators (PMA)	Mono-articular for hip joint (flexion, extension, abduction and adduction), knee joint (flexion and extension) and ankle joint (dorsiflexion and plantar-flexion)	Intelligent embedded control mechanism (a three-level PID joint torque control scheme)	[37]
Robotic gait trainer in water (RGTW)	2008	Over-ground gait trainers with orthosis	McKibben pneumatic muscle	Mono-articular for hip joint (flexion and extension) and knee joint (flexion and extension)	Position control system	[38]
Powered ankle-foot exoskeleton	2009	Foot orthoses	Pneumatic artificial muscle (PAM)	Mono-articular for ankle joint (dorsiflexion and plantar-flexion)	Electromyography (EMG) control with feed-forward algorithm	[39–42]
Powered knee-ankle-foot orthosis (KAFO)	2009	Knee and foot orthoses	McKibben pneumatic muscle	Mono-articular for knee joint (flexion and extension) and ankle joint (dorsiflexion and plantar-flexion)	Physiological-inspired controller using electromyography	[43]
Continuous passive motion (CPM)	2009	Stationary gait and ankle trainers	Pneumatic artificial muscle (PAM)	–	–	[44]

Table 1. *Cont.*

Comparison of Existing Pneumatic Muscle Actuated Lower-Limb Rehabilitation Orthosis Systems						
Orthosis system	**Time Scale**	**Robotic System Types**	**Actuators**	**Antagonistic Actuators**	**Control System**	**References**
Power-assist lower-limb orthosis	2010	Over-ground gait trainers (mobile)	McKibben pneumatic muscle	Mono-articular for knee joint (extension)	Inverse control and loop transfer recovery (LTR) feedback control	[45]
Active ankle-foot orthosis (AAFO)	2011	Foot orthoses	McKibben pneumatic muscle	Mono-articular for ankle joint (plantar-flexion)	Feedback control that utilizes a fuzzy logic gait phase detection system	[46]
Bio-inspired active soft orthotic device	2011	Foot orthoses	Pneumatic artificial muscle (PAM)	Mono-articular for ankle joint (dorsiflexion, inversion and eversion)	Feed-forward and feedback controllers	[47]
Active modular elastomer sleeve for soft wearable assistance robots	2012	Knee orthoses	Miniaturized McKibben pneumatic muscle	Mono-articular for knee joint (flexion and extension)	Through shape and rigidity control	[48]
Knee-ankle-foot orthosis (KAFO)	2012	Knee and foot orthoses	Pneumatic artificial muscle (PAM)	Mono-articular for hip joint (flexion and extension) and knee joint (flexion and extension)	–	[49]
Orthosis for walking assistant	2013	Hip orthoses	Straight fiber pneumatic artificial muscle (PMA)	Mono-articular for hip joint (flexion)	Dual pneumatic control system (DPCS) with a pulse-width modulation (PWM) signal	[50]
Six degree of freedom robotic orthosis for gait rehabilitation	2013	Treadmill gait trainers	McKibben pneumatic muscle	Mono-articular for hip joint (flexion and extension) and knee joint (flexion and extension)	Adaptive impedance control using boundary-layer-augmented sliding mode control (BASMC)	[51,52]

Table 1 shows the comparison of existing pneumatic muscle actuated lower-limb rehabilitation orthosis systems. Based on the evaluations of these systems for the past 10 years, it can be concluded that researchers' interests shifted to the implementation of the natural compliant-type actuators (*i.e.*, McKibben muscle, rubbertuators, air muscle, PAM, PMA, *etc.*). This was proven by the development of different types of assistive gait rehabilitation orthosis system prototypes, including foot orthoses, hip orthoses, knee-foot orthoses, stationary gait and ankle trainers, over-ground gait trainers with orthoses, mobile over-ground gait trainers and treadmill gait trainers [31–52]. In addition, the improvement of the control system implementation, since the year 2004 up until 2013, showed that researchers were gradually trying to improve the control of pneumatic muscle-actuated lower-limb orthoses, as illustrated in Table 1. In the beginning, only a simple angular position control was proposed to activate the control valves. Later, it was shifted to the implementation of proportional myoelectric control, intelligent embedded control, inverse control, feedback control (which utilized a fuzzy logic), rigidity control and, subsequently, adaptive impedance control. The exponential growth of these systems might also be due to the advantageous attributes of the pneumatic muscle actuator, as well as its nonlinear dynamic behavior. However, according to the evaluations of currently existing systems, it could be understood that suitable control schemes and strategies have yet to be found. Regardless, this only suggests that the space available for orthosis device improvements and enhancements, in either mechanical design or control scheme and strategy, is still boundless. This opportunity will attract researchers' interest in devising distinctive ideas and strategies to rectify previous methods or to discover new methods for the control system. Even though many different robotic system types for lower-limb rehabilitation orthoses have been developed, each prototype only implements the use of mono-articular muscles alone, either for hip, knee or ankle joints (*i.e.*, flexion, extension, abduction, adduction, plantar-flexion, dorsiflexion, inversion, eversion, *etc.*). However, no attempt has been made to introduce the implementation of bi-articular muscles, either to compensate for the lack of force/torque at the joints or to improve the control scheme and strategy performance.

3. Control Scheme and Strategy

The need for improved control strategies in handling the antagonistic actuator of pneumatic muscles will determine the progression of growth in lower-limb rehabilitation orthosis systems. Based on previous research, it is possible to utilize a standard PID controller in a feedback loop to control the joint angle of the assistive robotics within desired values. Nevertheless, without additional model paradigms or integrated controllers, it may not be able to accurately control a compliant robotic system, due to the complex and highly nonlinear dynamics of the pneumatic muscle. Thus, the resulting position control would be rather poor. For that reason, the implementation of conventional PID controllers should come with additional control strategies, such as additional model paradigms, auto-tuning, a nonlinear system, adaptive control, intelligent control (*i.e.*, neural network, fuzzy logic, genetic algorithm, *etc.*), robust control and stochastic control. A control scheme and strategy that enables a much simpler approach for the control system implementation in orthotic rehabilitation robotics is strongly desired. Therefore, in this review article, the implementation of co-contraction controls in manipulating the antagonistic actuators and the advantages will be discussed and elaborated thoroughly.

3.1. Pneumatic Muscle Actuators' Control System

Even though numerous control systems have been established for pneumatic actuators, especially pneumatic cylinders, only a fraction have been for artificial pneumatic muscles. From 1993 to 1995, some examples of well-known controllers that could be implemented, adopted by Caldwell *et al.*, were tested on a feed-forward PID regulator to develop an adaptive controller for a pneumatic artificial muscle (PAM) manipulator [53–55]. Likewise, in 1995, Gustavo *et al.* developed an adaptive position control for antagonistic pneumatic muscle actuators via adaptive pole-placement [56]. Furthermore, in 1995, Hamerlain *et al.* introduced a variable structure control that included a high robust performance, with respect to model errors, parameter variations and quick responses [57]. Within the same year, Iskarous *et al.* proposed intelligent control using a neuro-fuzzy network to control the complex dynamic properties of muscle actuators [58]. In 1996, van der Smagt *et al.*, introduced a neural network-based controller to a pneumatic robot arm; with complex, highly nonlinear dynamics that change over time, due to internal influences [59]. Additionally, in 1996, Cai and Yamaura presented a robust tracking control approach by implementing a sliding mode controller [60]. Within the same year, Colin *et al.* proposed position and PID controllers for force manipulation using adaptive pole-placement techniques [61].

Afterwards, in 1999, Repperger *et al.* handled the nonlinear factor with a nonlinear feedback controller, using a gain scheduling method [62]. Tondu and Lopez also employed a sliding-mode control approach in the year 2000 [22]. Contrarily, Carbonell *et al.* introduced nonlinear control of a pneumatic muscle actuator by using adaptive back-stepping and sliding-mode tracking controllers in 2001 [63,64]. In 2003, Folgheraiter *et al.* developed an adaptive controller based on a neural network for an artificial hand [65]. In the same year, Balasubramanian and Rattan proposed the feed-forward control of a nonlinear pneumatic muscle system using fuzzy logic [66]. From 2004 to 2006, Ahn and Tu proposed an intelligent switching control scheme by utilizing a learning vector quantization neural network and a nonlinear PID control to improve the control performance of a PAM manipulator using a neural network (NN) [67,68]. In 2008, Harald *et al.*, developed the cascade sliding mode (SM) control scheme for a high-speed linear axis pneumatic muscle [69]. Moreover, Seung *et al.* proposed a trajectory tracking control using a neural network based on PID control in 2009 [70]. In 2010, Xing *et al.* introduced the tracking control of pneumatic artificial muscle actuators based on a sliding-mode and non-linear disturbance observer (SMCBNDO) in order to improve the robustness and performance of the trajectory tracking control [71].

Unfortunately, applying a complicated control algorithm does not always indicate the best solution used to control pneumatic muscles. There is an argument in the field of rehabilitation robotics regarding what is the best control system of the orthotic problem for rehabilitation. It is preferred that control systems be simplified as much as possible; multiple sensors and impedances only increase the complexity of control systems. Rather than using a very complicated algorithm for a system, a much simpler approach may be proposed.

3.2. Co-Contraction of Antagonistic Muscle Control

An early study of the co-contraction of antagonist muscle control was carried out by Neville Hogan in 1984, which introduced the adaptive control of mechanical impedance by co-activation of antagonist muscles [72]. This research study focused on biomechanical modeling and the analysis of simultaneous co-activation of antagonist muscles by controlling the mechanical impedance. A dynamic optimization theory was used to obtain a prediction of antagonist co-activation, thus enabling a criterion function minimization, which represented the task of maintaining an upright posture. Based on the research findings, it was concluded that under normal psychological conditions, significant levels of the simultaneous activation of antagonist muscles were observed. In addition, the levels of antagonist muscles co-activation were also increased with the increment of gravitational torques. The modeled isometric muscle torque is represented in the following:

$$T_{biceps} = (T_o - K_{QS}\theta)u_{biceps} \tag{1}$$

$$T_{triceps} = -(T_o + K_{QS}\theta)u_{triceps} \tag{2}$$

$$(u) \text{ is the neural control} \begin{cases} 0 \leq u_{biceps} \leq 1 \\ 0 \leq u_{triceps} \leq 1 \end{cases} \tag{3}$$

Joint stiffness at maximum activation is:

$$\left(0 \leq K_{QS} \leq {}^{2T_o}\!/_{\pi}\right) \tag{4}$$

where (T_o) is the maximum isometric muscle torque.

Subsequently, in 1988, William R. Murray *et al.* carried on this research by implementing a simple model demonstrating the quasi-static behavior of skeletal muscles, in which the force generated by the muscle was the neural activation of the muscle and the bilinear function of the muscle length [73,74]. This muscle activation could be defined as the synchronized activation of agonist and antagonist muscle groups, acting in the same plane and crossing at the same joint. It was verified that the relationship between antagonistic actuators (*i.e.*, agonist and antagonist) could be linearly related in the occurrence of various fixed levels of co-contractions. In other words, the plane of agonist and antagonist muscle activity, the "equilibrium line" or the locus of all feasible levels of muscular activation, will be a straight line for which a particular equilibrium position is sustained. In addition, the intercepts and slopes of these equilibrium lines are such that the expected levels of muscular activation are counterintuitive. This explained why the anterior activation levels were higher than the posterior activation levels for all, regardless of how low the levels of muscular activity were.

Since then, numerous research studies have been implemented on the co-contraction of antagonistic muscle control, which have proven its ability to increase the stiffness and stability at the joints during volitional movements [75–86]. Based on these research studies, it was shown that by utilizing information from the antagonistic muscle co-contraction, muscular activation levels could be manipulated to control the movements of the joints. Recently, in 2013, Klauer *et al.* introduced the nonlinear joint-angle feedback control of electrical stimulated and λ-controlled antagonistic muscle pairs, in order to control the human limb movements in neural-prosthetic systems [87,88]. The desired recruitment levels, λ, of both muscles were estimated using the electrical stimulation-evoked electromyography (EMG) measurements. The proposed controller enabled the tracking of reference joint torques and predefined muscular co-contraction using exact linearization methods. Based on the outcomes of the result, the control system was able to rapidly compensate for muscle fatigue and then change the muscular thresholds. It could be said that this is a prerequisite for a neural-prosthetic system's practical application within clinical environments. The asymptotically stable system for the torques is depicted in the following:

$$T_i(k) = k_{s,i}\left(\theta_{max,i} - \theta(k)\right)\left(\frac{q^{-2}(1-a)}{1-aq^{-1}}\right)\left(\frac{1-b_i}{1-b_i q^{-1}}\right) r_{\lambda_i} \tag{5}$$

where (λ_i) is the muscular recruitment level, (r_{λ_i}) is the desired recruitment level, (q^{-1}) is the backward shift operator, (q^{-2}) is the delay of two sampling steps and (k) is the sampling index.

$$\theta \in \left[\theta_{max,1}, \theta_{max,2}\right] \tag{6}$$

$$a \in [0, 1] \tag{7}$$

$$b_i \in [0, 1] \tag{8}$$

3.3. Simulation of the Co-Contraction Model for Antagonistic Muscles

In recent years, plenty of research studies have been carried out on assistive robotics for rehabilitation, either using motors or pneumatic muscle actuators for the robotic system's source of power [5–8]. Consequently, these studies have become the basis for many findings. Famous researchers in this field, such as Daniel Ferris, have mentioned that powered orthoses could assist the task-specific practicing of the gait, with the long-term goal of improving patient's inherent locomotor capabilities [89]. According to Kalyan K. Mankala and Sunil K. Agrawal *et al.*, passive swing assistance was able to assist patients, with less than ordinary muscle strength, to attain better gait trajectories [90]. Furthermore, analyses on the implementation of mono- and bi-articular actuators for achieving the high muscle moment required at the joints and better gait trajectories were also taken into consideration in real practice [91–95]. The study of antagonistic muscle co-contraction suggested that the control of the orthosis, which implements these mono- and bi-articular actuators, could achieve good joint stiffness and stability [75–86]. The design was biologically inspired (by human muscles), as it employed two compliant elements to manipulate the joints. Usually, this type of orthosis system, implemented antagonistically, actuated joints using the pneumatic-type muscle actuators. In addition, the co-contraction activations were also able to reduce the kinematic variability; whereby, through the increment of co-contraction activations, the kinematic variability could be reduced with the exception of the low co-contraction activation levels [96]. Therefore, it could be concluded that the modeling of co-contraction to represent the movement of antagonistic actuators may be beneficial.

An early study of the co-contraction model was proposed by William K. Durfee *et al.* in 1989. They developed task-based methods for evaluating electrically-simulated antagonist muscle controllers in a novel animal model [97]. The stimulus activation levels of two antagonist muscles, which manipulated an anesthetized cat's intact ankle joint, were determined by the controller output. In this study, three types of controllers were evaluated: the first was open loop reciprocal control; the second was P-D closed loop reciprocal control; and the third was open loop co-contraction control (Figure 9). Based on the results of the analysis, it was shown that in the visual feedback, the performance of the open loop co-contraction control was comparable to the performance of the P-D closed loop control. This suggested that, in some cases of clinical neural prosthesis implementation, the feedback controller may not be required for good control system performance. In addition, these results also suggested the importance of co-contraction for position control tasks in neural prostheses. However, the disadvantage of this control scheme was that it required more than one input command for each degree of freedom of motion, which could cause premature muscle fatigue.

Figure 9. (**a**) Open loop reciprocal control; (**b**) P-D closed loop reciprocal control; and (**c**) open loop co-contraction control [97].

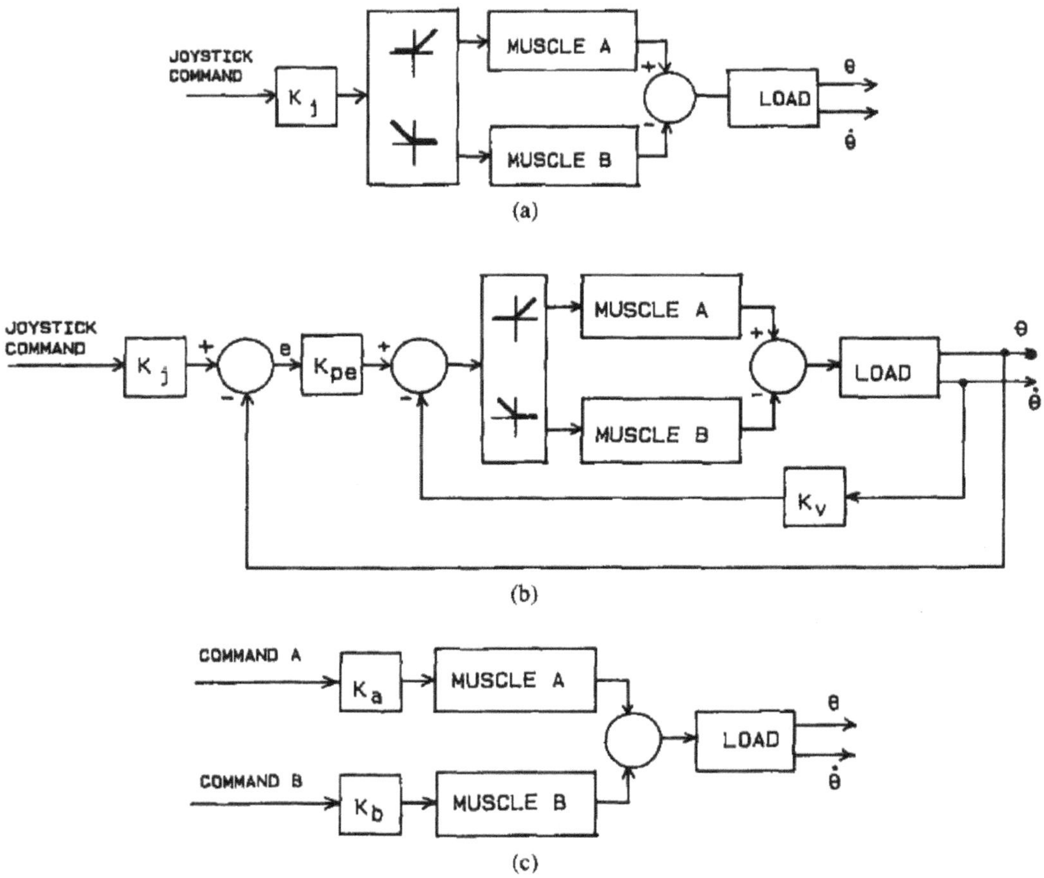

A simulation study of the co-contraction model control scheme for simultaneously manipulating antagonistic actuators was reinitiated by Mohammed *et al.* in 2005. It was mentioned in their study of a co-contraction muscle control strategy for paraplegics that the co-contraction of antagonistic muscle functions (basically, quadriceps and hamstrings) is not necessarily restricted to opposing motion, but may yield to increasing joint stiffness and stable movements [98]. The magnitude of antagonistic muscle co-contractions was first determined based on the optimization of the static linear constraints of muscle forces acting on the joint; whereby, the redundancy of two muscles in co-contraction (*i.e.*, agonist and antagonist) spanning the joint was resolved using the linear minimization of the total stress in the antagonistic muscles. Afterwards, the relationship between the amounts of muscle co-contractions and the maximum force for the antagonistic muscle actuators was computed by implementing weight factors. However, to ensure the robustness and the safe movement of the orthosis, due to the nonlinearity and the presence of a second order system, a high order sliding mode (HOSM) controller was implemented. In addition, Mohammed *et al.* continued their research in 2010 by introducing an inverse model that considered the muscular dynamic contraction of muscle actuators [99]. This dynamic contraction consisted of two main components: the first was activation dynamics, and the second was contraction dynamics (*i.e.*, force-length and force-velocity relationships). However, the activation dynamics was neglected, as its role was assumed to not be essential during the optimization. The inability of most optimization models to compute muscle co-contractions may be caused by the utilization of monotonous increment objective functions that penalize every additional increment of muscle force. The amount of co-contraction muscle forces (*i.e.*, quadriceps and hamstrings) was derived as follows:

$$F_q = \xi_q F_{max,q} + r_q F^2{}_{max,q} \left(\frac{M - \xi_q \sum_q (r_q F_{max,q})}{\sum_q (r_q F_{max,q})^2} \right) \tag{9}$$

$$F_h = \xi_h F_{max,h} + r_h F^2{}_{max,h} \left(\frac{M - \xi_h \sum_h (r_h F_{max,h})}{\sum_h (r_h F_{max,h})^2} \right) \tag{10}$$

The constraints are:

$$\begin{cases} 0 \le F_i \le F_{max,i} \\ \sum_i r_i F_i = M \end{cases} \quad (i = q, h) \tag{10}$$

(ξ_q) and (ξ_h) are the weight factors;
(F_{max}) is the maximum isometric muscle force;
(r) is the radius.

Subsequently, a simulation research study was instigated by Heitmann *et al.* in 2012 on the muscle co-contraction of a three-link biomechanical limb that modulates the damping and stability of the joints. This study was conducted to replicate the natural relationship, without information on anatomical detail, between the muscle activation and joint dynamics [100]. It was proven that the muscle co-contraction was able to alter the damping and the stiffness of the limb joint without altering the net joint torque, and its effect was incorporated into the model by attaching each manipulator joint with a pair of antagonist muscles. These muscles could be activated individually with each other using ideal mathematical forms of muscle co-contraction. This mathematical equation was derived from natural force-length-velocity relationships of contractile muscle tissue. From the simulation result and numerical stability analysis, it was proven that the damping in the biomechanical limb had increased consistently with the human motor control observation. Moreover, it was also revealed that under identical levels of muscle co-contraction, the bi-stable equilibrium positions could co-exist when the opponent muscles were configured with asymmetric contractile element force-length properties. There were two implications of these findings: the first was the practical implication for the nonlinear bio-mimetic actuator design; and the second was the theoretical implication of the biological motor control, which presumes that antagonist muscle systems are universally mono-stable.

In 2011, Kawai *et al.* had also instigated a simulation study for manipulating antagonistic mono- and bi-articular muscle actuators using a co-contraction-based model [101]. The purpose of this simulation study was to verify the proposed passivity-based control for two degrees of freedom (2 DOF) for human arm manipulators. The bi-articular manipulator dynamics for three muscle torques (*i.e.*, two pairs of mono-articular and a pair of bi-articular actuators) was constructed in order to design the control inputs for the system. The important property of passivity was used to examine the stability analysis of the proposed control law, even though the bi-articular manipulator dynamics passivity could not be determined based on the antagonistic bi-articular muscles. Afterwards, in 2012, Sano, K.; Kawai, H. *et al.* proposed a simulation study of the same 2 DOF manipulator systems using open loop control [102]. Compared to their previous simulation study, the Lyapunov method was used to examine the stability analysis of the proposed control law. However, the anticipated approach did not coincide with the bi-articular manipulator dynamic's uncertainties. This simulation study was then extended to a robust control method that enabled semi-global asymptotic tracking, using RISE control, due to an uncertain nonlinear model of the lower limb of the human body, in 2013 [103]. The results showed that the lower limb was able to be positioned in the desired trajectories in the presence of un-modeled bounded disturbances. However, the torque generated at the knee joint was less when compared to their previous method, due to the antagonistic bi-articular muscles. The contractile force of the flexor muscle (u_{fi}) and extensor muscle (u_{ei}) was derived as follows:

$$T_i = (u_{ei} - u_{fi})l_p - (u_{ei} + u_{fi})k_i {l_p}^2 q_i + (u_{e3} - u_{f3})l_p - (u_{e3} + u_{f3})k_3 {l_p}^2 (q_1 + q_2) \quad (11)$$

where ($i = 1, 2$), (l_p) and (k_i) are the radius of the joints;

(q_1) and (q_2) are the hip and knee joint angles;
(u_{e1}) and (u_{f1}) are the antagonistic mono-articular muscles for the hip joint;
(u_{e2}) and (u_{f2}) are the antagonistic mono-articular muscles for the knee joint;
(u_{e3}) and (u_{f3}) are the antagonistic bi-articular muscles.

Within the same year (2013), Kawai *et al.* also proposed a design of the co-contraction level of antagonistic muscles with muscle contraction dynamics for tracking the control of human lower limbs [104,105]. The manipulation of the antagonistic muscle co-contraction level was dependent on the angular velocity of human lower limbs. Based on the research findings, it could be verified that the co-contraction of antagonist muscles played an important role in the joint's stiffness and stability. In addition, the muscle co-contraction was not only useful for compensating for the joint's stiffness and stability, it was also able to maneuver the direction of the output force.

3.4. Co-Contraction Model for Antagonistic Actuators

Numerous studies have been investigated regarding the co-contraction movements of human antagonistic muscles. However, their model implementations in controlling the antagonistic muscle actuators of lower-limb orthoses have not been completely discovered. In addition, any research paper that focuses on the implementation of mono-articular and bi-articular muscle actuators using pneumatic muscles for the lower-limb rehabilitation orthosis has yet to be extensively investigated; thus, simply actuating the actuators may not give a good result for the joint's stiffness and the stability of the lower-limb leg orthosis and its position trajectory. Therefore, based on the evaluation and suggestion of the related research findings, the simultaneous co-contraction movements between the agonist and antagonist muscle actuators should be considered with respect to the control system.

4. Conclusions

In this review article, the evaluation and comparison of the developed lower-limb rehabilitation orthoses using the pneumatic muscle-type actuators, including the control algorithms and strategies intended to provide stiffness and stability with respect to the control system, were reviewed. Although a considerable amount of work is now complete, the field is still rapidly evolving. The issue of which is the most effective control algorithm is still widely open. However, randomized controlled trials are necessary for identifying suitable control algorithms, even though this is expensive and time consuming. In conclusion, a few remarks about suggestions for future research of pneumatic muscle-actuated gait trainer system are as follows: firstly, the pneumatic muscle actuators' arrangement in the lower-limb orthosis should be antagonistic; secondly, the co-contractive movement of the antagonistic pneumatic muscles should provide good stiffness and stability for the leg orthosis system; thirdly, a model paradigm is essential for generating adequate co-contractive input data for manipulating the antagonistic muscle actuators; and finally, the developed model should be managed by controllers to deal with the presence of dynamic properties and the nonlinearity behavior of the system.

Acknowledgments

This work was supported by KAKENHI: Grant-in-Aid for Scientific Research (B) 21300202.

Author Contributions

Mohd Azuwan Mat Dzahir reviewed the related journal and conference papers, summarized the literature reviews and prepared the manuscript; Shin-ichiroh Yamamoto re-evaluated the manuscript.

Conflicts of Interest

The authors declare no conflict of interests.

References

1. Wernig, A.; Muller, S.; Nanassy, A.; Cagol, E. Laufband therapy based on "rules of spinal locomotion" is effective in spinal cord injured persons. *Eur. J. Neurosci.* **1995**, *7*, 823–829.
2. Wernig, A.; Nanassy, A.; Muller, S. Maintenance of locomotor abilities following laufband (treadmill) therapy in para- and tetraplegic persons: Follow-up studies. *Spinal Cord* **1998**, *36*, 744–749.
3. Dietz, V.; Muller, R.; Colombo, G. Locomotor activity in spinal man: Significance of afferent input from joint and load receptors. *Brain* **2002**, *125*, 2626–2634.
4. Dietz, V.; Harkema, S.J. Locomotor activity in spinal cord-injured persons. *J. Appl. Physiol.* **2004**, *96*, 1954–1960.
5. Díaz, I.; Gil, J.J.; Sánchez, E. Lower-limb robotic rehabilitation: Literature review and challenges.

J. Robot. **2011**, *2011*, 759764:1–759764:11.

6. Marchal-Crespo, L.; Reinkensmeyer, D.J. Review of control strategies for robotic movement training after neurologic injury. *J. Neuro-Eng. Rehabil.* **2009**, *6*, 20:1–20:15.
7. Pennycott, A.; Wyss, D.; Vallery, H.; Klamroth-Marganska, V.; Riener, R. Towards more effective robotic gait training for stroke rehabilitation: A review. *J. Neuro-Eng. Rehabil.* **2012**, *9*, 1–13.
8. Dollar, A.M.; Herr, H. Lower extremity exoskeletons and active orthoses: Challenges and state-of-the-art. *IEEE Transact. Robot.* **2008**, *24*, 144–158.
9. Colombo, G.; Wirz, M.; Dietz, V. Driven gait orthosis for improvement of locomotor training in paraplegics patients. International Medical Society of Paraplegia. *Spinal Cord* **2001**, *39*, 252–255.
10. Jazernik, S.; Colombo, G.; Morari, M. Automatic gait pattern adaptation algorithms for rehabilitation with a 4-DOF robotic orthosis. *IEEE Transact. Robot. Autom.* **2004**, *20*, 574–582.
11. Lunenburger, L.; Colombo, G.; Riener, R. Biofeedback for robotic gait rehabilitation. *J. Neuro-Eng. Rehabil.* **2007**, *4*, 1:1–1:11.
12. Freivogel, S.; Schmalohr, D.; Mehrholz, J. Improved walking ability and reduced therapeutic stress with an electromechanical gait device. *J. Rehabil. Med.* **2009**, *41*, 734–739.
13. Fisher, S.; Lucas, L.; Thrasher, T.A. Robot-assisted gait training for patients with hemiparesis due to stroke. *Top Stroke Rehabil.* **2011**, *18*, 269–276.
14. West, R.G. Powered Gait Orthosis and Method of Utilizing SAME. U.S. Patent 6689075, 29 January 2004.
15. Veneman, J.; Kruidhof, R.; Hekman, E.; Ekkelenkamp, R.; van Asseldonk, E.; van der Kooij, H. Design and evaluation of the LOPES exoskeleton robot for interactive gait rehabilitation. *IEEE Trans. Neural Syst. Rehabil. Eng.* **2007**, *15*, 379–386.
16. Vallery, H.; Veneman, J.; van Asseldonk, E.; Ekkelenkamp, R.; Buss, M.; van der Kooij, H. Compliant actuation of rehabilitation robots. *IEEE Robot. Autom. Mag.* **2008**, *15*, 60–69.
17. Banala, S.K.; Kim, S.H.; Agrawal, S.K.; Scholz, J.P. Robot assisted gait training with active leg exoskeleton (ALEX). *IEEE Trans. Neural Syst. Rehabil. Eng.* **2009**, *17*, 2–8.
18. Banala, S.K.; Agrawal, S.K.; Kim, S.H.; Scholz, J.P. Novel gait adaptation and neuromotor training results using an active leg exoskeleton. *IEEE/ASME Trans. Mechatron.* **2010**, *15*, 216–225.
19. Monaco, V.; Galardi, G.; Coscia, M.; Martelli, D.; Micera, S. Design and evaluation of NEUROBike: A neuro-rehabilitative platform for bedridden post-strike patients. *IEEE Trans. Neural Syst. Rehabil. Eng.* **2012**, *20*, 845–852.
20. Taherifar, A.; Mousavi, M.; Rassaf, A.; Ghiasi, F.; Hadian, M.R. LOKOIRAN—A novel robot for rehabilitation of spinal cord injury sand stroke patients. In Proceedings of the RSI/ISM International Conference on Robotics and Mechatronics, Tehran, Iran, 12–14 February 2013.
21. Pietrusinski, M.; Cajigas, I.; Severini, G.; Bonato, P.; Mavroidis, C. Robotic gait rehabilitation trainer. *IEEE/ASME Trans. Mechatron.* **2014**, *19*, 490–499.
22. Tondu, B.; Lopez, P. Modelling and control of McKibben artificial muscle robot actuators. *IEEE Control Syst. Mag.* **2000**, *20*, 15–38.
23. Caldwell, D.G.; Razak, A.; Goodwin, M. Braided artificial muscle actuators. In Proceedings of the IFAC, Southampton, UK, 18–21 April 1993; pp. 507–512.
24. Caldwell, D.G.; Medrano-Cerda, G.A.; Goodwin, M. Characteristics and adaptive control of pneumatic muscle actuators for a robotic elbow. In Proceedings of the International Conference

Robotics Automation, San Diego, CA, USA, 8–13 May 1994; Volume 4, pp. 3558–3563.
25. Caldwell, D.G.; Medrano-Cerda, G.A.; Goodwin, M. Control of pneumatic muscle actuators. *IEEE Control Syst. Mag.* **1995**, *15*, 40–48.
26. Caldwell, D.G.; Tsagarakis, N.; Medrano-Cerda, G.A. Biomimetic actuators: Polymeric pseudo muscular actuators and pneumatic muscle actuators for biological emulation. *Mechatronics* **2000**, *10*, 499–530.
27. Davis, S.; Tsagarakis, N.; Canderle, J.; Caldwell, D.G. Enhanced modeling and performance in braided pneumatic muscle actuators. *Int. J. Robot. Res.* **2003**, *22*, 213–227.
28. Davis, S.; Caldwell, D.G. Braid effects on contractile range and friction modeling in pneumatic muscle actuators. *Int. J. Robot. Res.* **2006**, *25*, 359–369.
29. Chou, C.P.; Hannaford, B. Static and dynamic characteristics of McKibben pneumatic artificial muscles. In Proceedings of the IEEE International Conference Robotics Automation, San Diego, CA, USA, 8–13 May 1994; Volume 1, pp. 281–286.
30. Chou, C.P.; Hannaford, B. Measurement and modeling of McKibben pneumatic artificial muscles. *IEEE Trans. Robot. Autom.* **1996**, *12*, 90–102.
31. Nascimento, B.G.; Vimieiro, C.B.; Nagem, D.A.; Pinotti, M. Hip orthosis powered by pneumatic artificial muscle: Voluntary activation in absence of myoelectrical signal. *Artif. Organs* **2008**, *32*, 317–322.
32. Vimieiro, C.B.S.; do Nascimento, B.G.; Nagem, D.A.P.; Pinotti, M. Development of a hip orthosis using pneumatic artificial muscles. In Proceeding of TMSi, São Paulo, Spain, 18–19 July 2005.
33. Bharadwaj, K.; Sugar, T.G. Kinematics of a robotic gait trainer for stroke rehabilitation. In Proceedings of the IEEE International Conference on Robotics and Automation, Orlando, FL, USA, 15–19 May 2006.
34. Ferris, D.P.; Gordon, K.E.; Sawicki, G.S.; Peethambaran, A. An improved powered ankle-foot orthosis using proportional myoelectric control. *Gait Posture* **2006**, *23*, 425–428.
35. Ferris, D.P.; Czerniecki, J.M.; Hannaford, B. An ankle-foot orthosis powered by artificial pneumatic muscles. *J. Appl. Biomech.* **2005**, *21*, 189–197.
36. Gordon, K.E.; Sawicki, G.S.; Fessis, D.P. Mechanical performance of artificial pneumatic muscles to power an ankle-foot orthosis. *J. Biomech.* **2006**, *39*, 1832–1841.
37. Costa, N.; Bezdicek, M.; Brown, M.; Gray, J.O.; Caldwell, D.G. Joint motion control of a powered lower limb orthosis for rehabilitation. *Int. J. Autom. Comput.* **2006**, *3*, 271–281.
38. Miyoshi, T.; Hiramatsu, K.; Yamamoto, S.I.; Nakazawa, K.; Akai, M. Robotic gait trainer in water: Development of an underwater gait-training orthosis. *Disabil. Rehabil.* **2008**, *30*, 81–87.
39. Malcom, P.; Fiers, P.; Segers, V.; van Caekenberghe, I.; Lenoir, M.; de Clercq, D. Experimental study on the role of the ankle push off in the walk-to-run transition by means of a powered ankle-foot-exoskeleton. *Gait Posture* **2009**, *30*, 322–327.
40. Malcom, P.; Segers, V.; van Caekenberghe, I.; de Clercq, D. Experimental study of the influence of the m. tibialis anterior on the walk-to-run transition by means of a powered ankle-foot-exoskeleton. *Gait Posture* **2009**, *29*, 6–10.

41. Galle, S.; Malcom, P.; Derave, W.; de Clercq, D. Adaptation to walking with an exoskeleton that assists ankle extension. *Gait Posture* **2013**, *38*, 495–499.
42. Malcom, P.; Derave, W.; Galle, S.; de Clercq, D. A simple exoskeleton that assist plantarflexion can reduce the metabolic cost of human walking. *PLoS One* **2013**, *8*, 0056137.
43. Sawicki, G.S.; Fessis, D.P. A pneumatically powered knee-ankle-foot orthosis (KAFO) with myoelectric activation and inhibition. *J. Neuro-Eng. Rehabil.* **2009**, *6*, 23:1–23:16.
44. Deaconescu, T.T.; Deaconescu, A.I. Pneumatic muscle actuated equipment for continuous passive motion. *IAENG Trans. Eng. Technol.* **2009**, doi:10.1063/1.3256258.
45. Yeh, T.J.; Wu, M.J.; Lu, T.J.; Wu, F.K.; Huang, C.R. Control of McKibben pneumatic muscles for a power-assist, lower-limb orthosis. *Mechatronics* **2010**, *20*, 686–697.
46. Carberry, J.; Hinchly, G.; Buckerfield, J.; Taylor, E.; Burton, T.; Madgwick, S.; Vaidyanathan, R. Parametric design of an active ankle foot orthosis with passive compliance. In Proceedings of the Computer-Based Medical System (CBMS), Bristol, UK, 27–30 June 2011.
47. Park, Y.; Chen, B.; Young, D.; Stirling, L.; Wood, R.; Goldfield, E.; Nagpal, R. Bio-inspired Active Soft Orthotic Device for Ankle Foot Pathologies. In Proceedings of the International Conference on Robots and Systems (IROS), San Francisco, CA, USA, 25–30 September 2011.
48. Park, Y.; Chen, B.; Majidi, C.; Wood, R.; Nagpal, R.; Goldfield, E. Active Modular Elastomer Sleeve for Soft Wearable Assistance Robots. In Proceedings of the IEEE International Conference on Robots and Systems (IROS), Vilamoura, Portugal, 7–12 October 2012.
49. Teng, C.M.; Wong, Z.Y.; The, W.Y.; Chong, Y.Z. Design and development of inexpensive pneumatically-powered assisted knee-ankle-foot orthosis for gait rehabilitation-preliminary finding. In Proceedings of the International Conference on Biomedical Engineering (ICoBE), Penang, Malaysia, 27–28 February 2012.
50. Kawamura, T.; Takanaka, K. Development of an orthosis for walking assistance using pneumatic artificial muscle-a quantitative assessment of the effect of assistance. In Proceedings of the International Conference on Rehabilitation Robotics, Seattle, WA, USA, 24–26 June 2013.
51. Hussain, S.; Xie, S.Q.; Jamwal, P.K. Adaptive impedance control of a robotic orthosis for gait rehabilitation. *IEEE Trans. Cybern.* **2013**, *43*, 1025–1034
52. Hussain, S.; Xie, S.Q.; Jamwal, P.K. Robust nonlinear control of an intrinsically compliant robotic gait training orthosis. *IEEE Trans. Syst. Man Cybern.: Syst.* **2013**, *43*, 655–665.
53. Caldwell, D.G.; Medrano-Cerda, G.A.; Goodwin, M.J. Braided pneumatic actuator control of a multi jointed manipulator. In Proceedings of the International Conference on Systems, Man and Cybernetics, Le Touquet, France, 17–20 October 1993.
54. Caldwell, D.G.; Medrano-Cerda, G.A.; Goodwin, M. Characteristics and adaptive control of pneumatic muscle actuators for a robotic elbow. In Proceedings of the International Conference on Robotics and Automation, San Diego, CA, USA, 8–13 May 1994.
55. Caldwell, D.G.; Medrano-Cerda, G.A.; Goodwin, M. Control of a pneumatic muscle actuators. *IEEE Trans. Control Syst.* **1995**, *15*, 40–48.
56. Medrano-Cerda, G.A.; Bowler, C.J.; Caldwell, D.G. Adaptive position control of antagonistic pneumatic muscle actuators. In Proceedings of the International Conference on Intelligent Robots and Systems, Pittsburgh, PA, USA, 5–9 August 1995; Volume 1, pp. 378–383.

57. Hamerlain, M. An anthropomorphic robot arm driven by artificial muscles using a variable structure control. In Proceedings of the International Conference on Intelligent Robots and Systems, Pittsburgh, PA, USA, 5–9 August1995; Volume 1, pp. 550–555.
58. Iskarous, M.; Kawamura, K. Intelligent control using a neuro-fuzzy network. In Proceedings of the International Conference on Intelligent Robots and Systems, Pittsburgh, PA, USA, 5–9 August1995; Volume 3, pp. 350–355.
59. Van der Smagt, P.; Groen, F.; Schulten, K. Analysis and control of a rubbertuator arm. *Biol. Cybern.* **1996**, *75*, 433–440.
60. Cai, D.; Yamaura, H. A robust controller for manipulator driven by artificial muscle actuator. In Proceedings of the IEEE International Conference on Control Applications, Dearborn, MI, USA, 15–18 September 1996.
61. Bowler, C.J.; Caldwell, D.G.; Medrano-Cerda, G.A. Pneumatic muscle actuators: Musculature for an anthropomorphic robot arm. *IEEE Colloq. Actuator Technol.* **1996**, *8*, 1–6.
62. Repperger, D.W.; Johnson, K.R.; Phillips, C.A. Nonlinear feedback controller design of a pneumatic muscle actuator system. In Proceedings of the American Control Conference, San Diego, CA, USA, 2–4 June 1999.
63. Carbonell, P.; Jiang, Z.P.; Repperger, D.W. Nonlinear control of a pneumatic muscle actuator: Backstepping *vs.* Sliding-mode. In Proceedings of the IEEE International Conference on Control Applications, Mexico City, Mexico, 5–7 September 2001.
64. Carbonell, P.; Jiang, Z.P.; Repperger, D.W. A fuzzy backstepping controller for a pneumatic muscle actuator system. In Proceedings of the IEEE International Symposium on Intelligent Control, Mexico City, Mexico, 5–7 September 2001; pp. 353–358.
65. Folgheraiter, M.; Gini, G.; Perkowski, M.; Pivtoraiko, M. Adaptive reflex control for an artificial hand. In Proceedings of the SYROCO, Wrocław, Poland, 1–3 September 2003.
66. Balasubramanian, K.; Rattan, K.S. Feedforward control of a non-linear pneumatic muscle system using fuzzy logic. In Proceedings of the IEEE International Conference on Fuzzy Systems, St. Louis, MO, USA, 25–28 May 2003; Volume 1, pp. 272–277.
67. Ahn, K.K.; Thanh, T.C. Improvement of the control performance of pneumatic artificial muscle manipulators using an intelligent switching control method. *KSME Int. J.* **2004**, *18*, 1388–1400.
68. Ahn, K.K.; Thanh, T.C. Nonlinear PID control to improve the control performance of the pneumatic artificial muscle manipulator using neural network. *J. Mech. Sci. Technol.* **2005**, *19*, 106–116.
69. Aschemann, H.; Schindele, D. Sliding-mode control of a high-speed linear axis driven by pneumatic muscle actuators. *IEEE Trans. Ind. Electron.* **2008**, *55*, 3855–3864
70. Cho, S.H. Trajectory tracking control of a pneumatic c-y table using neural network based PID control. *Int. J. Precis. Eng. Manuf.* **2009**, *10*, 37–44.
71. Xing, K.; Huang, J.; Wang, Y.; Wu, J.; Xu, Q.; He, J. Tracking control of pneumatic artificial muscle actuators based on sliding mode and non-linear disturbance observer. *IET Control Theory Appl.* **2010**, *10*, 2058–2070.
72. Hogan, N. Adaptive control of mechanical impedance by coactivation of antagonist muscles. *IEEE Trans. Autom. Control* **1984**, *29*, 681–690.

73. Murray, W.R. Modelling elbow equilibrium in the presence of co-contraction. In Proceedings of the Bioengineering Conference, Durham, NH, USA, 10–11 March 1988; pp. 190–193.
74. Murray, W.R.; Hogan, N. Co-contraction of antagonist muscles: Prediction and observation. In Proceedings of the Joint Dynamics and Control, IEEE International Conference on Engineering in Medicine and Biology Society (EMBS), New Orleans, LA, USA, 4–7 November 1988; Voluem 4, 1926–1927.
75. Migliore, S.A.; Brown, E.A.; de Weerth, S.P. Novel nonlinear elastic actuators for passively controlling robotic joint compliance. *Trans. ASME* **2007**, *129*, 406–412.
76. Schepelmann, A.; Taylor, M.D.; Geyer, H. Development of a testbed for robotic neuromuscular controllers. In *Robotics: Science and Systems VIII*; University of Sydney: Sydney, Australia, 2013.
77. Laffranchi, M.; Tsagarakis, N.G.; Cannella, F.; Caldwell, D.G. Antagonistic and series elastic actuators: A comparative analysis on the energy consumption. In Proceedings of the IEEE/RSJ International Conference on Intelligent Robots and Systems, St. Louis, MO, USA, 10–15 October 2009.
78. Klauer, C.; Raisch, J.; Schauer, T. Advanced control strategies for neuro-prosthetic systems. In Proceedings of the Technically Assisted Rehabilitation (TAR), Berlin, Germany, 14–15 March 2013.
79. Mitrovic, D.; Klanke, S.; Osu, R.; Kawato, M.; Vijayakumar, S. A computational model of limb impedance control based on principles of internal model uncertainty. *PLoS One* **2010**, doi:10.1371/journal.pone.0013601.
80. Dierick, F.; Domicent, C.; Detrembleur, C. Relationship between antagonistic leg muscles co-contraction and body centre of gravity mechanics in different level gait disorders. *J. Electromyogr. Kinesiol.* **2002**, *12*, 59–66.
81. Milner, T.E.; Cloutier, C. The effect of antagonist muscle co-contraction on damping of the wrist joint during voluntary movement. In Proceedings of the IEEE 17th Annual Conference on Engineering in Medicine and Biology Society, Montreal, Canada, 20–23 September 1995; Volume 2, 1247–1248.
82. Hollerbach, K.; Ramos, C.F.; Kazerooni, H. Destabilizing effects of muscular co-contraction in human-machine interaction. In Proceedings of the American Control Conference, San Francisco, CA, USA, 2–4 June 1993.
83. Fu, C.; Wang, R. The influence of co-contraction to the arm impedance during free planar movement. In Proceedings of the International Conference on Bioinformatics and Biomedical Engineering (ICBBE), Beijing, China, 11–13 June 2009; pp. 1–3.
84. Lynch, C.L.; Sayenko, D.; Popovic, M.R. Co-contraction of antagonist muscles during knee extension against gravity: Insights for functional electrical simulation control design. In Proceedings of the IEEE International Conference on Engineering in Medicine and Biology Society (EMBS), San Diego, California, USA, 28 August–1 September 2012.
85. Odhner, L.; Asada, H. Equilibrium point control of artificial muscles using recruitment of many motor units. In Proceedings of the IEEE/RAS International Conference on Biomedical Robotics and Biomechatronics, Scottsdale, AZ, USA 19–22 October 2008.

86. Missenard, O.; Mottet, D.; Perrey, S. The role of cocontraction in the impairment of movement accuracy with fatigue. *Exp. Brain Res.* **2008**, *185*, 151–156.
87. Klauer, C.; Raisch, J.; Schauer, T. Nonlinear joint-angle feedback control of electrically simulated and λ-controlled antagonistic muscle pairs. In Proceedings of the European Control Conference (ECC), Zurich, Switzerland, 17–19 July 2013.
88. Spagnol, P.; Klauer, C.; Previdi, F.; Raisch, J.; Schauer, T. Modelling and online-identification of electrically stimulated antagonistic muscles for horizontal shoulder abduction and adduction. In Proceedings of the European Control Conference (ECC), Zurich, Switzerland, 17–19 July 2013.
89. Ferris, D.P.; Lewis, C.L. Robotic lower limb exoskeletons using proportional myoelectric control. In Proceedings of the IEEE International Conference on Engineering in Medicine and Biology Society (EMBS), Minneapolis, MN, USA, 2–6 September 2009.
90. Mankala, K.K.; Banala, S.K.; Agrawal, S.K. Passive swing assistive exoskeletons for motor-incomplete spinal cord injury patients. In Proceedings of the IEEE International Conference on Robotics and Automation, Roma, Italy, 10–14 April 2007.
91. Kumamoto, M.; Oshima, T.; Fujukawa, T. Control properties of a two-joint link mechanism equipped with mono- and bi-articular actuators. In Proceedings of the IEEE International Workshop on Robot and Human Interactive Communication, Osaka, Japan, 27–29 September 2000.
92. Shimizu, S.; Momose, N.; Oshima, T.; Koyanagi, K. Development of robot leg which provided with the bi-articular actuator for training techniques of rehabilitation. In Proceedings of the IEEE International Symposium on Robot and Human Interactive Communication, Toyama, Japan, 27 September–2 October 2009.
93. Salvucci, V.; Oh, S.; Hori, Y. Infinity norm approach for precise force control of manipulators driven by bi-articular actuators. In Proceedings of the Annual Conference on IEEE Industrial Electronics Society, Glendale, AZ, USA, 7–10 November 2010; pp. 1908–1913.
94. Salvucci, V.; Oh, S.; Hori, Y.; Kimura, Y. Disturbance rejection improvement in non-redundant robot arms using bi-articular actuators. In Proceedings of the Industrial Electronics (ISIE), Gdańsk, Poland, 27–30 June 2011; pp. 2159–2164.
95. Salvucci, V.; Kimura, Y.; Oh, Y.; Hori, Y. Experimental verification of infinity norm approach for force maximization of manipulators driven by bi-articular actuators. In Proceedings of the American Control Conference on O'Farrell Street, San Francisco, CA, USA, 29 June–1 July 2011.
96. Selen, L.P.J.; Beek, P.J.; van Dieen, J.H. Can co-activation reduce kinematics variability? A simulation study. *Biol. Cybern.* **2005**, *93*, 373–381.
97. Durfee, W.K. Task-based methods for evaluating electrically stimulated antagonist muscle controllers. *IEEE Trans. Biomed. Eng.* **1989**, *36*, 309–321.
98. Mohammed, S.; Fraisse, P.; Guiraud, D.; Poignet, P.; el Makssoud, H. Towards a co-contraction muscle control strategy for paraplegics. In Proceedings of the 44th IEEE Conference on Decision and Control, Seville, Spain, 12–15 December 2005; pp. 7428–7433.

99. Mohammed, S.; Poignet, P.; Fraisse, P.; Guiraud, D. Optimal stimulation patterns for knee joint movement restoration during co-contraction of antagonist muscles. In Proceedings of the International Conference on Biomedical Robotics and Biomechatronics, Tokyo, Japan, 26–29 September 2010; pp. 678–692.
100. Heitmann, S.; Ferns, N.; Breakspear, M. Muscle co-contraction modulates damping and joint stability in a three-link biomechanical limb. *Front. Neurorobot.* **2012**, *5*, 5:1–5:14.
101. Kawai, H.; Murao, T.; Sato, R.; Fujita, M. Passivity-based control for 2DOF robot manipulators with antagonistic bi-articular muscles. In Proceedings of the IEEE International Conference on Control Applications (CCA), Denver, CO, USA, 28–30 September 2011; pp. 1451–1456.
102. Kawai, K.S.H.; Murao, T.; Fujita, M. Open-loop control for 2DOF robot manipulators with antagonistic bi-articular muscles. In Proceedings of the IEEE International Conference on Control Applications (CCA), Dubrovnik, Croatia, 3–5 October 2012; pp. 1346–1351.
103. Kawai, Y.; Kawai, H.; Fujita, M. RISE control for 2DOF human lower limb with antagonistic bi-articular muscles. In Proceedings of the IEEE International Conference on Control Applications (CCA), Hyderabad, India, 28–30 August 2013; pp. 109–114.
104. Kawai, Y.; Downey, R.J.; Kawai, H.; Dixon, W.E. Co-Contraction of Antagonist Bi-Articular Muscles for Tracking Control of Human Limb. Available online: http://wwwr.kanazawa-it.ac.jp/kawai/research/2013/CDC13KaDoKaDi.pdf (accessed on 12 March 2014).
105. Kawai, Y. A Design of Co-Contraction Level of Antagonist Muscles with Muscle Contraction Dynamics for Tracking Control of Human Limb. Available online: http://k-lab.e.ishikawa-nct.ac.jp/paper/2013/SICE13_0447_FI.pdf (accessed on 12 March 2014).

4

Computationally Efficient Adaptive Type-2 Fuzzy Control of Flexible-Joint Manipulators

Hicham Chaoui [1], **Wail Gueaieb** [1,*], **Mohammad Biglarbegian** [2] **and Mustapha C. E. Yagoub** [1]

[1] School of Electrical Engineering and Computer Science, University of Ottawa, 800 King Edward Avenue, Ottawa, ON K1N 6N5, Canada; E-Mails: h.chaoui@uottawa.ca (H.C.); myagoub@eecs.uottawa.ca (M.C.E.Y.)

[2] School of Engineering, University of Guelph, Guelph, ON N1G 2W1, Canada; E-Mail: mbiglarb@uoguelph.ca

* Author to whom correspondence should be addressed; E-Mail: wgueaieb@eecs.uottawa.ca

---

Abstract: In this paper, we introduce an adaptive type-2 fuzzy logic controller (FLC) for flexible-joint manipulators with structured and unstructured dynamical uncertainties. Simplified interval fuzzy sets are used for real-time efficiency, and internal stability is enhanced by adopting a trade-off strategy between the manipulator's and the actuators' velocities. Furthermore, the control scheme is independent of the computationally expensive noisy torque and acceleration signals. The controller is validated through a set of numerical simulations and by comparing it against its type-1 counterpart. The ability of the adaptive type-2 FLC in coping with large magnitudes of uncertainties yields an improved performance. The stability of the proposed control system is guaranteed using Lyapunov stability theory.

Keywords: type-2 fuzzy control; uncertain systems; robot manipulators; flexible structures; adaptive control

---

1. Introduction

Flexible-joint manipulators offer several advantages with respect to their rigid counterpart, such as light weight, lower cost, smaller actuators, larger work volume, better manoeuvrability and transportability, higher operational speed, power efficiency, and larger number of applications. Thus, they are often required to operate at high speed to yield high productivity. The conflicting requirements between high speed and high accuracy make the robotic control task a challenging research problem. Reducing the weight of the arms and/or increasing the operation speed make many industrial flexible-joint manipulators face arm vibration problems, particularly in high speed motion, because of the low stiffness. This can be resolved by increasing the stiffness. However, it increases the mass, depleting the advantages listed above. Typical challenges include severe friction nonlinearities, coupling stemming from the manipulator's flexibility, varying operating conditions, structured and unstructured dynamical uncertainties, and external disturbances.

Flexibility and nonlinear friction may have some destabilizing effects when failing to compensate for modeling uncertainties in controlling flexible structures. These phenomena have been thoroughly studied in many control systems for high quality servomechanisms. Several studies show these negative consequences, such as severe tracking errors, limit cycles, chattering, and excessive noise [1,2]. Many control laws have been proposed for flexible joints [1,3,4], including classical, robust and adaptive control laws, using techniques such as singular perturbations and energy methods [5], but they generally consider (structured) parametric uncertainties only. Several models and compensation schemes have been proposed. Adaptive control techniques have been regarded among the most promising solutions to such type of problems. However, not much has been achieved yet for systems that exhibit both flexibility and severe nonlinearities.

Flexible-joint manipulators are governed by complex dynamics and hence controlling them depends on their dynamic models. There are many modeling techniques for mechanical systems, such as Lagrangian approach, Hamilton's principle, and Kane method. Yet, the system is inevitably subjected to the ubiquitous presence of high, particularly unstructured, modeling nonlinearities, such as Coulomb friction and external disturbances. The presence of such uncertainties on a manipulator driven through a flexible joint significantly changes the system's dynamics as opposed to when the load is driven with a rigid joint [6,7]. In this case, solving the inverse dynamics of the system is not realizable since the motor position is not uniquely defined at standstill. This last condition also illustrates that the actuator's state cannot be observed continuously from the load output. Henceforth, only an approximate inverse model can be realized. Therefore, modeling the system's dynamics based on presumably accurate mathematical models cannot be applied efficiently in this case. This raises the urgency to consider alternative approaches for the control of this type of manipulator systems to keep up with their increasingly demanding design requirements.

Various control techniques were proposed over the years to control flexible-joint manipulators [8–10]. De Luca *et al.* [11] and Khorasani [12] proposed feedback linearization-based controllers. However, these controllers depend on excessively noisy joint acceleration and jerk signal measurements and are hence unreliable in most real-world robotic systems. On the other hand, C. de Wit [13] proposed a robust control scheme for friction overcompensation due to uncertainties in friction models. Even as such, the suggested controllers require the full *a priori* knowledge of the system's dynamics. This problem has been partially overcome by several adaptive control schemes [14–16]. Most of these control techniques capitalize on the singular perturbation theory to extend the adaptive control theory developed for rigid bodies to flexible ones [17–20]. M. Spong [21] reduced the flexible-joint manipulators model to the standard rigid manipulators model as the joint stiffness tends to infinity. This model has been widely used by many researchers to achieve better tracking performance. For example, F. Ghorbel *et al.* [5] used a rigid manipulator's conventional method as slow controller and a fast feedback control law to damp out the oscillations of the joint flexibility modes. In a similar way, K. Khorasani *et al.* [22] illustrated how standard adaptive control schemes for rigid robots may be generalized for flexible-joint manipulators under a certain set of assumptions. Although many of these controllers are shown to be quite performant in theory, they failed to address important issues that might stand against their practical implementation, like basing the control laws on joint torques and their derivative [23,24], for instance, which are well known to be extremely noisy in real-life applications.

Moreover, such type of control algorithms uses online continuous estimation through well-defined adaptation laws of a set of the plant's physical parameters to approximate the system's dynamics. For it to provide a satisfactory performance, a typical adaptive control algorithm assumes that the dynamic model is perfectly known and free of significant external (unmodeled) disturbances. In other words, the controller is only robust to parametric or structured (also called modeled) uncertainties and possibly to minor unstructured uncertainties. Moreover, the unknown physical parameters must have constant or slowly varying nominal values. An explicit linear parameterization of the uncertain dynamics parameters also has to exist, and even if it does, it might not be trivial to derive, especially with complex dynamic systems. Although the latter condition is guaranteed for robotic systems, it might not be the case for many other dynamic models. Although some conventional adaptive control techniques, proposed in the literature, did indeed tackle external disturbance attenuation in addition to the compensation for parametric uncertainties, they did not take into consideration the effects of modeling uncertainties [14].

There is an increasing interest in developing new adaptive control schemes for robot manipulators [25,26]. For example, in [25] an adaptive neural network based sliding mode control was developed. The authors in [27] developed a neural network and an estimator to estimate the external perturbation on flexible robot manipulators. This work does not consider uncertainty in the robot parameters and was validated in simulations only. In another work, an adaptive based output-feedback control strategy for global position stabilization was developed in [28]. Although this work guarantees the adaptive regulation objective without the need for velocity feedback, it does not consider parameter uncertainties. A voltage-based adaptive control methodology for flexible joint manipulators was developed in [29,30]. The advantage of this work is that it is not torque-based control. However, it does not consider dynamic parameter uncertainties and was validated only in simulations.

In the literature, several papers on the robust control of robot manipulators have been developed, most of which use sliding mode techniques [31–33]. In [31] a multiple model/control strategy was proposed. It uses sliding mode control to reduce the high gain control for robot manipulators with large parameter uncertainty. In another work, a digital sliding mode controller for manipulators with three joints was developed by Corradini *et al.* [32]. This controller is not adaptive and uses sliding mode structure, which is prone to chattering.

On the other hand, computational intelligence tools, such as artificial neural networks and fuzzy logic controllers, have been credited in various applications as powerful tools capable of providing robust controllers for mathematically ill-defined systems that may be subjected to structured and unstructured uncertainties [34,35]. The universal approximation theorem has been the main driving force behind the increasing popularity of such methods as it shows that they are theoretically capable of uniformly approximating any continuous real function to any degree of accuracy. This has led to the recent advances in the area of intelligent control [36,37]. Various neural network and fuzzy logic models have been applied in the control of flexible joint manipulators, which have led to a satisfactory performance [9,10]. H. Chaoui *et al.* [38–41] used a neural network based adaptive control approach inspired by sliding mode control to learn the system's dynamics. A time-delay neurofuzzy network was suggested in [42], where a linear observer was used to estimate the joint velocity signals and eliminated the need to measure them explicitly. Subudhi *et al.* [43] presented a hybrid architecture composed of a neural network to control the slow dynamic subsystem and an H_∞ to control the fast subsystem. A feedback linearization technique using a Takagi–Sugeno neurofuzzy engine was adopted in [44]. Despite the success witnessed by neural network-based control systems, they remain incapable of incorporating any human-like expertise already acquired about the dynamics of the system in hand, which is considered one of the main weaknesses of such soft computing methodologies. In another work, an adaptive neural network based sliding mode control was developed in citeSun-2011. However, only simulations were used to verify the results. Type-1 FLCs have also been developed for robot manipulators. Some of these control techniques have hybridized type-1 FLCs with sliding mode control to achieve good robustness; e.g., Li and Huang [45] developed an MIMO adaptive sliding mode based manipulator. However, only simulations were performed to validate the theory.

As type-2 FLCs can handle uncertainties more effectively than type-1 FLCs, they are becoming a more viable tool for the control of uncertain systems [46]. Melin and Castillo [47,48] reviewed genetic algorithms, particle swarm and ant colony optimization methods in the design of optimal type-2 fuzzy systems for different applications. The authors concluded that although genetic algorithms have been used more frequently, the other two methods are rapidly gaining ground for the design of optimum type-2 FLCs. The same authors adopted the chemical reaction algorithm (CRA) to tune the parameters of type-1 and type-2 FLCs for a unicycle tracking applications [49].

While several type-1 FLCs have been developed for robot manipulators [50], very few interval type-2 FLCs have been devised for this purpose [51–53]. In [52], a type-2 fuzzy controller was developed and tested on a parallel robot. Most recently, in [53], a type-2 TSK fuzzy controller was used to control a modular robot. Chen developed a sliding mode based type-2 FLC for trajectory tracking. The computational complexity of the hybrid controller is rather significant because of its learning structure. In all of these works, it was concluded that type-2 FLCs can outperform their type-1 counterparts despite noise and uncertainty.

The present work contributes to the merits and the latest developments of type-2 fuzzy logic theory for the design and implementation of an adaptive type-2 FLC for the control of flexible-joint robot manipulators with uncertain dynamics. The combination of high, particularly unstructured, nonlinearities, such as in the form of Coulomb friction, and manipulator's joint elasticity changes significantly the system's dynamics. In this case, the system's inverse dynamic model cannot be found and only an approximation can be made. Therefore, modeling the system's dynamics based on presumably accurate mathematical models cannot be applied efficiently in this case. Conventional type-1 fuzzy logic systems (FLSs) can be used to identify the behavior of this highly nonlinear system with various types of uncertainties. However, type-1 fuzzy sets may not fully capture the uncertainties in the system due to membership functions and knowledge base imprecision. Hence, higher types of fuzzy sets have to be considered. However, the computational complexity of operations on fuzzy sets increases with the increasing type of the fuzzy set. Therefore, interval type-2 fuzzy sets are adopted in this work for their simplicity and efficiency to capture the severe nonlinearities of flexible-joint robot manipulators. Thus far, type-2 FLSs have been used in very few control applications, such as nonlinear control and mobile robot navigation [54]. The work presented in this manuscript has two main contributions: (i) it devises a novel adaptive control law for type-2 FLCs; and (ii) it proposes a computationally efficient inference mechanism for such type of controllers, which are reputed for their typically high computational complexity, to make them more suitable for real-time applications. This work represents one of the scarce attempts in developing adaptive type-2 FLC to control flexible-joint manipulators with uncertain dynamics. To the best of our knowledge, adaptive type-2 FLCs with a Mamdani structure have not yet been developed for the control of robot manipulators. As shall be detailed later, and unlike other types of control systems proposed in the literature, the proposed adaptive type-2 FLC is proven to be stable by Lyapunov stability theory and does not depend on the excessively noisy acceleration signals or joint torque measures [23,24]. We also present a comparative study between the proposed adaptive type-2 FLC and its type-1 counterpart to better assess their respective performances in various operating conditions.

The rest of the paper is organized as follows: Section 2 outlines the dynamical model of a typical flexible-joint manipulator. In Section 3, an overview of type-2 fuzzy logic systems is presented. In Section 4, we present interval type-2 FLSs and describe the functionality of a type-2 fuzzy inference engine. The design of the proposed controller is detailed in Section 5. In Section 6, simulation results are reported and discussed before concluding the paper with a few remarks about this important, yet complex, control problem in Section 7.

2. Flexible-Joint Manipulator Dynamics

2.1. Modeling of a Flexible-Joint Manipulator

The schematic representation for the ith flexible-joint in a multi-joint manipulator is shown in Figure 1. The actuator is coupled to a flexible transmission through an r:1 reduction gear. The transmission is dynamically simplified as a linear torsional spring linked directly to the load (e.g., manipulator link.)

Figure 1. Flexible-joint model.

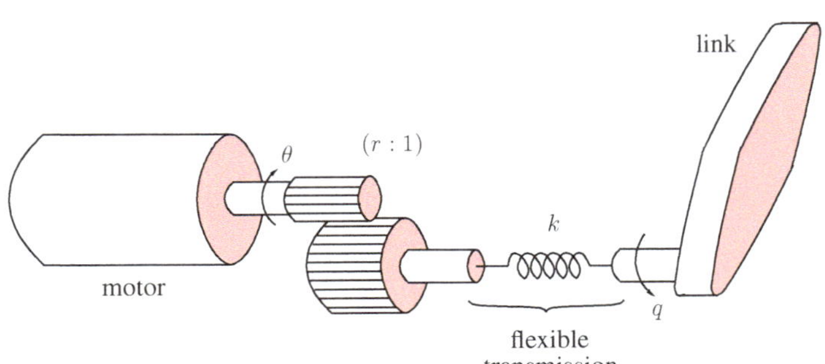

Consider a robot manipulator with n revolute flexible joints. Using Euler–Lagrange formulation, the dynamic equations of the manipulator can be written as:

$$M(q)\ddot{q} + C(q,\dot{q})\dot{q} + G(q) = \tau_t - \tau_{fl} - \tau_{dl} \quad (1a)$$

$$J_m\ddot{\theta} = \tau_m - \frac{1}{r}\tau_t - \tau_{fm} - \tau_{dm} \quad (1b)$$

$$\tau_t = K(\frac{\theta}{r} - q) \quad (1c)$$

where,

$q \in \mathbb{R}^n$:vector of links' positions
$\theta \in \mathbb{R}^n$:vector of motors' positions
$M(q) \in \mathbb{R}^{n\times n}$:manipulator's positive definite inertial matrix
$C(q,\dot{q}) \in \mathbb{R}^{n\times n}$:matrix of Coriolis and centrifugal terms
$G(q) \in \mathbb{R}^n$:vector of gravitational torques
$J_m \in \mathbb{R}^{n\times n}$:motors' diagonal inertial matrix
$\tau_t \in \mathbb{R}^n$:vector of transmission torques
$\tau_m \in \mathbb{R}^n$:motors' generalized torque vector (control input)
$\tau_{fl} \in \mathbb{R}^n$:load friction vector
$\tau_{fm} \in \mathbb{R}^n$:motors' friction vector
$\tau_{dl} \in \mathbb{R}^n$:load's unmodeled dynamics and external disturbance vector
$\tau_{dm} \in \mathbb{R}^n$:motors' unmodeled dynamics and external disturbance vector
$K \in \mathbb{R}^{n\times n}$:diagonal matrix of joints' stiffness coefficients
$r \in \mathbb{R}$:gear ratio

The dynamics of a robotic manipulator is characterized by the following properties:

Property 1 *The inertia matrix $M(q)$ is characterized by the following properties.*
(1) Positive Definite Symmetric (PDS), i.e., $M^T(q) = M(q)$ and $x^T M(q)\, x > 0$ for any non-null vector x.
(2) Upper and lower bounded, i.e., there exist two scalars $\alpha_1(q)$ and $\alpha_2(q)$ such that $\alpha_1(q)I \leq M(q) \leq \alpha_2(q)I$, where I is the identity matrix.

Property 2 *The Coriolis and centripetal term $C(q,\dot{q})$ has the following properties.*
(1) Matrix $\dot{M}(q) - 2C(q,\dot{q})$ is skew symmetric, i.e.,

$$x^T(\dot{M}(q) - 2C(q,\dot{q}))\, x = 0 \quad \forall\, x \in \mathbb{R}^n$$

(2) $C(q,\dot{q})\dot{q}$ is quadratic in $\dot{q}$ and bounded, i.e., there exists a scalar $\alpha_3(q)$ such that $||C(q,\dot{q})\dot{q}|| \leq \alpha_3(q)||\dot{q}||^2$.

Property 3 *The gravity vector $G(q)$ is bounded, i.e., $||G(q)|| \leq \alpha_4(q)$, for a scalar $\alpha_4(q)$.*

Before we proceed further, we introduce the following realistic assumption.

Assumption 1 *The norm of the unknown disturbance τ_d is upper bounded by a scalar b_d, i.e., $||\tau_d|| \leq b_d$.*

2.2. Friction Modeling

Friction is highly nonlinear and it is therefore important to capture the essence of the friction phenomena with models of reasonable complexity. The behavior of friction has been extensively examined lately, a good accurate representation was introduced in [4] as the sum of Coulomb, viscous, and static friction terms. The model of such a memoryless linear or angular friction $F \in \mathbb{R}^n$ operating along a linear or angular displacement rate vector $\dot{x} = (\dot{x}_1, \ldots, \dot{x}_n)^T \in \mathbb{R}^n$ can be expressed as

$$F(\dot{x}) = F_c \,\text{sign}(\dot{x}) + F_v \,\dot{x} + F_s \,\Upsilon(\dot{x})\,\text{sign}(\dot{x}) \quad (2)$$

where $F_c = \text{diag}(F_{c_1}, \ldots, F_{c_n})$, $F_v = \text{diag}(F_{v_1}, \ldots, F_{v_n})$ and $F_s = \text{diag}(F_{s_1}, \ldots, F_{s_n})$ are the Coulomb, viscous and static friction positive definite diagonal matrices, respectively, and $\Upsilon(\dot{x}) = \text{diag}(e^{-(\dot{x}_1/\eta_{s_1})^2}, \ldots, e^{-(\dot{x}_n/\eta_{s_n})^2})$ is a positive definite diagonal matrix representing the rate of decay of the static friction for some decay rate scalars $\eta_{s_1}, \ldots, \eta_{s_n}$, along the n degrees of freedom. The term $\text{sign}(\dot{x}) \in \mathbb{R}^n$ is the vector $(\text{sign}(\dot{x}_1), \ldots, \text{sign}(\dot{x}_n))^T$ defined by

$$\text{sign}(\dot{x}_i) = \begin{cases} 1 & , \text{if } \dot{x}_i \geq 0 \\ -1 & , \text{if } \dot{x}_i < 0 \end{cases} \quad , \quad i = 1, \ldots, n$$

Note the friction's high sensitivity and nonlinearity in the vicinity of $\dot{x} = 0$, as illustrated in Figure 2. At very low velocity, manipulators are likely to exhibit undesirable stick-slip, a cycle of stop and motion. Friction will act as a constraint and not as a motion generator [55]. τ is the control input and ϵ is a very small constant.

Figure 2. Friction model.

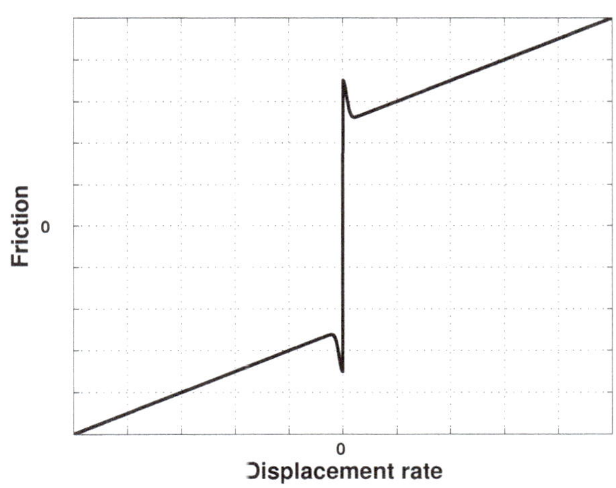

In a single stage speed reduction system, a lumped flexibility model can be used if the dominant flexion appears in the gear teeth. In the following, the input gears inertia are combined with the actuators inertia J_m while the output gears inertia is lumped with the load inertia $M(q)$. As shown in [38] and the references therein, this model reduction method has been used in multi-stage reduction systems, such as planetary gears, and in multi-mass flexibility models, like harmonic drives, for instance.

2.3. Problem Statement

Given the desired trajectories q_d and $\dot{q}_d$, the aim is to design a control law τ_m which ensures that the manipulator's position q and velocity $\dot{q}$ track their desired trajectories under unknown or uncertain dynamics and in the presence of external disturbances. The proposed controller uses q, $\dot{q}$, and $\dot{\theta}$ as the system's measurable states. The manipulator's parameters, $M(q)$, $C(q, \dot{q})$, $G(q)$, J_m, τ_{fl}, τ_{fm}, τ_{dl}, τ_{dm} are assumed to be unknown or uncertain.

3. Type-2 FLSs

A type-2 FLS is comprised of five components: fuzzifier, rule base, fuzzy inference engine, type-reducer and defuzzifier. A block diagram of a typical type-2 FLS is depicted in Figure 3.

Figure 3. Block diagram of a type-2 FLS.

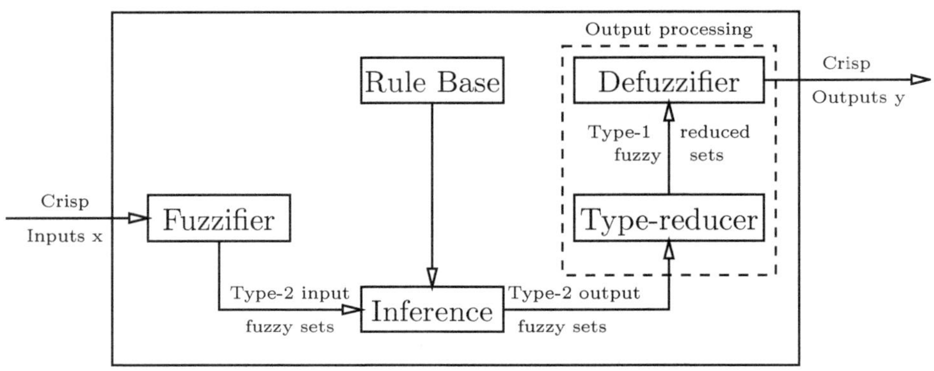

A type-2 fuzzy set is defined by a fuzzy membership function, where the membership value or grade for each element of this set is a fuzzy set in the interval $[0, 1]$ rather than a crisp value. As such, the footprint of uncertainty (FOU) provides type-2 FLSs with additional degrees of freedom, making the membership functions of type-2 fuzzy sets three dimensional functions. Therefore, type-2 fuzzy sets can handle more types of uncertainties with higher magnitudes using a smaller rule base than their type-1 counterparts.

A type-2 fuzzy set, denoted as $\tilde{A}$, is characterized by a type-2 membership function $\mu_{\tilde{A}}(x, u)$, where $x \in X$ and $u \in J_x \subseteq [0, 1]$, *i.e.*,

$$\tilde{A} = \{((x, u), \mu_{\tilde{A}}(x, u) \mid \quad \forall x \in X, \quad \forall u \in J_x \subseteq [0, 1]\}$$

in which $0 \leq \mu_{\tilde{A}}(x, u) \leq 1$. For a continuous universe of discourse, $\tilde{A}$ can be expressed as

$$\tilde{A} = \int_{x \in X} \int_{u \in J_x} \mu_{\tilde{A}}(x, u)/(x, u) \quad J_x \subseteq [0, 1]\}$$

where J_x is referred to as the primary membership of x. As in type-1 fuzzy logic, discrete fuzzy sets are represented by the symbol $\sum$ instead of $\int$. The secondary membership function associated to $x = x'$, for a given $x' \in X$, is the type-1 membership function defined by $\mu_{\tilde{A}}(x = x', u)$, $\forall u \in J_x$. The uncertainty in the primary membership of a type-2 fuzzy set $\tilde{A}$ is represented by the FOU and is illustrated in Figure 4. Note that the FOU is also the union of all primary memberships.

$$FOU(\tilde{A}) = \bigcup_{x \in X} J_x$$

Figure 4. Type-2 fuzzy logic membership function.

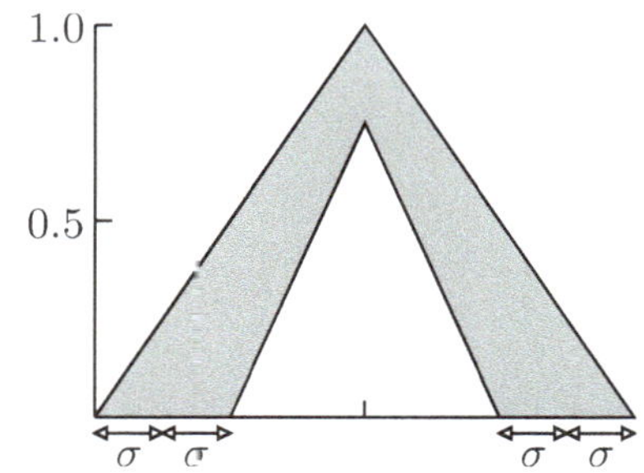

The upper and lower membership functions, denoted by $\overline{\mu}_{\tilde{A}}(x)$ and $\underline{\mu}_{\tilde{A}}(x)$, respectively, are two type-1 membership functions that represent the upper and lower bounds for the footprint of uncertainty of an interval type-2 membership function $\mu_{\tilde{A}}(x, u)$, respectively [56].

The structure of rules in the type-2 FLS and its inference engine is similar to those in type-1 FLS. The inference engine combines rules and provides a mapping from input type-2 fuzzy sets to output type-2 fuzzy sets. In this process, unions and intersections of type-2 sets as well as compositions of type-2 relations are used. The output of the type-2 inference engine is a type-2 set. Using extension principle, type-1 defuzzification can derive a crisp output from type-1 fuzzy set; similarly, for a higher type set as type-2, this operation derives the type-2 sets to a type-1 set. This process is the so called "type-reduction".

4. Interval Type-2 FLSs

The computation intensity of type-2 FLSs has been behind the development of interval fuzzy sets [56,57] as it provides a simplified and efficient way to compute meet and join operations and perform type-reduction for FLSs. It distributes the uncertainty evenly among all admissible primary memberships and offers a balanced trade-off between performance and complexity. Before proceeding further, a few concepts need to be introduced. An FLS with an interval singleton type-2 fuzzifier and product or minimum t-norm satisfies the following properties [57]:

1. the firing strength of the lth fuzzy rule is an interval type-1 fuzzy set defined as

$$F^l(x') \equiv \sqcap_{i=1}^{p} \mu_{\tilde{F}_i^l}(x'_i) = [\underline{f}^l(x'), \overline{f}^l(x')] \equiv [\underline{f}^l, \overline{f}^l]$$

where

$$\underline{f}^l(x') = \underline{\mu}_{\tilde{F}_1^l}(x'_1) \star \cdots \star \underline{\mu}_{\tilde{F}_p^l}(x'_p) \tag{3}$$

$$\overline{f}^l(x') = \overline{\mu}_{\tilde{F}_1^l}(x'_1) \star \cdots \star \overline{\mu}_{\tilde{F}_p^l}(x'_p) \tag{4}$$

with the t-norm operator denoted by "$\star$".

2. the fired output consequent set of the lth rule is a type-1 fuzzy set characterized by a membership function

$$\mu_{\tilde{B}^l}(y) = \int_{b^l \in [\underline{f}^l \star \underline{\mu}_{\tilde{G}^l}(y), \overline{f}^l \star \overline{\mu}_{\tilde{G}^l}(y)]} 1/b^l \qquad \forall y \in Y$$

with $\underline{\mu}_{\tilde{G}^l}(y)$ and $\overline{\mu}_{\tilde{G}^l}(y)$ being the lower and upper membership grades of $\mu_{\tilde{G}^l}(y)$.

3. if N out of a total of L fuzzy rules in the FLS fire, where $N \leq L$, then the overall aggregated output fuzzy set is defined by a type-1 membership function $\mu_{\tilde{B}}(y)$ obtained by combining the fired output consequent sets into one. In other words, $\mu_{\tilde{B}}(y) = \sqcup_{l=1}^{N} \mu_{\tilde{B}^l}(y)$, where $\mu_{\tilde{B}^l}(y)$ is defined in Equation (5).

The following is a brief description of the different stages of the type-2 fuzzy logic inference engine.

$$\mu_{\tilde{B}^l}(y) = \int_{b \in \left[[\underline{f}^1 \star \underline{\mu}_{\tilde{G}^1}(y)] \vee \cdots \vee [\underline{f}^N \star \underline{\mu}_{\tilde{G}^N}(y)], [\overline{f}^1 \star \overline{\mu}_{\tilde{G}^1}(y)] \vee \cdots \vee [\overline{f}^N \star \overline{\mu}_{\tilde{G}^N}(y)]\right]} 1/b^l, \quad \forall y \in Y \tag{5}$$

4.1. Type-2 Fuzzification

In the fuzzification stage, the crisp input vector with n elements $x = (x_1, \ldots, x_n)^T$ in the universe of discourse $X_1 \times X_2 \times \ldots \times X_n$ is mapped into type-2 fuzzy sets [56,57]. The type-2 fuzzification process is schematically depicted in Figure 5. The upper and lower membership functions are computed for each point of the universe of discourse, resulting in an interval type-1 set $[\underline{f}^l, \overline{f}^l]$ for each rule l.

Figure 5. Interval type-2 inference process.

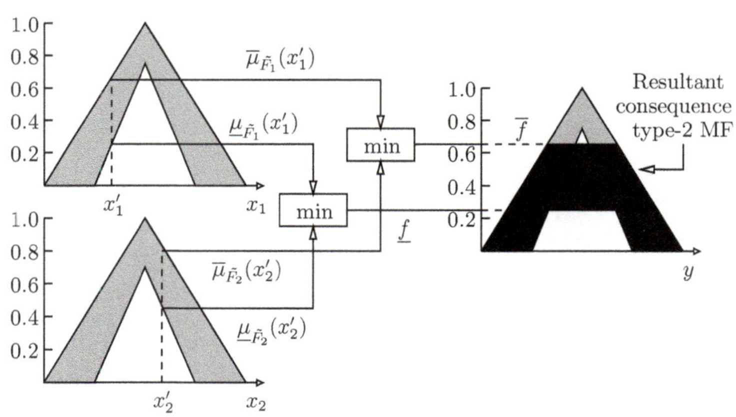

4.2. Type-2 Fuzzy Rule Base

Type-2 IF-THEN rules have the same structure as their type-1 counterpart. However, the antecedents and the consequents are represented by interval type-2 fuzzy sets. Thus, the lth rule is of the form:

$$R^l : \text{IF } x_1 \text{ is } \tilde{F}_1^l \text{ and } x_2 \text{ is } \tilde{F}_2^l \text{ and } \ldots \text{ and } x_n \text{ is } \tilde{F}_n^l$$
$$\text{THEN } y_1 \text{ is } \tilde{G}_1^l \text{ and } y_2 \text{ is } \tilde{G}_2^l \text{ and } \ldots \text{ and } y_m \text{ is } \tilde{G}_m^l$$

where $\tilde{\bullet}$ is type-2 fuzzy set.

4.3. Type-2 Fuzzy Inference Engine

A type-2 fuzzy inference engine provides a mapping from the input type-2 fuzzy sets to the output ones. Each rule l in the knowledge base is interpreted as a type-2 fuzzy implication that, when aggregated with the fuzzified inputs, infers a type-2 fuzzy set $\tilde{B}^l$ such that:

$$\mu_{\tilde{B}^l}(y) = \sqcup_{\mathbf{x}\in X}[\mu_{\tilde{A}_x}(x) \sqcap \mu_{R^l}(x, y)]$$

The t-norm and t-conorm used for the type-2 FLC herein are the "minimum" and "maximum" operators, respectively. These operators have been accredited in the literature for their computational efficiency and satisfactory performance.

4.4. Type Reduction

Type-reduction is an "extended version" of type-1 defuzzification methods (using the extension principle) because this operation reduces a type-2 output fuzzy set to a type-1 fuzzy set, which is called the "type-reduced set". Several type-reduction methods have been suggested in the literature, such as the center-of-sums, the height, the modified height and the center-of-sets. The calculation of type-reduced sets is performed in two stages. First, the centroids of the type-2 interval consequent sets of the fuzzy rules are computed. This is conducted ahead of time and is not part of the control cycle. In the second stage, the type-reduced sets are computed at each control cycle before being defuzzified [54].

Calculation of the Rule Consequents Centroids The centroid of the tth output fuzzy set y_k^t is a type-1 interval set determined by its left and right most points, y_{lk}^t and y_{rk}^t, respectively, which are expressed by [57]:

$$y_k^t = [y_{lk}^t, y_{rk}^t] = y(\theta_1, \ldots, \theta_z) = \frac{\sum_{z=1}^{Z} y_z\theta_z}{\sum_{z=1}^{Z} \theta_z} \tag{6}$$

Algorithm 1 describes the iterative procedures for computing y_{rk}^t and y_{lk}^t [54]. Without loss of generality, we assume that y_z, $z = 1, \ldots, Z$, are arranged in an ascending order; *i.e.*, $y_1 \leq y_2 \leq \ldots \leq y_Z$.

Note that both algorithms are guaranteed to converge in at most Z iterations. However, as mentioned earlier, these procedures are conducted ahead of time and are not part of the control cycle. Such a property helps alleviate the heavy computational burden that is usually associated to computational intelligence-based controllers in general.

4.5. Calculation of the Type-Reduced Set

Using the centroid method, the center-of-sets type reduction reduces the resulting type-2 fuzzy sets to an interval type-1 fuzzy set $[y_{lk}^i, y_{rk}^i]$ for each rule i. The inferred interval type-1 fuzzy set is then defined by $[y_{lk}, y_{rk}]$, such as:

$$y_{lk} = \frac{\sum_{i=1}^{L} f_l^i y_{lk}^i}{\sum_{i=1}^{L} f_l^i} \tag{7}$$

$$y_{rk} = \frac{\sum_{i=1}^{L} f_r^i y_{rk}^i}{\sum_{i=1}^{L} f_r^i} \tag{8}$$

where f_l^i, f_r^i are the firing strengths corresponding to y_{lk}^i and y_{rk}^i of rule i, to minimize y_{lk}^i and maximize y_{rk}^i. Algorithms 2 and 3 reveal the iterative procedures to compute y_{lk} and y_{rk}, respectively [57]. Without loss of generality, we here assume that the pre-computed y_{lk}^i, $i = 1, \ldots, L$, are arranged in an ascending order; *i.e.*, $y_{lk}^1 \leq y_{lk}^2 \leq \ldots \leq y_{lk}^L$. Hence, y_{lk} and y_{rk} can be mathematically expressed as:

$$y_{lk} = \frac{\sum_{u=1}^{Q} \overline{f}^u y_{lk}^u + \sum_{v=Q+1}^{L} \underline{f}^v y_{lk}^v}{\sum_{u=1}^{Q} \overline{f}^u + \sum_{v=Q+1}^{L} \underline{f}^v}$$

$$y_{rk} = \frac{\sum_{u=1}^{R} \underline{f}^u y_{rk}^u + \sum_{v=R+1}^{L} \overline{f}^v y_{rk}^v}{\sum_{u=1}^{R} \underline{f}^u + \sum_{v=R+1}^{L} \overline{f}^v}$$

It is worth pointing out that both procedures are proven to converge in no more than L iterations, where L is the total number of rules [57].

Algorithm 1: Computing y_{rk}^t or y_{lk}^t.

---

begin
- Set $\theta_z = h_z$ for $z = 1, \ldots, Z$;
- Compute $y' = y(h_1, \ldots, h_z)$ using Equation (6);
- Set Stop = False;
- **while** *Stop = False* **do**
  - Find e, where $1 \leq e \leq Z - 1$, such that $y_e \leq y' \leq y_{e+1}$;
  - **if** y_{rk}^t *is to be computed* **then**
    - Set $\theta_z = h_z - \Delta_z$ for $z \leq e$;
    - Set $\theta_z = h_z + \Delta_z$ for $z > e + 1$;
    - Compute $y'' = y(h_1 - \Delta_1, \ldots, h_e - \Delta_e, h_{e+1} + \Delta_{e+1}, \ldots, h_z + \Delta_z)$ using Equation (6);
  - **else if** y_{lk}^t *is to be computed* **then**
    - Set $\theta_z = h_z + \Delta_z$ for $z \leq e$;
    - Set $\theta_z = h_z - \Delta_z$ for $z > e + 1$;
    - Compute $y'' = y(h_1 + \Delta_1, \ldots, h_e + \Delta_e, h_{e+1} - \Delta_{e+1}, \ldots, h_z - \Delta_z)$ using Equation (6);
  - **if** $y'' = y'$ **then**
    - Stop = True;
    - **if** y_{rk}^t *is to be computed* **then**
      - Set $y_{rk}^t = y''$ (y'' is the maximum value of $y(\theta_1, \ldots, \theta_z)$);
    - **else if** y_{lk}^t *is to be computed* **then**
      - Set $y_{lk}^t = y''$;
  - **else**
    - Set $y' = y''$;

end

Algorithm 2: Computing y_{lk}.

begin
- Set $f_r^i = (\underline{f}^i + \overline{f}^i)/2$, for $i = 1, \ldots, L$, with $\underline{f}^i$ and $\overline{f}^i$ as defined in Equations (3) and (4);
- Set $y'_{lk} = y_{lk}$;
- Set Stop = False;
- **while** *Stop = False* **do**
  - Find Q $(1 \leq Q \leq L-1)$ such that $y_{lk}^Q \leq y'_{lk} \leq y_{lk}^{Q+1}$;
  - Compute y_{lk} as in Equation (7), using $f_l^i = \overline{f}^i$ for $i \leq Q$ and $f_l^i = \underline{f}^i$ for $i > Q$;
  - Let $y''_{lk} = y_{lk}$;
  - **if** $y''_{lk} = y'_{lk}$ **then**
    - Stop = True;
    - Set $y''_{lk} = y_{lk}$;
  - **else**
    - Set $y'_{lk} = y''_{lk}$;

end

Algorithm 3: Computing y_{rk}.

begin
- Set $f_r^i = (\underline{f}^i + \overline{f}^i)/2$, for $i = 1, \ldots, L$, with $\underline{f}^i$ and $\overline{f}^i$ as defined in Equations (3) and (4);
- Set $y'_{rk} = y_{rk}$;
- Set Stop = False;
- **while** *Stop = False* **do**
  - Find R $(1 \leq R \leq L-1)$ such that $y_{rk}^R \leq y'_{rk} \leq y_{rk}^{R+1}$;
  - Compute y_{lk} as in Equation (8), using $f_r^i = \overline{f}^i$ for $i \leq R$ and $f_r^i = \underline{f}^i$ for $i > R$;
  - Let $y''_{rk} = y_{rk}$;
  - **if** $y''_{rk} = y'_{rk}$ **then**
    - Stop = True;
    - Set $y''_{rk} = y_{rk}$;
  - **else**
    - Set $y'_{rk} = y''_{rk}$;

end

4.6. Type-2 Defuzzification

The type-reduced set $Y_{cos}(X)_k$ determined by its left most and right most points, y_{lk} and y_{rk}, respectively, is defuzzified using the interval set average formula to get a crisp output value. As such, the defuzzified crisp output for each output k is formulated as [56]:

$$Y_k(x) = \frac{y_{lk} + y_{rk}}{2}$$

In this work, we are concerned with adaptive type-2 FLC, and hence, Algorithms 1–3 will be replaced with the adaptive mechanism explained in the next section. As such, we overcome the heavy computational burden traditionally associated to computational intelligence-based controllers.

5. Control Strategy

Let $\Delta q = q_d - q$ and $\Delta\theta = \theta_d - \theta$ denote the links' and actuators' position errors, respectively, with θ_d being the unknown desired time-dependent motor position vector. The control strategy is based on the design of an adaptive controller that not only leads to a precise tracking of the system's nominal desired signals but also improves the motors' internal stability. Should the motors' desired position θ_d have been available, the control strategy would be based on tracking Δq and $\Delta\theta$ to zero. Since that is not the case, we define the following compounded velocity error signal:

$$\Delta\dot{\epsilon}_r = \dot{q}_d - \left(\lambda\dot{q} + (1-\lambda)\frac{1}{r}\dot{\theta}\right) \tag{9}$$

for a diagonal matrix $\lambda = \text{diag}(\lambda_1, \lambda_2, \ldots, \lambda_n)$ with $\lambda_i \in [0, 1]$, $i = 1, \ldots, n$. The feedback gain λ is introduced to provide a trade-off between the link tracking performance and internal stability, due to the high nonlinear coupling between the two. Note how a choice of $\lambda_i = 1, \forall\, i \in \{1, \ldots, n\}$, completely annihilates the motor's internal stability factor. The fuzzy control strategy is based on a human operator experience to interpret a situation and initiate its control action. A block diagram for the fuzzy controller is illustrated in Figure 6. Given the desired control signals q_d and $\dot{q}_d$, the link's position error Δq and the compounded velocity error $\Delta\dot{\epsilon}_r$ are computed. The FLC takes these two inputs and provides a control action τ_{m} that is proportional to the input values. These signals are quantized into 5 levels represented by a set of linguistic variables: Negative Large (NL), Negative Small (NS), Zero (Z), Positive Small (PS), and Positive Large (PL). In this study, triangular membership functions are used, mainly due to their high computational and performance efficiencies [58]. To assess the performance of both types of controllers, the proposed fuzzy controller is implemented in two different ways: the first is based on a type-1 fuzzy control scheme while the second is based on a type-2. The input membership functions adopted by both types of control systems are shown in Figure 7. In order to perform a fair comparison between type-1 and type-2 FLCs, the type-2 membership functions have been designed such that they are very similar to those of type-1 and have small blurbs in their FOUs. The fuzzy rules of the two control techniques are the same, they were chosen heuristically and can be refined by an expert (see Table 1). These rules are based on three hypotheses: (i) when the input signals are far from their respective nominal zero-valued surfaces, then the FLC's output assumes a high value; (ii) when the inputs are approaching the nominal zero-valued surfaces, the output is adjusted to a smaller value for a smoother approach; (iii) once the inputs are on the nominal zero-valued surfaces, then the output is set to zero. This way, the FLC forces the link position error Δq and the compounded velocity error $\Delta\dot{\epsilon}_r$ to approach zero. It is worth mentioning that an empirical study was conducted beforehand to tune the input membership functions. The center of area method is used for defuzzification.

Figure 6. Block diagram of the proposed fuzzy control scheme.

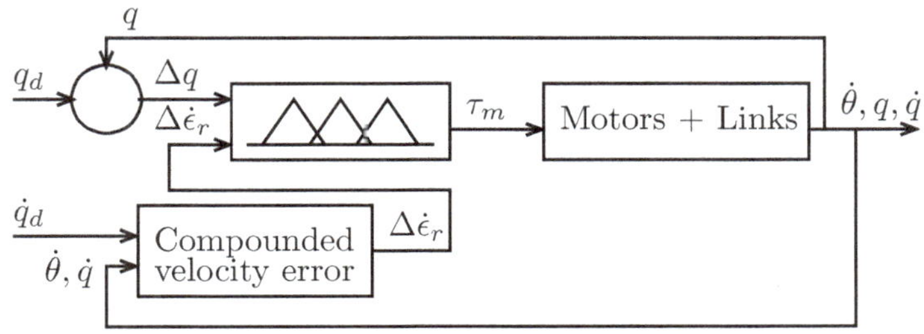

Figure 7. Fuzzy membership functions: (**a**) type-1; and (**b**) type-2.

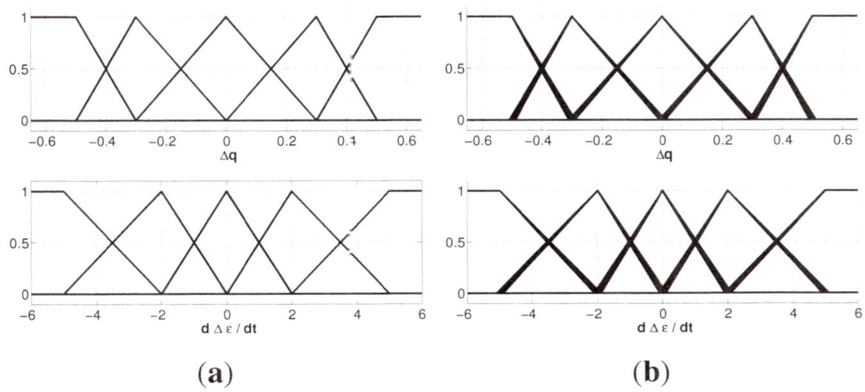

Table 1. Fuzzy rules for type-1 and type-2 FLCs.

	Δq				
$\Delta\dot{\epsilon}_r$	**NL**	**NS**	**Z**	**PS**	**PL**
PL	Z	PL	PL	PL	PL
PS	NS	Z	PS	PS	PL
Z	NL	NS	Z	PS	PL
NS	NL	NS	NS	Z	PS
NL	NL	NL	NL	NL	Z

5.1. Adaptive Type-2 FLC

The adaptive type-2 FLC structure is depicted in Figure 8. It consists of four layers. Input nodes and type-2 fuzzification nodes are shown in layer 1 and layer 2, respectively, forming the antecedent part of the fuzzy rules. Consequent parts are represented by layers 3 and 4, which are constructed with fuzzy rule nodes and output nodes. They are linked by interval weighting factors $[w_{lz}^m \; w_{rz}^m]$.

The adaptive type-2 FLC's output can be written as:

$$Y = \Phi^T w + \epsilon = \Phi^T \hat{w} \tag{10}$$

where $w \in \mathbb{R}^{z \times m}$ is a weight matrix and Φ is a m-dimensional vector of known functions (regressor).

Figure 8. Adaptive Type-2 fuzzy logic control structure.

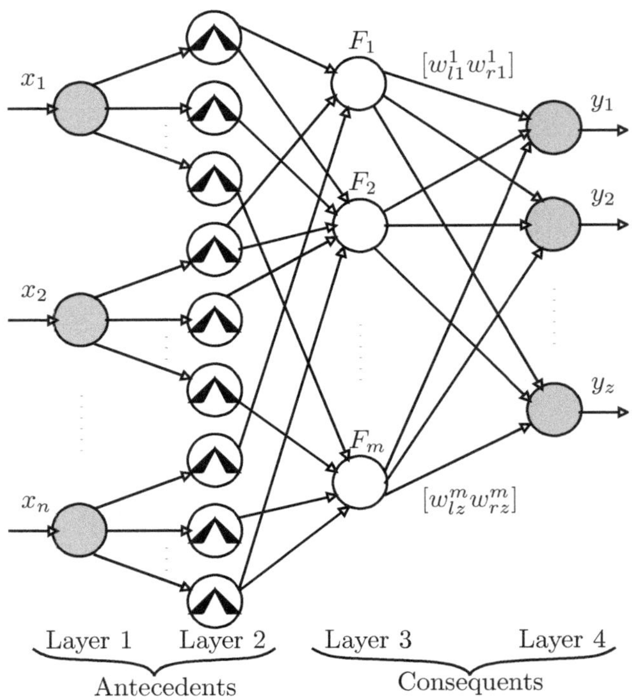

Define the following signals:

$$s = \Delta\dot{e}_r + \Psi\Delta q = \dot{q}_r - \Lambda\dot{q} - (1-\Lambda)\frac{1}{r}\dot{\theta} \tag{11}$$

$$\dot{q}_r = \dot{q}_d + \Psi\Delta q \tag{12}$$

where $\Psi = [\psi_1, \psi_2, \ldots, \psi_n]$ with ψ_i being a positive constant, $i = 1, \ldots, n$.

Recall Euler–Lagrange formulation (1):

$$M(q)\ddot{q} + C(q,\dot{q})\dot{q} + G(q) + \tau_{fl} + \tau_{dl} - \tau_t = 0$$
$$J_m\ddot{\theta} + \tau_{fm} + \tau_{dm} - \tau_m = -\frac{1}{r}\tau_t$$

where q, $\dot{q}$, $\ddot{q}$, $\ddot{\theta} \in \mathbb{R}^n$. Substituting for τ_t, and letting the stiffness constant K tend to infinity (singular perturbation), we obtain the following rigid model [16]:

$$M_t(q)\ddot{q} + C(q,\dot{q})\dot{q} + G(q) + \tau_{Frd} = \tau_m \tag{13}$$

where,

$$M_t(q) = J_m r + M(q)$$
$$\tau_{Frd} = \tau_{fl} + \tau_{dl} + \tau_{fm}r + \tau_{dm}r + (1-r)\tau_m$$

Using the linear in parameters property of the manipulator dynamics, we can approximate the model with a linear regression:

$$M_t(q)\ddot{q}_r + C(q,\dot{q})\dot{q}_r + G(q) + \tau_{Frd} = \Phi(\ddot{q}_r, \dot{q}_r, \dot{q}, q)^T w$$

The control law is:

$$\tau_m = \Phi^T \hat{w} + K_D s \tag{14}$$

where K_D is a positive diagonal matrix gain and the sign $\hat{\bullet}$ denotes the parameter estimate vector.

Theorem 1 *Consider the nonlinear system in Equation (1) with reference signal (11) and control law (14). The adaptive control law is asymptotically stable and the tracking error converges to zero with the following adaptation law:*

$$\dot{\hat{w}} = -\Gamma\ \Phi\ s$$

where $\Gamma = diag(\gamma_1, \gamma_2, \ldots, \gamma_n)$ *and* γ_i *is a positive constant,* $i = 1, \ldots, n$.

Proof: Taking the derivative of the error signal s yields

$$\dot{s} = \ddot{q}_r - \Lambda\ddot{q} - (1-\Lambda)\frac{1}{r}\ddot{\theta}$$

Let $K \to \infty$, *i.e.*, displacement $(\theta - q) \to 0$. We get:

$$\dot{s} = \ddot{q}_r - \ddot{q}$$

$$M_t(q)\dot{s} = M_t(q)\ddot{q}_r - M_t(q)\ddot{q}$$

Substituting $M\ddot{q}$ from Equation (13):

$$M_t(q)\dot{s} = M_t(q)\ddot{q}_r + C(q,\dot{q})\dot{q}_r + G(q) + \tau_{Frd} - C(q,\dot{q})s - \tau_m$$

The linear in parameters property yields

$$M_t(q)\dot{s} = \Phi^T w - C(q,\dot{q})s - \tau_m \tag{15}$$

Set

$$\tau_m = \Phi^T \hat{w} + K_D s \tag{16}$$

Equation (15) becomes

$$M_t(q)\dot{s} = \Phi^T \tilde{w} - C(q,\dot{q})s - K_D s \tag{17}$$

where $\tilde{w} = w - \hat{w}$.

Choose the following Lyapunov candidate:

$$V = \frac{1}{2}\left\{\ s^T M_t(q) s\ +\ \tilde{w}^T\ \Gamma^{-1}\ \tilde{w}\ \right\}$$

Taking the derivative of V:

$$\dot{V} = s^T M_t(q)\dot{s} + \frac{1}{2} s^T \dot{M}_t(q) s + \tilde{w}^T \Gamma^{-1} \dot{\tilde{w}}$$

Substituting for $M_t(q)\dot{s}$:

$$\dot{V} = s^T \Phi^T \tilde{w} + s^T\{\frac{1}{2}(\dot{M}_t(q) - 2C(q,\dot{q}))\}s + \tilde{w}^T \Gamma^{-1}\dot{\tilde{w}} - s^T K_D s$$

$s^T\{(\dot{M}_t(q) - 2C(q,\dot{q}))\}s = 0$ due to the skew-symmetry property. Hence,

$$\dot{V} = s^T \Phi^T \tilde{w} + \tilde{w}^T \Gamma^{-1} \dot{\hat{w}} - s^T K_D s$$

Setting the adaptation law as

$$\dot{\hat{w}} = -\Gamma\, \Phi\, s$$

leads to

$$\dot{V} = -s^T K_D\, s \leq 0$$

Therefore V, and so s, $\tilde{w}$ and $\hat{w}$, are bounded and converge to finite values. It follows from Equation (16) that τ_m is bounded, which implies that all the terms in Equation (17), including $\dot{s}$, are bounded. Thus, $\ddot{V} = -2s^T K_D \dot{s}$ is also bounded. Hence, from Barbalet's Lemma, it implies that $\lim_{t\to\infty} \dot{V} = 0$. Therefore, $\lim_{t\to\infty} s = 0$.

6. Simulation Results and Discussion

6.1. Simulation Setup

To demonstrate the performance of the proposed controller, two numerical simulations are carried out on a single link flexible-joint manipulator. Table 2 summarizes the manipulator's physical parameters along with their respective values. The stiffness coefficient and gear ratio are set to be $K = 7$ N·m/rad and $r = 1$, respectively. The link's mass and length are taken as $m = 0.21$ Kg and $l = 0.3$ m, respectively. The manipulator's dynamics in terms of its physical parameters is defined by: $M(q) = I$, $C(q,\dot{q}) = 0$, and $G(q) = mgl\sin(q)$, where $g = 9.8$ m/s^2 is the gravitational constant and I is the link's rotational inertia given in Table 2. The nonlinear friction model described in Section 2 is considered to model the actuator's and load's mechanical frictions, τ_{fm} and τ_{fl}, respectively.

Table 2. Manipulators physical parameters.

Parameter	**Link**	**Motor**
rotational inertia (kg·m^2)	$I = 5.05 \times 10^{-2}$	$J_m = 4 \times 10^{-3}$
viscous friction coefficient (N·m·s/rad)	$F_{vl} = 4 \times 10^{-3}$	$F_{vm} = 3 \times 10^{-3}$
Coulomb friction coefficient (N·m)	$F_{cl} = 1 \times 10^{-2}$	$F_{cm} = 4 \times 10^{-3}$
static friction coefficient (N·m)	$F_{sl} = 2 \times 10^{-3}$	$F_{sm} = 2 \times 10^{-3}$
static friction decreasing rate (rad/s)	$\eta_{sl} = 7 \times 10^{-2}$	$\eta_{sm} = 5 \times 10^{-2}$

The manipulator's desired position trajectory is taken as the step response of a critically damped second order system with a natural frequency of 3 rad/s, as shown in Figure 9. The feedback gain λ in Equation (9) is set to 0.8. The control structure scheme and the system's model are implemented in Simulink™ while the fuzzy control engines are programmed in C. Both controllers are set to operate at a bandwidth of 100 Hz.

Figure 9. Manipulator's position and velocity reference signals.

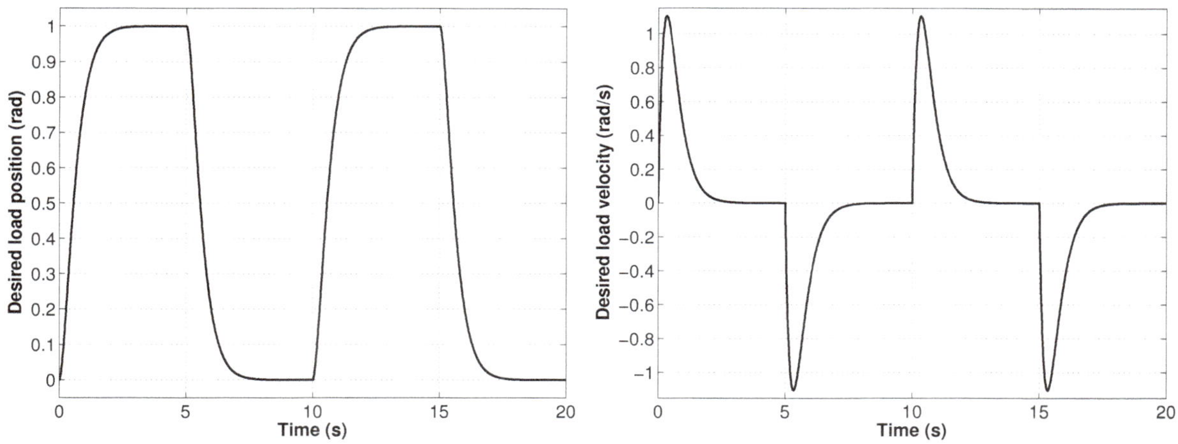

6.2. Numerical Simulations and Results

Two simulations are carried out to highlight the proposed type-2 FLC as opposed to type-1 in tolerating a higher degree of parametric and modeling uncertainties. In both simulations, the system's dynamics is assumed to be *a priori* unknown. For each simulation, the system's response is studied taking into account the manipulator's position and velocity errors, the joint's internal stability, and the controller's output torque, τ_m.

The first simulation is meant to study the controller's ability to sustain various types and magnitudes of load uncertainties. For this purpose, the load's inertia and the link's mass are both doubled abruptly at 5 s and returned back to their original values at 15 s of the simulation. The results are shown in Figure 10. A slight increase is noticed in the manipulator's position and velocity errors due to a heavier load. However, the error signals and the controller's output under type-1 FLC are fairly fluctuating as opposed to a smooth and steady convergence behavior with the type-2 FLC. It is quite important to notice here the degradation in the actuator's internal stability (Figure 10(e)) under type-1 FLC despite the settling of the load's velocity. The superiority of the adaptive type-2 FLC in compensating for such a type of uncertainty is manifested with a better load position, speed accuracy and control effort performance over its type-1 counterpart.

In the second simulation, the elastic joint's stiffness coefficient is changed abruptly to $K = 5$ N·m/rad at 5 s and returned back to its original value ($K = 7$ N·m/rad) at 15 s of the simulation. The controller's performance under such conditions is revealed in Figure 11. As in simulation 1, adaptive type-1 FLC is able to maintain bounded error signals but fails to make them converge smoothly. This is especially clear with the load's position and velocity errors (Figure 11(a) and 11(c)) and the motor's internal stability (Figure 11(e)). On the other hand, it is clear that the adaptive type-2 FLC does indeed maintain these signals smoother and with less control effort than its type-1 counterpart.

Figure 10. System's response to varying load's mass and inertia: (**a**), (**b**) manipulator's position error; (**c**), (**d**) manipulator's velocity error; (**e**), (**f**) motor's velocity *vs.* manipulator's velocity; and (**g**), (**h**) controller's output torque (τ_m).

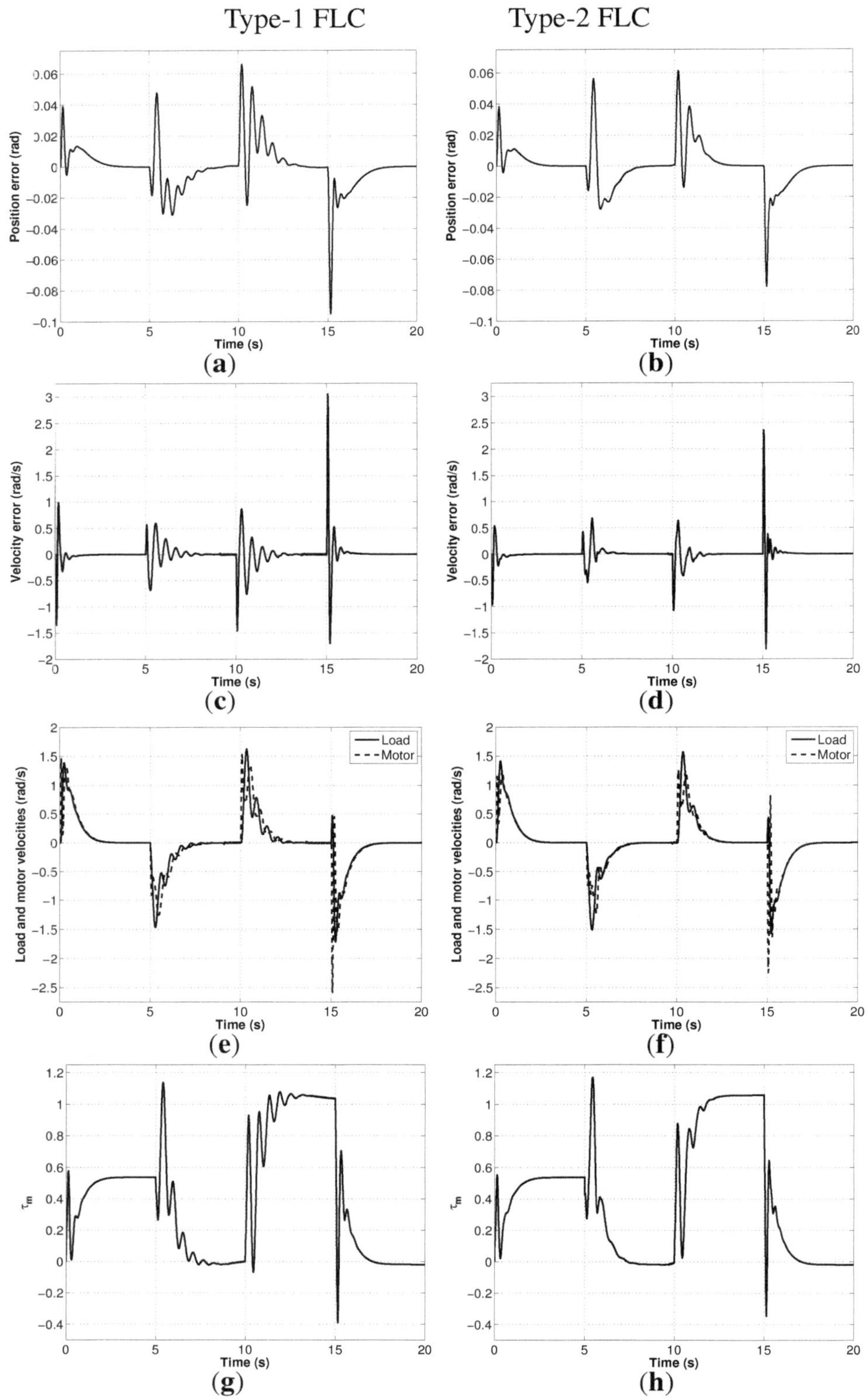

Figure 11. System's response to varying stiffness coefficient: (**a**), (**b**) manipulator's position error; (**c**), (**d**) manipulator's velocity error; (**e**), (**f**) motor's velocity *vs.* manipulator's velocity; and (**g**), (**h**) controller's output torque (τ_m).

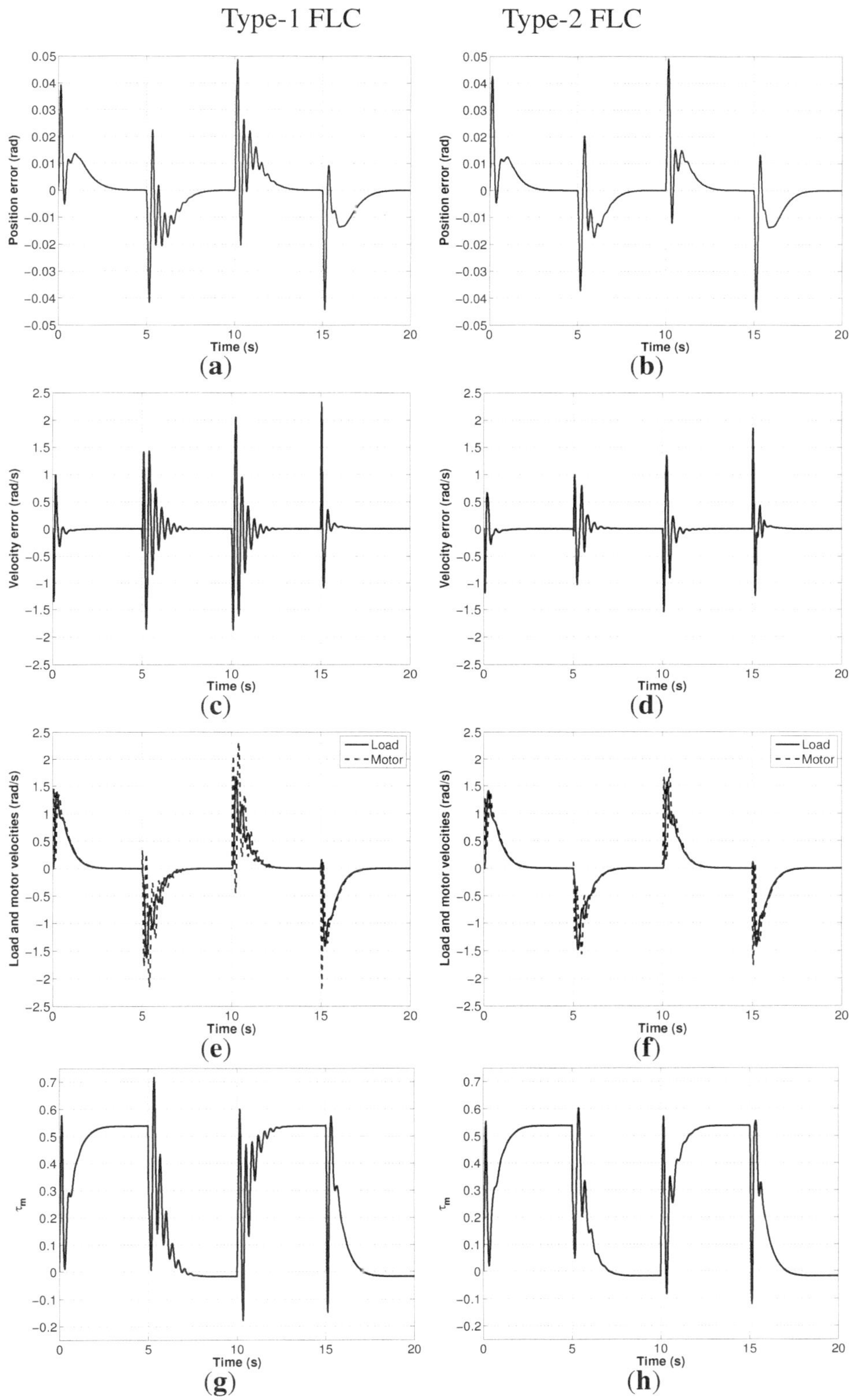

7. Conclusions

In this paper, an adaptive type-2 FLC with a Mamdani inference engine has been proposed for flexible-joint manipulators in the presence of dynamical modeling and parametric uncertainties of various magnitudes. The control strategy takes into account the actuators' relative stability by introducing a trade-off criterion between the actuators' internal stability and the links' position. A comparison of the proposed controller and its type-1 counterpart is performed under similar operating conditions. The simulations show the superiority of the adaptive type-2 FLC in compensating for high magnitude of uncertainties, which confirms the theoretical credentials associated to type-2 FLSs.

Acknowledgements

The authors would like to acknowledge Natural Sciences and Engineering Research Council of Canada (NSERC) for partially supporting this work through the Discovery program.

Conflict of Interest

The authors declare no conflict of interest.

References

1. Olsson, H.; Astrom, K.; de Wit, C.C.; Gafvert, M.; Lischinsky, P. Friction models and friction compensation. *Eur. J. Control* **1998**, *4*, 176–195.
2. Seidl, D.R.; Lam, S.L.; Putman, J.A.; Lorenz, R.D. Neural network compensation of gear backlash hysteresis in position-controlled mechanisms. *IEEE Trans. Ind. Appl.* **1995**, *31*, 1475–1483.
3. Sweet, L.; Good, M. Redefinition of the robot motion-control problem. *Control Syst. Mag. IEEE* **1985**, *5*, 18–25.
4. Armstrong, B.; de Wit, C.C. Friction modeling and compensation. *Control Handb.* **1996**, *77*, 1369–1382.
5. Ghorbel, F.; Hung, J.; Spong, M. Adaptive control of flexible-joint manipulators. *Control Syst. Mag. IEEE* **1989**, *9*, 9–13.
6. Seidl, D.; Lam, S.L.; Putman, J.; Lorenz, R. Neural network compensation of gear backlash hysteresis in position-controlled mechanisms. *IEEE Trans. Ind. Appl.* **1995**, *31*, 1475–1483.
7. Seidl, D.; Reineking, T.; Lorenz, R. Neural Network Compensation of Gear Backlash Hysteresis in Position-Controlled Mechanisms. In Proceedings of Conference Record of the IEEE Industry Applications Society Annual Meeting, Houston, TX, USA, October 1992; Volume 2, pp. 1937–1944.
8. Benosman, M.; Vey, G.L. Control of flexible manipulators: A survey. *Robotica* **2004**, *22*, 533–545.
9. Chaoui, H.; Gueaieb, W. Type-2 fuzzy logic control of a flexible-joint manipulator. *J. Intell. Robot. Syst.* **2008**, *51*, 159–186.
10. Chaoui, H.; Sicard, P.; Gueaieb, W. ANN-based adaptive control of robotic manipulators with friction and joint elasticity. *IEEE Trans. Ind. Electron.* **2009**, *56*, 3174–3187.

11. Luca, A.D.; Isidori, A.; Nicolo, F. Control of Robot Arm with Elastic Joints via Nonlinear Dynamic Feedback. In Proceedings of the IEEE Conference on Decision and Control Including the Symposium on Adaptive Processes, Ft. Lauderdale, FL, USA, 11–13 December 1985; pp. 1671–1679.
12. Khorasani, K. Nonlinear feedback control of flexible joint manipulators: A single link case study. *IEEE Trans. Autom. Control* **1990**, *35*, 1145–1149.
13. De Wit, C. Robust control for servo-mechanisms under inexact friction compensation. *Automatica* **1993**, *29*, 757–761.
14. Ott, C.; Albu-Schaffer, A.; Hirzinger, G. Comparison of Adaptive and Nonadaptive Tracking Control Laws for a Flexible Joint Manipulator. In Proceedings of the IEEE International Conference on Intelligent Robots and Systems, Lausanne, Switzerland, 30 September–4 October 2002; Volume 2, pp. 2018–2024.
15. Al-Ashoor, R.; Patel, R.; Khorasani, K. Robust adaptive controller design and stability analysis for flexible-joint manipulators. *IEEE Trans. Syst. Man Cybern.* **1993**, *23*, 589–602.
16. Ghorbel, F.; Spong, M.W. Adaptive Integral Manifold Control of Flexible Joint Robot Manipulators. In Proceedings of the IEEE International Conference on Robotics and Automation, Nice, France, 12–14 May 1992; Volume 1, pp. 707–714.
17. Spong, M.W. Modeling AND control of elastic joint robots. *J. Dyn. Syst. Meas. Control* **1987**, *109*, 310–319.
18. Ge, S.S.; Postlethwaite, I. Adaptive neural network controller design for flexible joint robots using singular perturbation technique. *Trans. Inst. Meas. Control* **1995**, *17*, 120–131.
19. Taghirad, H.; Khosravi, M. Design and Simulation of Robust Composite Controllers for Flexible Joint Robots. In Proceedings of the IEEE International Conference on Robotics and Automation, Taipei, Taiwan, 14–19 September 2003; Volume 3, pp. 3108–3113.
20. Huang, L.; Ge, S.; Lee, T. Adaptive Position/force Control of an Uncertain Constrained Flexible Joint Robots-Singular Perturbation Approach. In Proceedings of the SICE Annual Conference, Sapporo, Japan, 4–6 August 2004; pp. 1693–1698.
21. Spong, M. Modeling and control of elastic joint robots. *J. Dyn. Syst. Meas. Control* **1987**, *109*, 310–319.
22. Khorasani, K. Adaptive control of flexible joint robots. *IEEE Trans. Robot. Autom.* **1992**, *8*, 250–267.
23. Ott, C.; Albu-Schaffer, A.; Kugi, A.; Stramigioli, S.; Hirzinger, G. A Passivity Based Cartesian Impedance Controller for Flexible Joint Robots-Part I: Torque Feedback and Gravity Compensation. In Proceedings of the IEEE International Conference on Robotics and Automation, New Orleans, LA, USA, 26 April–1 May 2004; pp. 2659–2665.
24. Tian, L.; Goldenberg, A. Robust Adaptive Control of Flexible Joint Robots with Joint Torque Feedback. In Proceedings of the IEEE International Conference on Robotics and Automation, Nagoya, Japan, May 1995; Volume 1, pp. 1229–1234.
25. Sun, T.; Pei, H.; Pan, Y.; Zhou, H.; Zhang, C. Neural network-based sliding mode adaptive control for robot manipulators. *Neurocomputing* **2011**, *74*, 2377–2384.

26. Li, Y.; Tong, S.; Li, T. Adaptive fuzzy output feedback control for a single-link flexible robot manipulator driven DC motor via backstepping. *Nonlinear Anal. R. World Appl.* **2013**, *14*, 483–494.
27. Yen, H.M.; Li, T.H.S.; Chang, Y.C. Adaptive neural network based tracking control for electrically driven flexible-joint robots without velocity measurements. *Comput. Math. Appl.* **2012**, *64*, 1022–1032.
28. Lopez-Araujo, D.J.; Zavala-Rio, A.; Santibanez, V.; Reyes, F. Output-feedback adaptive control for the global regulation of robot manipulators with bounded inputs. *Int. J. Control Autom. Syst.* **2013**, *11*, 816–822.
29. Fateh, M.M. Nonlinear control of electrical flexible-joint robots. *Nonlinear Dyn.* **2011**, *67*, 2549–2559.
30. Fateh, M.M.; Khorashadizadeh, S. Robust control of electrically driven robots by adaptive fuzzy estimation of uncertainty. *Nonlinear Dyn.* **2012**, *69*, 1465–1477.
31. Islam, S.; Liu, P.X. Robust sliding mode control for robot manipulators. *IEEE Trans. Ind. Electron.* **2011**, *58*, 2444–2453.
32. Corradini, M.L.; Fossi, V.; Giantomassi, A.; Ippoliti, G.; Longhi, S.; Orlando, G. Discrete time sliding mode control of robotic manipulators: Development and experimental validation. *Control Eng. Pract.* **2012**, *20*, 816–822.
33. Islam, S.; Liu, P.X. Robust adaptive fuzzy output feedback control system for robot manipulators. *ASME/IEEE Trans. Mechatron.* **2011**, *16*, 288–296.
34. Karray, F.; Gueaieb, W.; Al-Sharhan, S. The hierarchical expert tuning of PID controllers using tools of soft computing. *IEEE Trans. Syst. Man Cybern.* **2002**, *32*, 77–90.
35. De Silva, C.W. *Intelligent Control Fuzzy Logic Applications*; CRC Press: Boca Raton, FL, USA, 1995.
36. Kim, E. Output feedback tracking control of robot manipulators with model uncertainty via adaptive fuzzy logic. *IEEE Trans. Fuzzy Syst.* **2004**, *12*, 368–378.
37. Gueaieb, W.; Karray, F.; Al-Sharhan, S. A robust adaptive fuzzy position/force control scheme for cooperative manipulators. *IEEE Trans. Control Syst. Technol.* **2003**, *11*, 516–528.
38. Chaoui, H.; Sicard, P.; Lakhsasi, A.; Schwartz, H. Neural Network Based Model Reference Adaptive Control Structure for a Flexible Joint with Hard Nonlinearities. In Proceedings of the IEEE International Symposium on Industrial Electronics, Ajaccio, France, 4–7 May 2004; Volume 1, pp. 271–276.
39. Chaoui, H.; Sicard, P.; Lakhsasi, A. Reference Model Supervisory Loop for Neural Network Based Adaptive Control of a Flexible Joint with Hard Nonlinearities. In Proceedings of the IEEE Canadian Conference on Electrical and Computer Engineering, Niagara Falls, ON, Canada, 2–5 May 2004; Volume 4, pp. 2029–2034.
40. Chaoui, H.; Gueaieb, W.; Yagoub, M.; Sicard, P. Hybrid Neural Fuzzy Sliding Mode Control of Flexible-Joint Manipulators with Unknown Dynamics. In Proceedings of the 32nd Annual Conference of the IEEE Industrial Electronics Society (IECON-2006), Paris, France, 6–10 November 2006; pp. 4082–4087.

41. Chaoui, H.; Gueaieb, W.; Yagoub, M.C. Artificial Neural Network Control of a Flexible-Joint Manipulator under Unstructured Dynamic Uncertainties. In Proceedings of the IEEE International Workshop on Robotic and Sensors Environments, Ottawa, ON, Canada, 12–13 October 2007.
42. Hui, D.; Fuchun, S.; Zengqi, S. Observer-based adaptive controller design of flexible manipulators using time-delay neuro-fuzzy networks. *J. Intell. Robot. Syst. Theory Appl.* **2002**, *34*, 453–466.
43. Subudhi, B.; Morris, A. Singular perturbation based neuro-H infinity control scheme for a manipulator with flexible links and joints. *Robotica* **2006**, *24*, 151–161.
44. Park, C.W. Robust stable fuzzy control via fuzzy modeling and feedback linearization with its applications to controlling uncertain single-link flexible joint manipulators. *J. Intell. Robot. Syst. Theory Appl.* **2004**, *39*, 131–147.
45. Li, T.H.S.; Huang, Y.C. MIMO adaptive fuzzy terminal sliding-mode controller for robotic manipulators. *Inform. Sci.* **2011**, *180*, 4641–4660.
46. Cazarez-Castro, N.R.; Aguilar, L.T.; Castillo, O. Designing type-1 and type-2 fuzzy logic controllers via fuzzy lyapunov synthesis for nonsmooth mechanical systems. *Eng. Appl. Artif. Intell.* **2012**, *25*, 971–979.
47. Castillo, O.; Melin, P. Optimization of type-2 fuzzy systems based on bio-inspired methods: A concise review. *Inform. Sci.* **2012**, *205*, 1–19.
48. Castillo, O.; Martinez-Marroquin, R.; Melin, P.; Valdez, F.; Soria, J. Comparative study of bio-inspired algorithms applied to the optimization of type-1 and type-2 fuzzy controllers for an autonomous mobile robot. *Eng. Appl. Artif. Intell.* **2012**, *192*, 19–38.
49. Melin, P.; Astudillo, L.; Castillo, O.; Valdez, F.; Garcia, M. Optimal design of type-2 and type-1 fuzzy tracking controllers for autonomous mobile robots under perturbed torques using a new chemical optimization paradigm. *Expert Syst. Appl.* **2013**, *40*, 3185–3195.
50. Feng, G. A survey on analysis and design of model-based fuzzy control systems. *IEEE Trans. Fuzzy Sets Syst.* **2006**, *14*, 676–697.
51. Chen, C.S. Supervisory adaptive tracking control of robot manipulators using interval type-2 TSK fuzzy logic system. *IET Control Theory Appl.* **2011**, *5*, 1796–1807.
52. Linda, O.; Manic, M. Uncertainty-robust design of interval type-2 fuzzy logic controller for delta parallel robot. *IEEE Trans. Ind. Inf.* **2011**, *57*, 661–670.
53. Biglarbegian, M.; Melek, W.W.; Mendel. J.M. Design of novel interval type-2 fuzzy controllers for modular and reconfigurable robots: Theory and experiments. *IEEE Trans. Ind. Electron.* **2011**, *58*, 1371–1384.
54. Hagras, H.A. A hierarchical type-2 fuzzy logic control architecture for autonomous mobile robots. *IEEE Trans. Fuzzy Syst.* **2004**, *12*, 524–539.
55. Sicard, P. Trajectory Tracking of Flexible Joint Manipulators with Passivity Based Controller. Ph.D. Thesis, Rensselaer Polytechnic Institute, Troy, NY, USA, 1993.
56. Liang, Q.; Mendel, J.M. Interval type-2 fuzzy logic systems: Theory and design. *IEEE Trans. Fuzzy Syst.* **2000**, *8*, 535–550.
57. Mendel, J.M. *Uncertain Rule-Based Fuzzy Logic Systems: Introduction and New Directions*; Prentice-Hall: Upper Saddle River, NJ, USA, 2001.

58. Pedryez, W. Why triangular membership functions? *Fuzzy Sets Syst.* **1994**, *64*, 21–30.

5

A Review of Camera Viewpoint Automation in Robotic and Laparoscopic Surgery

Abhilash Pandya [1,*], Luke A. Reisner [1], Brady King [2], Nathan Lucas [1], Anthony Composto [1], Michael Klein [2] and Richard Darin Ellis [3]

[1] Department of Electrical and Computer Engineering, Wayne State University, Detroit, MI 48202, USA; E-Mails: lreisner@wayne.edu (L.A.R.); nathan.lucas@wayne.edu (N.L.); aj6568@wayne.edu (A.C.)

[2] Department of Pediatric Surgery, Children's Hospital of Michigan, Detroit, MI 48201, USA; E-Mails: bwking@wayne.edu (B.K.); mklein@dmc.org (M.K.)

[3] Department of Industrial and Systems Engineering, Wayne State University, Detroit, MI 48202, USA; E-Mail: rdellis@wayne.edu

* Author to whom correspondence should be addressed; E-Mail: apandya@ece.eng.wayne.edu

Abstract: Complex teleoperative tasks, such as surgery, generally require human control. However, teleoperating a robot using indirect visual information poses many technical challenges because the user is expected to control the movement(s) of the camera(s) in addition to the robot's arms and other elements. For humans, camera positioning is difficult, error-prone, and a drain on the user's available resources and attention. This paper reviews the state of the art of autonomous camera control with a focus on surgical applications. We also propose potential avenues of research in this field that will support the transition from direct slaved control to truly autonomous robotic camera systems.

Keywords: robotic surgery; laparoscopic surgery; autonomous camera control; teleoperation; path planning

1. Introduction

In current robot-assisted surgery, the robotic system does not function with any amount of autonomy. Instead, a surgeon remotely controls one or more slaved mechanical arms using a master controller. This is known as teleoperation, which does not represent a truly intelligent/autonomous robotic system. Moreover, the surgeon is encumbered by the need to manipulate the camera as well as the surgical instruments. In this paper, we review autonomous camera control in the context of the next generation of surgical robotics that goes beyond direct slaved control. We envision surgical robotic systems that understand the current task and can evaluate various inputs to intelligently generate movements that aid the surgeon.

Viewpoint control is a key aspect of teleoperation and is conducive for automation. Due to the physical separation of the surgeon from the patient, an indirect view (from a camera), and the indirect manipulation of the remote site, some significant challenges arise that could be aided by automation. The user is removed from the extensive sensory experience of the remote site, which means there is limited visual, auditory, and haptic feedback. This results in a reduced situation awareness and a high mental workload [1–3]. Furthermore, it is difficult for the user to obtain optimal camera viewpoints in a dynamic environment or to react effectively to irregular events in the scene due to task overload, latency issues, and complex camera positioning issues. The surgeon has to continually manage the camera position to achieve effective viewpoints through manual control or verbal communication with a second operator. This can lead to errors as well as longer operating times. Therefore, we focus our attention on a robot-controlled camera.

"A Roadmap for U.S. Robotics: From Internet to Robotics" [4] states that "designing effective and intuitive interfaces for remote users to control telepresence robots is an open challenge." In addition, the Roadmap points out the need for research that will allow "robots to learn new skills [from observation of humans]." Teleoperating a surgical robot requires the use of indirect visual information that is difficult to manage, which poses many technical and human performance challenges. In particular, the user is expected to control the movement of the camera in addition to the robot's arms. Camera positioning is a difficult task that can be error-prone and a drain on the user's resources [1]. Although camera control research is applicable in many different teleoperation domains, this paper reviews the current state of the art in camera control for telerobotic surgery. It also provides a discussion of potential avenues of research to advance this field and fulfill the unmet need for improved, intelligent camera control [5].

1.1. Current Surgical Robots

Despite being called "robots," the robotic surgery systems that have been available for clinical use have lacked any real autonomy. In general, today's surgical robots act as slaves under the control of the surgeon. For example, the da Vinci Surgical System (Intuitive Surgical, Sunnyvale, CA, USA) is used to perform a variety of minimally invasive procedures, such as prostatectomies. The da Vinci is entirely controlled by the surgeon, who uses hand controls and foot pedals to manipulate a camera arm and several instrument arms.

There are also a number of robots designed specifically to manipulate a laparoscope, such as the AESOP (Intuitive Surgical), EndoAssist [6] (Armstrong Healthcare, High Wycombe, UK), FreeHand (Freehand 2010 Ltd., Guildford, UK), and ViKY EP (EndoControl, Grenoble, France). These devices employ various forms of control, such as hand control, voice recognition, foot control, and head tracking. Despite the various forms of control, there is still a one-to-one correspondence between the surgeon's actions and movements of the laparoscope.

Another group of surgical robots is used to execute preprogrammed motions rather than follow the real-time control input from the surgeon. Examples include the neuromate (Renishaw, Gloucestershire, UK), a neurosurgery robot; the ROBODOC (Curexo Technology Corporation, Fremont, CA, USA), an orthopedic robot; and the CyberKnife (Accuray, Sunnyvale, CA, USA), a radiosurgery robot. Although these robots may seem to act independently, they are actually slaves to their rigidly preprogrammed behaviors. Therefore, like the other robots, they lack the intelligence and awareness to be considered autonomous. The surgeon must expend considerable effort to control all of these robots.

1.2. Motivation for Autonomous Camera Research

Open surgery, where the surgical site is directly viewed, provides a clear 3D view and easier access. Surgeons can see the surgical area and surrounding anatomy from the natural viewpoint of their eyes. Open surgery, however, comes with various disadvantages. This includes a larger incision, sometimes more bleeding, more pain for the patient, and sometimes other complications (e.g., adhesions for certain operations) that may be severe enough to require reoperation [7].

In laparoscopic or robotic surgery, the surgeon views the site through the use of a tool (a laparoscope) that is inserted through a small opening in the patient's body. The image from the laparoscope is displayed on a monitor located somewhere in the operating room. The laparoscopic view is often quite different from the view during open surgery. The laparoscope provides a magnified image, and the viewing angle varies due to the angle of insertion and the angle of the scope's lens.

In many cases, the tighter field of view and alternative angle provided by the laparoscope can be better for the surgeon and the patient. For example, during a laparoscopic fundoplication (a surgery that treats acid reflux disease), the scope can easily look directly down on the gastroesophageal junction under the liver, avoiding the need to mobilize the liver. Other advantages of laparoscopic surgery include a smaller incision, less pain, reduced hospital stay, and often a reduced chance of needing a reoperation. Some drawbacks include typically worse depth perception, a smaller field of view (due to the magnification), and the fact that the surgeons have added camera control (either direct or indirect) to their already intensive workload. Note that a stereo/3D laparoscope is sometimes used to improve the depth perception, particularly in robotic surgery.

In a typical laparoscopic procedure, an assistant is needed to control the camera arm. Directing an assistant to control the camera can be challenging (especially with angled scopes) and lead to increased time for accomplishing the operation. Expert camera holders (*i.e.*, those who are also expert surgeons) improve the performance of the surgeon by generally providing a desirable view. In some cases, the camera positioner may even help guide the steps of the operation. This occurs when an attending (teaching) surgeon holds the camera for a resident who is learning to do the procedure. In most cases, however, it is the expert who performs the operation and often struggles because the less experienced and skilled camera holder does not completely understand which view to provide at a given time.

Robotic surgery adds the advantages of motion scaling, tremor filtration, a comfortable interface, the lack of a fulcrum effect, and a wrist at the end of the surgical tool with a greater range of motion than the human wrist. However, it still retains the disadvantages of a reduced field of view and the additional task of controlling a camera arm. Because there is no assistant for the camera arm, the surgeon controls both the tools and the camera. Operations do not flow as smoothly when the surgeon must frequently stop momentarily to change the camera's viewpoint. Due to the work involved, the surgeon sometimes settles for a suboptimal camera view. The surgeon may also allow the tools to temporarily leave the field of view (due to the effort required to reposition the camera), which can lead to injury to organs outside the field of view.

The aforementioned limitations are the motivations for research and development in autonomous camera navigation systems for minimally invasive surgery (both laparoscopic and robotic) and for remote teleoperation in general. With an autonomous camera system, the human camera assistant used in a typical laparoscopy can be eliminated, and the surgeon will not have to manually move the camera arm in a robotic surgery. Figure 1 demonstrates how an autonomous camera system would differ from current minimally invasive approaches.

Figure 1. Comparison among the mechanisms used to move the camera in traditional laparoscopy, current robotic surgery, and envisioned future procedures with an autonomous camera control system.

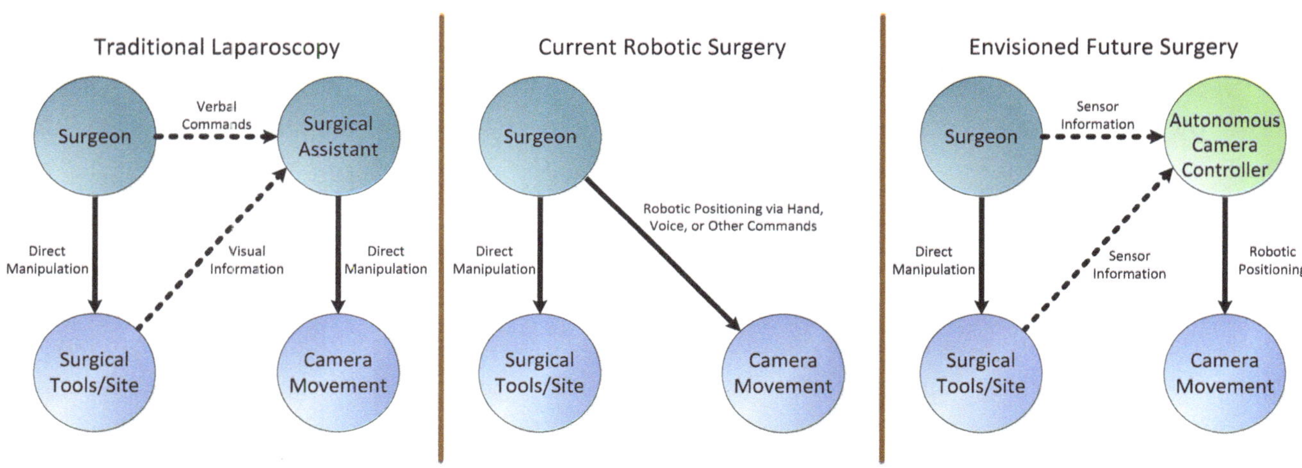

Some work on optimal visualization for medical robotics was done to ascertain the effects of camera views on surgical tasks using multiple cameras. It was found that using two distinct camera views on different monitors (one zoomed in and one zoomed out) helps both surgeons and non-surgeons perform complex tasks with fewer errors. While not the same as an autonomous camera system, the research showed that efficient camera views are important for surgical performance [8].

2. Current Research Approaches

Camera navigation can be approached with a reactive, proactive, or combined control strategy. In a reactive system, sensors track data such as surgeon eye gaze or tool position within the field of view, and the camera is moved in direct response to changes in these inputs. With a proactive strategy, a prediction-based approach is used to determine the best camera viewpoint. In this strategy, the system determines the current state of the procedure that is being undertaken and adjusts the camera's behavior by incorporating preexisting knowledge about the visualization requirements and types of movements needed for the procedure. The knowledge can be represented in different forms, such as simple heuristics or complex statistical models.

Several techniques [9–13] for autonomous camera control have been studied in the fields of virtual reality and animation. Some methods rely on reactive techniques where the behavior is related to the dynamic nature of the scene. Higher-level methods of control try to capture expert user knowledge to come up with general-purpose solutions that are often based on simple rules. In these systems, the parameters of the camera movement can be very application- or context-dependent. For example, an expert camera operator might say: "always keep the tools in the camera view." This information could be translated into a rule-based algorithm, although this rule may not apply to all situations. Currently, these kinds of techniques [9–13] are generally designed for computer graphics applications, which are very different from robotic systems operating in the real world. There, the camera views are constrained by physical limitations and are subject to real conditions.

2.1. Techniques for Tracking and Determining Operator Intent

In general, the challenges of teleoperation have been well studied. Dragan *et al.* [14] address the fundamental problem of teleoperation interfaces being limited by the inherent indirectness/remoteness of these systems. Their report discusses intelligent and customizable solutions to the negative aspects of remote operations. They state that the decision on what assistance should be provided to operators must be contextual and dependent upon the prediction of the user's intent. Their main recommendation is that a robot should learn certain strategies based on examples or even measurements of the movements of a human operator.

2.1.1. Eye Gaze Tracking

Eye tracking technology has its origins in pre-twentieth century studies of basic eye movements, leading to applied research in reading, scene perception, visual search, and other information processing (see Duchowski [15] for past and contemporary applications). The development of non-invasive techniques such as video-based eye tracking (Morimoto and Mimica [16], Hansen and Ji [17]), along with improvements in automatic fixation and saccade detection [18], have enabled applications for human–computer interaction (see Jacob and Karn [19] for a review).

In a human–computer interaction context, the user's point of gaze can be used as an input. For example, it is commonly used as a pointing device (to augment or replace a computer mouse) or to make inferences about the user's attention or intent. Eye movement has been employed for both direct and indirect input to interactively and automatically change computer interface and camera views. Goldberg and Schryver [20] developed an offline method for predicting a user's intended camera zoom level (*i.e.*, magnification/reduction). An accuracy of 65% was reported when predicting zooming in, zooming out, or no zoom change for simple shapes in a controlled interface environment. Less constrained scenarios, such as complex renderings or video of real scenes, would likely decrease the accuracy. In addition, the algorithm relied on previously recorded eye movement data and could not determine zoom intent in real time, limiting its use to non-interactive evaluations. However, advances in automatic identification of fixations and saccades (e.g., Salvucci and Goldberg [18]) might enable real-time adaptations.

Hansen *et al.* [21] used eye tracking as direct input to control planar translation and zoom of a gaze-based typing interface. Users fixated on a desired letter to type, and the interface responded by moving that letter toward the center of the screen. Smooth pursuit eye movement to visually track the letter confirmed the user's intent. Continued tracking as the interface zoomed in on the letter increased system confidence. This sequence culminated in selection of the letter once a threshold zoom level was reached. Events leading to selection of the wrong letter result in the user automatically looking away toward the intended letter. This natural response was utilized to correct errors prior to selection of the wrong letter.

Latif, Sherkat, and Lotfi [22,23] proposed a gaze-based interface to both drive a robot and change the on-board camera view. Although the design permitted hands-free teleoperation, a comparison of joystick control, gaze input with dwell-time activation, and gaze input with foot clutch found no statistically significant difference in task completion time, goal achievement, or composite NASA TLX (task load index) scores. Zhu, Gedeon, and Taylor [24] developed gaze-based automatic camera pan and tilt control that continually repositioned the viewpoint to bring the user's point of gaze to the center of the video screen. The simple proportional control algorithm only responded to where the user is looking. Both [22,23] and [24] demonstrated how eye movement can be used to control camera viewpoint in real-time, but neither sought to account for how task conditions and scene dynamics may influence optimal viewpoint.

Eye tracking has also been explored for surgical applications. Kwok *et al.* [25] used simultaneous displays of two users' point of gaze to indicate their respective intents during simulated robotic surgery tasks. Ali *et al.* [26] used eye tracking to automatically center a laparoscopic camera viewpoint at the user's point of gaze (see Figure 2). Like [24], their system did not adapt to the surgical task being performed, although intent prediction [20] and more sophisticated interaction schemes [21] using similar eye movement data hint at the possibility of more robust gaze-based automated camera systems. Mylonas *et al.* [27] used a binocular eye-tracking device that was directly integrated into a surgical console for 3D gaze localization. They demonstrated that accurate 3D gaze data can be captured and used to alter scene characteristics.

Figure 2. Diagram showing how a robotic camera arm can center a viewpoint based on eye gaze tracking [26]. If the eye gaze in (**a**) is focused at the top of the image, the camera zooms to that area in image (**b**). If the gaze is at the bottom of the image, the camera zooms to the area in image (**c**).

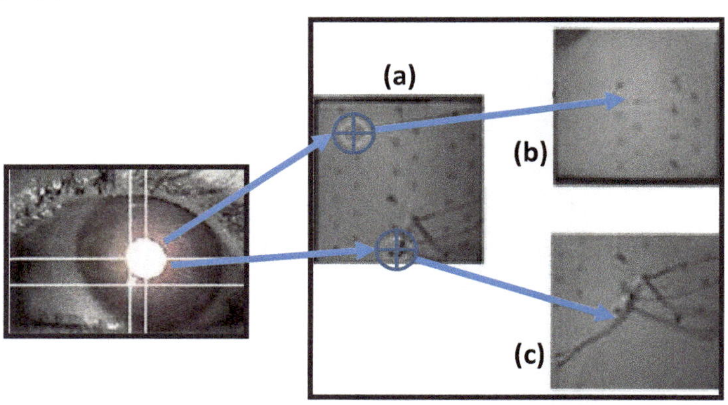

One weakness of relying solely on eye gaze is that tracking technology remains relatively unreliable. Variations in tracking performance in different environments and between individual users continue to be a challenge for applications. Eye gaze might be useful as one of the components of a system, but researchers should consider multimodal or redundant methods to increase the robustness of the interface.

2.1.2. Instrument Tracking

The teleoperator's intent is also expressed through any instruments he or she may be using. The operator manipulates instruments in order to achieve desired goals, which indicates that the operator's intent is closely related to the manner in which the instruments are manipulated. If an intelligent camera control system can ascertain and utilize the dynamic state of the instruments, it can better cooperate with the user.

The state of an instrument includes general properties such as its position, orientation, speed, and acceleration. It also includes instrument-specific properties, such as whether a gripper is open or closed. There are various mechanisms by which a control system can determine these properties. The following subsections describe some of these mechanisms.

2.1.3. Kinematic Tracking

If a system with robotic instrument control is being used, such as the da Vinci Surgical System, kinematic tracking [28] can generally be used to determine various properties of the instruments. A typical robotic system includes one or more jointed arms that manipulate instruments on behalf of the user. The robot arm includes sensors (like encoders or potentiometers) that can accurately determine the state of each joint. If the fixed properties of the physical structure of the robot arm are known (lengths of links, twists, *etc.*), they can be combined with the dynamic joint values to form a mathematical model of the robot arm [29]. Desired instrument properties, such as the position and orientation of the end-effector, can be computed from this model.

The positional information resulting from kinematic tracking is generally expressed in terms of some coordinate system that is specific to the robot. In order to express this information with respect to some other coordinate system, such as one associated with the view from the camera, a transformation that maps between the two systems must be determined. This transformation can be determined through a registration process. A typical registration process involves finding a number of features (such as predefined reference locations) in both coordinate systems and using an algorithm to compute a transformation between the systems [28].

2.1.4. Image-Based Tracking

In some applications, such as traditional laparoscopic surgery, there is no robotic instrument manipulation system. In this case, the properties of the instruments must be determined though other mechanisms. Since the systems addressed in this paper would all have a camera, it is logical to use the camera to assess instrument properties [30].

In addition, there is value in tracking objects in the environment besides the instruments. A basic autonomous camera system could use this information to avoid obstacles or focus on points of interest.

A more advanced autonomous system might evaluate how the user interacts with different objects to ascertain the user's intent or identify the task that is currently being performed.

There are many techniques for tracking objects in a camera image. Objects of interest can be tracked by identifying inherent, distinguishing characteristics of the objects. For example, this could include the shape, color, texture, and movement of the objects. To enhance the tracking, the objects can be modified to make them more recognizable in the camera image. For instance, colored markers or tracking patterns can be affixed to instruments to aid in detection by a computer algorithm [31–33].

Depending on the application and environment, image-based tracking techniques may not be sufficient. The resolution of the camera limits accuracy, and most images only contain 2D information. Objects may be occluded, insufficiently lit, or simply not have sufficiently distinguishable features. Moreover, the amount of computer processing required for some image analysis may be prohibitive for real-time applications. For these reasons, it may be necessary to add additional sensing techniques to track the instruments.

2.1.5. Other Tracking Technologies

There are many different types of sensors that can be used to track an instrument or other objects. Optical techniques, such as an infrared tracking system, can locate objects that have infrared markers attached to them. The objects being tracked do not require any wires, but the line of sight from the tracking system to the tracked objects must be kept clear.

A magnetic tracking system can also be used to locate instruments or other objects of interest. Magnetic sensors are affixed to the objects, and a magnetic transmitter emits a field that the sensors can detect. Unfortunately, the presence of objects in the operating room that affect or are affected by magnetic fields may make this technology infeasible for surgery [34–36].

Another type of tracking technology is called an inertial measurement unit (IMU). An IMU incorporates multiple sensors, such accelerometers, gyroscopes, and magnetometers, to track the orientation, position, or velocity of an object. An IMU can be relatively small and be designed to transmit data wirelessly. However, these devices experience increasing error over time (especially in position), and some of the sensors may be sensitive to interference from other devices in an operating room.

In summary, beyond the algorithmic and knowledge-based challenges posed, there are some hardware technological barriers that need to be overcome. Better tracking methods that are suitable for use in the operating room are required. As described above, there are limitations to the use of optical, magnetic, and inertial sensors in operating room.

2.2. Gesture/Task Classification

Building on the tracking technologies discussed above, progress has been made in relating basic tool-level data (e.g., tool position, orientation, and motion/dynamics) to more abstract knowledge-based constructs at the surgical task level [37]. For example, Speidel *et al.* [38] analyzed image data collected from an endoscope, highlighting visual features and key points to identify movement characteristics. Using this technique, they isolated and identified the trajectories, velocities, and shapes of surgical instruments, and then they used a Bayes classifier to predict the next surgical subtask.

Using recorded motions from a da Vinci Surgical System, Murphy [39] applied linear discriminant analysis to automatically classify seven surgical subtasks. Next, Murphy used this information to characterize skill based on the absence, presence, or order of the surgical gestures. In other work with the da Vinci platform, Lin *et al.* [40] applied an artificial neural network to classify signals into eight ad hoc surgeon gestures. Their study also aimed to automatically determine surgeon skill level. Reiley *et al.* [41] extended the research of Lin *et al.* [40], adding three gestures to Lin's breakdown (based on surgery observations) for a total of eleven gestures. However, five of their motions were consolidated into one gesture class for processing and analysis.

This body of work shows promise. There were generally acceptable classification accuracies and meaningful outputs. However, further research is required. The employed task breakdowns are ad hoc and potentially idiosyncratic. Future work should start from generally accepted and externally validated task analyses and surgery process models. Regardless, this work is important because it could bridge the gap from low-level signal processing approaches to knowledge-based approaches for camera control. Research is still needed, however, to determine the specific visualization requirements on a task-by-task basis.

In a related topic, Weede *et al.* [42] describe a cognitive robotic system based on closed-loop control. They focus on knowledge acquisition and surgical workflow analysis, and they relate their work to autonomous camera control. They make a clear distinction between automatic systems and autonomous systems. They contend that autonomous systems (unlike automatic systems) must take the environment and task context into account when making decisions. Another fertile area that may lead to insight into cognitive robotics is surgical skills acquisition. Reiley *et al.*'s [43] review paper discusses descriptive statistics based on time and motion analyses for surgical training. They state that developed skill models can also be useful in training robots for building human–robot collaborative systems.

2.3. Path Planning Related to Camera Motion Control

Motion control of virtual cameras (in the field of computer graphics) uses path-planning techniques and is related to the area of laparoscopic camera navigation. It is a PSPACE-hard problem with a complexity that is exponential in the number of degrees of freedom [10]. For a laparoscopic camera, a path-planning technique could, for example, use inputs such as the user's gaze, tool locations, object importance levels, and the surgical task timeline to generate an optimal trajectory for the camera (within kinematic, collision, and visual obstruction constraints). In robotic path planning, there are several ways to represent spatial knowledge for optimally planning solutions that would be useful for the autonomous camera problem. Note that traditional robotic path planning, which tells a robot how to move between two points, is not the same as viewpoint determination/generation, which defines where the camera should be pointed. That topic is discussed in a later section.

2.3.1. Search Algorithm Approaches

The simplest approach involves a grid-based search, which overlays a grid on an operational space. There are many probabilistic path planning algorithms that find paths and trajectories, such as the A* and D* search algorithms, which are based on Dijkstra's [44] algorithm for node traversal. In addition, visibility graphs can be used to create a set of nodes that are only connected where there are no obstacles. In a variety of applications, robots gather knowledge about their surroundings using sensors and build a map of their environment as they move.

2.3.2. Vector Field and Potential Field Approaches

Other approaches involve vector fields, where each point on a grid represents the flow of some quantity. The advantage of these methods is that computations such as gradients, curls, and line integrals can be performed to ascertain quantities such as rates of change, rotations, and the amount work done. An extension of vector fields is the concept of potential fields. In this paradigm, obstacles have a repulsive charge and goals have an attractive charge. It is possible to produce a potential field-based system for robotic path planning that has a low computational load.

There is a large body of research concerning potential fields [45–47]. Most notable is the work of Khatib [48], which proposes the use of artificial potential fields using a Lagrangian approach to model the potential energy/operational force vectors. In this approach, the manipulator moves in a field of forces where attractive poles are defined for the end-effector and targets and repulsive forces are defined for the manipulator and obstacles. Since this paper was published, the use of artificial potential fields for path planning has become ubiquitous. In another key paper by Barraquand *et al.* [49], a graph is built to connect the local minima of a potential field, and the graph is searched until a goal robot configuration is reached. Their main contribution lies in making the algorithm tractable. Only a local search is performed, which allows the robot to navigate a global path in near real time. A number of techniques to implement this approach are presented. Their distributed approach allowed the authors to construct potential fields numerically rather than analytically, which in turn allowed them to create a usable implementation.

2.4. Existing Methods for Viewpoint Generation in Computer Graphics

While there are existing methods that attempt to address the camera-control problem, they are mostly related to computer graphics/computer gaming [45,46,50,51]. In a key review paper [10] of camera control in computer graphics, the authors review a number of state-of-the-art techniques for the yet open research question of camera control. They include constraint-based, potential-field-based, and optimization-based approaches. A very important first conclusion they make is that camera control is difficult, requiring great deal of dexterity. It necessitates the manipulation of seven degrees of freedom: x-position, y-position, z-position, roll, pitch, yaw, and zoom. In addition, the movement speeds and accelerations must be controlled.

Due to the complexity of camera placement, the authors recommend that the next generation of camera control systems not only provide efficient implementations, but also include empirical knowledge gained from experts to extract higher-level properties that will facilitate view aesthetics. Based on this recommendation, we believe that strategies employed by expert users of a telerobotic system also need to be studied in order to incorporate the nuances of their specific application domain. The authors also outline key steps that are taken to reduce the computational complexity.

In a series of papers [9,47,52,53], Burelli *et al.* describe the use of an artificial potential field implementation to control a camera view for virtual/gaming environments. They describe a method that uses particles suspended in an artificial potential field to control both camera position and orientation. A scene's geometry and three different constraints for each object (visibility, projection size, and viewing angle) are modeled into the system using the notion of potential fields. The papers also discuss the case of under-constrained systems in which several camera poses can satisfy the same requirements. Their solution involves an optimization that allows convergence to the nearest configuration that meets all the constraints. Their implementation, however, is for a gaming environment where the state space of every object in the scene is known.

Hornung [54] provides a nomenclature that is defined relative to the type of camera view. The author formalizes cinematographic concepts and shows a methodology to transfer this knowledge to the autonomous camera movement domain using techniques in neural networks and artificial intelligence.

Several papers [13,55] describe a camera planning algorithm for computer graphics. The developed algorithm allows the user to select certain generic goals, strategies, and presentation methods. The system then uses a genetic algorithm to optimize the camera viewpoints. The authors claim that genetic algorithms work well with ambiguous data values found in camera planning.

Lastly, another active field in which camera path-planning techniques play an important role is virtual endoscopy. Hong *et al.* [56] describe a method that avoids collisions during virtual-reality-guided navigation of the colon. The surfaces of the colon and the centerline of the colon are modeled with potential fields, with the walls being repulsive and the centerline being attractive. This is a promising approach for a real-time viewpoint-generation methodology.

2.5. Existing Autonomous Camera Systems

Various autonomous camera systems have been created for minimally invasive surgery [26,30,31,33,57–60]. For tracking, most of these systems used image processing to identify the position of the tool tips relative to the camera. To move the camera, the systems generally used a limited number of rules to set the camera's target position and zoom level. For instance, a commonly used rule sets the camera's target position to the midpoint of two tracked tool tips and alters the zoom level as necessary to keep the tool tips in the camera's view [61]. These systems generally are implemented with laparoscopic instruments using image processing to automatically move a camera with no explicit human direction. A camera control algorithm automatically keeps the tools in the view. A system might for instance, zoom in when the tools are brought close together, and zoom out to always keep the tools in the field of view. The advantage of this approach is that no sensors are necessary; making adoption to existing laparoscopic systems easier (see Figure 3). In general, the testing of the surgical systems was limited to subjective feedback about their performance.

Other methods of tracking the tool tips include using the kinematics of the robot. In these types of systems, the 3D poses of the tools are calculated using the known/measured joint angles, the DH parameters of the system [29], and a keyhole-based (due to the trocar point of surgical insertion) forward kinematics algorithm [28]. Once the tools are located in the working region, the camera arm moves to keep the tools within the view of the camera.

Some more recent work has also focused the development of autonomous laparoscopic camera systems. Weede *et al.* [42,62] have developed a test system that applies a Markov model to predict the motions of the tools so the camera can follow them. The system is trained using data from previous surgical interventions so it can operate more like an expert camera operator. Yu *et al.* [63] propose algorithms for determining how to move the laparoscope from one viewing location to another using kinematic models of a robotic surgery system.

Figure 3. A test platform for automatic laparoscopic camera movement that tracks the tools via colored markers [61]. The system automatically centers the camera's view on the two tools and alters the zoom level (based on the distance between the tools) to ensure both tools are in view at all times.

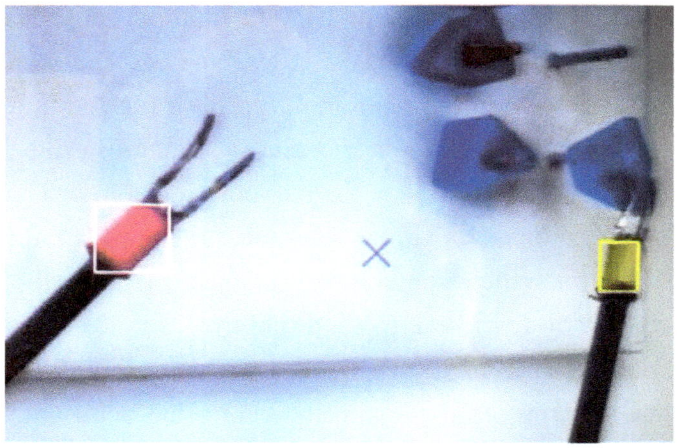

3. Potential Future Research Avenues

We envision that successful completion of the several research avenues discussed herein will provide the foundation for intelligent robots that perform automatic camera control in real time. This type of co-robotic system can reduce the cognitive workload of operators, allowing their focus to remain on the primary task. In addition, this technology can reduce the amount of time and the number of personnel required for teleoperations, producing a quantifiable cost savings to society in numerous applications. Moreover, the proposed ideas should help nurture the adoption of medical telerobotic systems in the marketplace by enhancing ease-of-use while reducing costs and errors.

As previously discussed, there are existing methods that attempt to address the camera-control problem. Most are based on reactive techniques, and others are related to computer graphics/computer gaming. Those methods are primarily based on simple virtual environments. They are generally focused on a camera's position for cinematographic effects and incorporate very little expert knowledge for camera operation.

To advance the field of camera control for surgical applications, there is a need for systems that are (A) capable of generating camera movements that match the quality of domain-specific expert camera operators to assist human partners; (B) able to learn from the knowledge and performance of expert human camera operators; and (C) generalizable for effective application in many different surgical and potentially non-surgical domains (military operations, space exploration, surveillance, border patrol, first response teams, *etc.*). We propose that the following are important research areas that will help meet those needs: (1) improved tracking methods; (2) algorithms for skills acquisition; (3) knowledge representation; (4) user intent modeling; and (5) methods for evaluation and testing. These areas are summarized in the following subsections.

3.1. Tracking

While there are noteworthy attempts to track surgical instruments through visual, kinematic, and other methods, these techniques give little information about the state of the rest of the scene. Examples of important entities include tissues that are bleeding, a foreign object, *etc.* It is conceivable that future tracking methods should not only track the instruments, but also be capable of tracking arbitrary objects defined by the user. In addition, as mentioned earlier, current forms of tracking (optical, magnetic, kinematic, *etc.*) have various limitations. Other methods of tracking need to be developed that are better suited to the operating room and will allow robust tracking of many (potentially ill-defined) objects.

3.2. Algorithms for Skills Acquisition

Gaining a deeper understanding of expert users of a system is essential. Research should be focused on forming a link between expert camera operators and parameters that define how the camera should be moved. There is a potential to use human factors methods, such as cognitive task analysis [64], to analyze how experts control laparoscopic cameras. This research could involve interviewing expert surgeons to discover and document salient varieties of camera control and their related task contexts. The results could be used to develop better robotic control algorithms that will be more responsive to the surgeon's needs.

Desired camera movement is dependent on the specific application, which is why research is needed to create fundamental movement algorithms and behaviors that can be generalized among robotic platforms. A generic method of training a robot for autonomous camera movement would be applicable to many different domains (in both surgery and general telerobotics). For instance, camera behavior parameters for a particular laparoscopic surgery could be derived from expert surgical camera operators from that domain. More specifically, the kinematic nuances of the domain could be derived from recordings of movements performed during user operations by a system equipped with various tracking sensors. The recorded movement data could be processed by various statistical and learning algorithms to develop models of how the camera should be moved.

For example, Reiley *et al.* and Lin *et al.* [40,41,43] describe the automation of skill evaluation, modeling, and capture. In another paper, van den Berg *et al.* demonstrated surgical task automation by collecting and learning from human operator data [65]. In their approach, trajectories of specific movements like knot tying are recorded, learned, and then autonomously executed with enhanced performance in terms of speed and smoothness.

3.3. Knowledge Representation

Research is needed to find effective ways to represent the contents of the working envelope for the camera control algorithms. Generically, this might consist of objects or regions of interest (tools, organs, *etc.*) and their properties (size, velocity, *etc.*). The representation should allow the encapsulation of not only the position and orientation of different objects, but also their relative levels of importance. The use of these representations should allow remote viewpoints to be automatically optimized based on a given set of input parameters. This could be accomplished by simulating how the camera would see the objects from different viewpoints and using mathematical optimization to select an effective viewpoint (camera orientation, position, and zoom level). The technical approaches could involve the development of control systems that build on a combination of optimization techniques and a system of representation, such as potential field theory.

In addition, these intelligent systems should allow data and knowledge from experts to be translated into spatial and other descriptive parameters in order to capture the nuances of domain experts. Such a framework should allow different entities have different types of influence on the camera's viewpoint. By adjusting the parameters of the various entities in the environment, the system's behavior can be altered to accommodate specific teleoperation applications. Relevant variables, such as object positions/dynamics, sensor inputs, user intentions, and changes in the image could be weighted (based on user inputs) to affect the camera's movement in domain-specific ways. The goal would be to embed the knowledge of experts into a model that can be used to generate effective camera trajectories for a particular application or task.

3.4. Detecting and Interpreting Operator Intent

To be most effective, an intelligent camera system requires the knowledge of what task(s) the operator is currently performing. Within all of the major research areas described above, one point is clear: moving a camera requires information about what is currently happening in the environment. This information can be used to optimize camera behaviors based on the context and better determine the operator's intent. It may also help the camera control algorithms handle unfamiliar situations.

Task analysis, a common technique to help chart and organize actions, can be used to categorize and organize data for use in an operator intent classification algorithm. More research into these methods is needed to efficiently establish implementations at the right level of detail to infer operator intent for telerobotic applications. There has been research that detects and segments the intent of a surgeon during robotic surgery [37,40], but this work has mainly focused on detecting surgical motions/tasks using complex tracking hardware. An extension of this work would be to use more natural methods of input, such as the video feed. Research that could extract the operator intentions from video would be much more advantageous because it would not require any additional tracking hardware or access to the robot's hardware. Moreover, it could be more easily applied to laparoscopic surgery, which is much more common than robotic surgery.

3.5. Evaluation and Testing

For the adoption of autonomous camera techniques, especially in the surgical arena, the systems developed for any application area must be thoroughly evaluated. A robust system of evaluating the selected viewpoints should be implemented, as well as ways to assess operator performance on particular tasks. Steinfeld *et al.* [66] provide an overview and analysis of metrics for human–robot interaction performance, and many of these could be easily adapted to measure task performance in robot-assisted camera control in surgical settings. These include aspects of overall task performance time, task performance efficiency, and the mental workload of the human member of the human–robot team. In evaluating the system's performance, the same tasks should be performed using both an expert human camera operator and the automated system to determine how the system compares to expert human operators. The analysis should consider unbiased metrics and potentially other factors deemed important by the expert users, such as: "are all the important actors in the scene?", "how much irrelevant information is present?", and "what level of detail is visible in the target area?"

Finally, clinical trials that test these systems in animal models and human surgical cases should be systematically performed. The U.S. Food and Drug Administration has determined that all technology must be validated for safe and effective use, and surgical equipment is no exception. Recent guidance from the FDA requires that human factors methods be applied during the design phase [67].

4. Conclusions

Successful completion of the proposed research topics will help provide the foundation for intelligent robots that perform autonomous camera control in real time. We also need standard, usable, and programmable surgical platforms that are sufficiently affordable for use in research and translate well to actual surgical platforms. This will allow mature research to have a streamlined path towards making a direct impact in real surgical robotic applications. The da Vinci Research Kit (DVRK), which is being developed by a consortium of universities and an industry partner (Intuitive Surgical), is an excellent starting point for this kind of research [68]. In addition, there are other potentially useful platforms, such as the RAVEN I and RAVEN II [69,70].

Although the focus of this paper has been on surgical robotics, the gains in this field will also affect remote operations in general. For instance, an astronaut can have eight or more camera views to simultaneously manage in space station operations. Similarly, manual camera control for unmanned military systems can reduce situational awareness under high workload and increase crew size requirements. For each of these examples, camera-related tasks are considerable sources of distraction, diverting users away from what should be the focus of human attention: on-orbit tasks, surveillance, military operations, *etc.* Autonomous camera control has the potential to enhance the performance of these tasks in ways that are similar to the improvements in surgical applications.

Acknowledgments

This work was supported in part by the Endowment for Surgical Research (ENSURE) and the Arvin I. Philippart Endowed Chair in Pediatric Surgical Research at Children's Hospital of Michigan.

Conflicts of Interest

The authors declare no conflict of interest.

References

1. Berguer, R.; Forkey, D.; Smith, W. The effect of laparoscopic instrument working angle on surgeons' upper extremity workload. *Surg. Endosc.* **2001**, *15*, 1027–1029.
2. Keehner, M.M.; Tendick, F.; Meng, M.V.; Anwar, H.P.; Hegarty, M.; Stoller, M.L.; Duh, Q.-Y. Spatial ability, experience, and skill in laparoscopic surgery. *Am. J. Surg.* **2004**, *188*, 71–75.
3. Zheng, B.; Cassera, M.A.; Martinec, D.V.; Spaun, G.O.; Swanström, L.L. Measuring mental workload during the performance of advanced laparoscopic tasks. *Surg. Endosc.* **2010**, *24*, 45–50.

4. Christensen, H.I.; Goldberg, K.; Kumar, V.; Messina, E. *A Roadmap for U.S. Robotics: From Internet to Robotics*; Robotics-Virtual Organization: Atlanta, GA, USA, 2013.
5. Kranzfelder, M.; Staub, C.; Fiolka, A.; Schneider, A.; Gillen, S.; Wilhelm, D.; Friess, H.; Knoll, A.; Feussner, H. Toward increased autonomy in the surgical OR: Needs, requests, and expectations. *Surg. Endosc.* **2013**, *27*, 1681–1688.
6. Aiono, S.; Gilbert, J.; Soin, B.; Finlay, P.; Gordan, A. Controlled trial of the introduction of a robotic camera assistant (endoassist) for laparoscopic cholecystectomy. *Surg. Endosc. Interv. Tech.* **2002**, *16*, 1267–1270.
7. Schäfer, M.; Krähenbühl, L.; Büchler, M. Comparison of adhesion formation in open and laparoscopic surgery. *Dig. Surg.* **1998**, *15*, 148–152.
8. Shah, R.D.; Cao, A.; Golenberg, L.; Ellis, R.D.; Auner, G.W.; Pandya, A.K.; Klein, M.D. Performance of basic manipulation and intracorporeal suturing tasks in a robotic surgical system: Single- *versus* dual-monitor views. *Surg. Endosc.* **2009**, *23*, 727–733.
9. Burelli, P.; di Gaspero, L.; Ermetici, A.; Ranon, R. Virtual Camera Composition with Particle Swarm Optimization. In Proceedings of the 8th International Symposium on Smart Graphics, Rennes, France, 27–29 August 2008; pp. 130–141.
10. Christie, M.; Olivier, P.; Normand, J.-M. Camera control in computer graphics. *Comput. Graph. Forum* **2008**, *27*, 2197–2218.
11. Drucker, S.M.; Zeltzer, D. Intelligent Camera Control in a Virtual Environment. In Proceedings of the Graphics Interface, Banff, AB, Canada, 18–20 May 1994; p. 190.
12. Gleicher, M.; Witkin, A. Through-the-Lens Camera Control. In Proceedings of the 19th Annual Conference on Computer Graphics and Interactive Techniques, Chicago, IL, USA, 26–31 July 1992; pp. 331–340.
13. Halper, N.; Olivier, P. Camplan: A Camera Planning Agent. In Proceedings of the AAAI Spring Symposium on Smart Graphics, Stanford, CA, USA, 20–22 March 2000; pp. 92–100.
14. Dragan, A.; Lee, K.C.T.; Srinivasa, S. Teleoperation with intelligent and customizable interfaces. *J. Hum.-Robot Interact.* **2013**, *2*, 33–79.
15. Duchowski, A.T. A breadth-first survey of eye-tracking applications. *Behav. Res. Methods Instrum. Comput.* **2002**, *34*, 455–470.
16. Morimoto, C.H.; Mimica, M.R. Eye gaze tracking techniques for interactive applications. *Comput. Vis. Image Underst.* **2005**, *98*, 4–24.
17. Hansen, D.W.; Ji, Q. In the eye of the beholder: A survey of models for eyes and gaze. *IEEE Trans. Pattern Anal. Mach. Intell.* **2010**, *32*, 478–500.
18. Salvucci, D.D.; Goldberg, J.H. Identifying Fixations and Saccades in Eye-Tracking Protocols. In Proceedings of the Symposium on Eye Tracking Research & Applications, Palm Beach Gardens, FL, USA, 6–8 November 2000; pp. 71–78.
19. Jacob, R.J.; Karn, K.S. Eye tracking in human-computer interaction and usability research: Ready to deliver the promises. In *The Mind's Eyes: Cognitive and Applied Aspects of Eye Movements*; Hyona, J., Radach, R., Deubel, H., Eds.; Elsevier Science: Amsterdam, The Netherlands, 2003; p. 573.

20. Goldberg, J.H.; Schryver, J.C. Eye-gaze-contingent control of the computer interface: Methodology and example for zoom detection. *Behav. Res. Methods Instrum. Comput.* **1995**, *27*, 338–350.
21. Hansen, D.W.; Skovsgaard, H.H.; Hansen, J.P.; Møllenbach, E. Noise Tolerant Selection by Gaze-Controlled Pan and Zoom in 3D. In Proceedings of the Symposium on Eye Tracking Research & Applications, Savannah, GA, USA, 26–28 March 2008; pp. 205–212.
22. Latif, H.O.; Sherkat, N.; Lotfi, A. Telegaze: Teleoperation through Eye Gaze. In Proceedings of the 7th IEEE International Conference on Cybernetic Intelligent Systems, London, UK, 9–10 September 2008; pp. 1–6.
23. Latif, H.O.; Sherkat, N.; Lotfi, A. Teleoperation through Eye Gaze (Telegaze): A Multimodal Approach. In Proceedings of the IEEE International Conference on Robotics and Biomimetics, Guilin, China, 19–23 December 2009; pp. 711–716.
24. Zhu, D.; Gedeon, T.; Taylor, K. "Moving to the centre": A gaze-driven remote camera control for teleoperation. *Interact. Comput.* **2011**, *23*, 85–95.
25. Kwok, K.-W.; Sun, L.-W.; Mylonas, G.P.; James, D.R.C.; Orihuela-Espina, F.; Yang, G.-Z. Collaborative gaze channelling for improved cooperation during robotic assisted surgery. *Ann. Biomed. Eng.* **2012**, *40*, 2156–2167.
26. Ali, S.M.; Reisner, L.A.; King, B.; Cao, A.; Auner, G.; Klein, M.; Pandya, A.K. Eye gaze tracking for endoscopic camera positioning: An application of a hardware/software interface developed to automate aesop. *Stud. Health Technol. Inform.* **2007**, *132*, 4–7.
27. Mylonas, G.P.; Darzi, A.; Zhong Yang, G. Gaze-contingent control for minimally invasive robotic surgery. *Comput. Aided Surg.* **2006**, *11*, 256–266.
28. Mudunuri, A.V. Autonomous Camera Control System for Surgical Robots. M.S. Thesis, Wayne State University, Detroit, MI, USA, 2011.
29. Denavit, J.; Hartenberg, R.S. A kinematic notation for lower-pair mechanisms based on matrices. *ASME J. Appl. Mech.* **1955**, *22*, 215–221.
30. Lee, C.; Wang, Y.-F.; Uecker, D.R.; Wang, Y. Image Analysis for Automated Tracking in Robot-Assisted Endoscopic Surgery. In Proceedings of the 12th IAPR International Conference on Pattern Recognition, Jerusalem, Israel, 9–13 October 1994; pp. 88–92.
31. Wei, G.-Q.; Arbter, K.; Hirzinger, G. Real-time visual servoing for laparoscopic surgery: Controlling robot motion with color image segmentation. *IEEE Eng. Med. Biol. Mag.* **1997**, *16*, 40–45.
32. Omote, K.; Feussner, H.; Ungeheuer, A.; Arbter, K.; Wei, G.-Q.; Siewert, J.R.; Hirzinger, G. Self-guided robotic camera control for laparoscopic surgery compared with human camera control. *Am. J. Surg.* **1999**, *177*, 321–324.
33. Fortney, D.R. *Real-Time Color Image Guidance System*; University of California at Santa Barbara: Santa Barbara, CA, USA, 2000.
34. Hastenteufel, M.; Vetter, M.; Meinzer, H.-P.; Wolf, I. Effect of 3D ultrasound probes on the accuracy of electromagnetic tracking systems. *Ultrasound Med. Biol.* **2006**, *32*, 1359–1368.
35. LaScalza, S.; Arico, J.; Hughes, R. Effect of metal and sampling rate on accuracy of flock of birds electromagnetic tracking system. *J. Biomech.* **2003**, *36*, 141–144.

36. King, D.L. Errors as a result of metal in the near environment when using an electromagnetic locator with freehand three-dimensional echocardiography. *J. Am. Soc. Echocardiogr.* **2002**, *15*, 731–735.
37. Golenberg, L.P. Task Analysis, Modeling, and Automatic Identification of Elemental Tasks in Robot-Assisted Laparoscopic Surgery. Ph.D. Thesis, Wayne State University, Detroit, MI, USA, 2010.
38. Speidel, S.; Delles, M.; Gutt, C.; Dillmann, R. Tracking of instruments in minimally invasive surgery for surgical skill analysis. In *Medical Imaging and Augmented Reality*; Springer: Berlin, Germany, 2006; pp. 148–155.
39. Murphy, T.E. Towards Objective Surgical Skill Evaluation with Hidden Markov Model-Based Motion Recognition. M.S. Thesis, Johns Hopkins University, Baltimore, MD, USA, 2005.
40. Lin, H.C.; Shafran, I.; Yuh, D.; Hager, G.D. Towards automatic skill evaluation: Detection and segmentation of robot-assisted surgical motions. *Comput. Aided Surg.* **2006**, *11*, 220–230.
41. Reiley, C.E.; Lin, H.C.; Varadarajan, B.; Vagvolgyi, B.; Khudanpur, S.; Yuh, D.; Hager, G. Automatic recognition of surgical motions using statistical modeling for capturing variability. *Stud. Health Technol. Inform.* **2008**, *132*, 396.
42. Weede, O.; Bihlmaier, A.; Hutzl, J.; Müller-Stich, B.P.; Wörn, H. Towards Cognitive Medical Robotics in Minimal Invasive Surgery. In Proceedings of the Conference on Advances in Robotics, Pune, India, 4–6 July 2013; pp. 1–8.
43. Reiley, C.E.; Lin, H.C.; Yuh, D.D.; Hager, G.D. Review of methods for objective surgical skill evaluation. *Surg. Endosc.* **2011**, *25*, 356–366.
44. Alexopoulos, C.; Griffin, P.M. Path planning for a mobile robot. *IEEE Trans. Syst. Man Cybern.* **1992**, *22*, 318–322.
45. Beckhaus, S.; Ritter, F.; Strothotte, T. Cubicalpath-Dynamic Potential Fields for Guided Exploration in Virtual Environments. In Proceedings of the Eighth Pacific Conference on Computer Graphics and Applications, Hong Kong, China, 3–5 October 2000; pp. 387–459.
46. Beckhaus, S.; Ritter, F.; Strothotte, T. Guided exploration with dynamic potential fields: The cubical path system. *Comput. Graph. Forum* **2001**, *20*, 201–211.
47. Burelli, P.; Jhala, A. Dynamic Artificial Potential Fields for Autonomous Camera Control. In Proceedings of the Fifth Artificial Intelligence and Interactive Digital Entertainment, Palo Alto, CA, USA, 14–16 October 2009.
48. Khatib, O. Real-time obstacle avoidance for manipulators and mobile robots. *Int. J. Robot. Res.* **1986**, *5*, 90–98.
49. Barraquand, J.; Langlois, B.; Latombe, J.-C. Numerical potential field techniques for robot path planning. *IEEE Trans. Syst. Man Cybern.* **1992**, *22*, 224–241.
50. Bares, W.H.; Grégoire, J.P.; Lester, J.C. Realtime Constraint-Based Cinematography for Complex Interactive 3D Worlds. In Proceedings of the Tenth Conference on Innovative Applications of Artificial Intelligence, Madison, WI, USA, 26–30 July 1998; pp. 1101–1106.
51. He, L.-W.; Cohen, M.F.; Salesin, D.H. The Virtual Cinematographer: A Paradigm for Automatic Real-Time Camera Control and Directing. In Proceedings of the 23rd Annual Conference on Computer Graphics and Interactive Techniques, New Orleans, LA, USA, 4–9 August 1996; pp. 217–224.

52. Burelli, P. Interactive Virtual Cinematography. Ph.D. Thesis, IT University Of Copenhagen, Copenhagen, Denmark, 2012.
53. Burelli, P.; Yannakakis, G.N. Global Search for Occlusion Minimisation in Virtual Camera Control. In Proceedings of the IEEE Congress on Evolutionary Computation, Barcelona, Spain, 18–23 July 2010; pp. 1–8.
54. Hornung, A.; Lakemeyer, G.; Trogemann, G. An autonomous real-time camera agent for interactive narratives and games. In *Intelligent Virtual Agents*; Rist, T., Aylett, R., Ballin, D., Rickel, J., Eds.; Springer: Berlin/Heidelberg, Germany: 2003; Volume 2792, pp. 236–243.
55. Jardillier, F.; Languénou, E. Screen-space constraints for camera movements: The virtual cameraman. *Comput. Graph. Forum* **1998**, *17*, 175–186.
56. Hong, L.; Muraki, S.; Kaufman, A.; Bartz, D.; He, T. Virtual Voyage: Interactive Navigation in the Human Colon. In Proceedings of the 24th Annual Conference on Computer Graphics and Interactive Techniques, Los Angeles, CA, USA, 3–8 August 1997; pp. 27–34.
57. Ko, S.-Y.; Kim, J.; Lee, W.-J.; Kwon, D.-S. Compact laparoscopic assistant robot using a bending mechanism. *Adv. Robot.* **2007**, *21*, 689–709.
58. Ko, S.-Y.; Kwon, D.-S. A Surgical Knowledge Based Interaction Method for A Laparoscopic Assistant Robot. In Proceedings of the 13th IEEE International Workshop on Robot and Human Interactive Communication, Kurashiki, Okayama, Japan, 20–22 September 2004; pp. 313–318.
59. Kai-Tai, S.; Chun-Ju, C. Autonomous and Stable Tracking of Endoscope Instrument Tools with Monocular Camera. In Proceedings of the IEEE/ASME International Conference on Advanced Intelligent Mechatronics, Kachsiung, Taiwan, 11–14 July 2012; pp. 39–44.
60. Azizian, M.; Khoshnam, M.; Najmaei, N.; Patel, R.V. Visual servoing in medical robotics: A survey. Part I: Endoscopic and direct vision imaging—Techniques and applications. *Int. J. Med. Robot. Comput. Assist. Surg.* **2013**, doi:10.1002/rcs.1531.
61. King, B.W.; Reisner, L.A.; Pandya, A.K.; Composto, A.M.; Ellis, R.D.; Klein, M.D. Towards an autonomous robot for camera control during laparoscopic surgery. *J. Laparoendosc. Adv. Surg. Tech.* **2013**, *23*, 1027–1030.
62. Weede, O.; Monnich, H.; Muller, B.; Worn, H. An Intelligent and Autonomous Endoscopic Guidance System for Minimally Invasive Surgery. In Proceedings of the IEEE International Conference on Robotics and Automation, Shanghai, China, 9–13 May 2011; pp. 5762–5768.
63. Yu, L.; Wang, Z.; Sun, L.; Wang, W.; Wang, T. A Kinematics Method of Automatic Visual Window for Laparoscopic Minimally Invasive Surgical Robotic System. In Proceedings of the IEEE International Conference Mechatronics and Automation, Takamatsu, Japan, 4–7 August 2013; pp. 997–1002.
64. Crandall, B.; Klein, G.A.; Hoffman, R.R. *Working Minds: A Practitioner's Guide to Cognitive Task Analysis*; MIT Press: Cambridge, MA, USA, 2006.
65. Van den Berg, J.; Miller, S.; Duckworth, D.; Hu, H.; Wan, A.; Fu, X.-Y.; Goldberg, K.; Abbeel, P. Superhuman Performance of Surgical Tasks by Robots Using Iterative Learning from Human-Guided Demonstrations. In Proceedings of the IEEE International Conference on Robotics and Automation, Anchorage, AK, USA, 3–7 May 2010; pp. 2074–2081.

66. Steinfeld, A.; Fong, T.; Kaber, D.; Lewis, M.; Scholtz, J.; Schultz, A.; Goodrich, M. Common Metrics for Human-Robot Interaction. In Proceedings of the 1st ACM SIGCHI/SIGART Conference on Human-Robot Interaction, Salt Lake City, UT, USA, 2–3 March 2006; pp. 33–40.
67. Food and Drug Administration. Draft Guidance for Industry and Food and Drug Administration Staff—Applying Human Factors and Usability Engineering to Optimize Medical Device Design. Available online: http://www.fda.gov/medicaldevices/deviceregulationandguidance/guidancedocuments/ucm259748.htm (accessed on 17 July 2014).
68. Chen, Z.; Deguet, A.; Taylor, R.; DiMaio, S.; Fischer, G.; Kazanzides, P. An Open-Source Hardware and Software Platform for Telesurgical Robotics Research. In Proceedings of the MICCAI Workshop on Systems and Architecture for Computer Assisted Interventions, Nagoya, Japan, 22–26 September 2013.
69. Lum, M.J.; Friedman, D.C.; Sankaranarayanan, G.; King, H.; Fodero, K.; Leuschke, R.; Hannaford, B.; Rosen, J.; Sinanan, M.N. The RAVEN: Design and validation of a telesurgery system. *Int. J. Robot. Res.* **2009**, *28*, 1183–1197.
70. Hannaford, B.; Rosen, J.; Friedman, D.W.; King, H.; Roan, P.; Cheng, L.; Glozman, D.; Ma, J.; Kosari, S.N.; White, L. Raven-II: An open platform for surgical robotics research. *IEEE Trans. Biomed. Eng.* **2013**, *60*, 954–959.

6

IMU and Multiple RGB-D Camera Fusion for Assisting Indoor Stop-and-Go 3D Terrestrial Laser Scanning

Jacky C.K. Chow [1,*], Derek D. Lichti [1], Jeroen D. Hol [2], Giovanni Bellusci [2] and Henk Luinge [2]

[1] Department of Geomatics Engineering, Schulich School of Engineering, University of Calgary, 2500 University Drive N.W., Calgary, AB T2N 1N4, Canada; E-Mail: ddlichti@ucalgary.ca

[2] Xsens Technologies B.V., Pantheon 6a, Enschede 7521 PR, The Netherlands; E-Mails: jeroen.hol@xsens.com (J.D.H.); giovanni.bellusci@xsens.com (G.B.); henk.luinge@xsens.com (H.L.)

* Author to whom correspondence should be addressed; E-Mail: jckchow@ucalgary.ca

---

Abstract: Autonomous Simultaneous Localization and Mapping (SLAM) is an important topic in many engineering fields. Since stop-and-go systems are typically slow and full-kinematic systems may lack accuracy and integrity, this paper presents a novel hybrid "continuous stop-and-go" mobile mapping system called Scannect. A 3D terrestrial LiDAR system is integrated with a MEMS IMU and two Microsoft Kinect sensors to map indoor urban environments. The Kinects' depth maps were processed using a new point-to-plane ICP that minimizes the reprojection error of the infrared camera and projector pair in an implicit iterative extended Kalman filter (IEKF). A new formulation of the 5-point visual odometry method is tightly coupled in the implicit IEKF without increasing the dimensions of the state space. The Scannect can map and navigate in areas with textureless walls and provides an effective means for mapping large areas with lots of occlusions. Mapping long corridors (total travel distance of 120 m) took approximately 30 minutes and achieved a Mean Radial Spherical Error of 17 cm before smoothing or global optimization.

Keywords: SLAM; Kinect; IMU; LiDAR; ICP; Kalman filter; visual odometry; monocular vision; tightly-coupled; navigation

---

1. Introduction

Visual Simultaneous Localization and Mapping (V-SLAM) is an important topic for autonomous robot navigation. It is often branded as Structure from Motion (SFM) in computer vision and photogrammetric bundle adjustment in photogrammetry; two disciplines that also study the 3D reconstruction of an object/scene while inferring the 6 degrees-of-freedom movement of the optical system, albeit for different purposes. Traditionally the robotics community had been adopting computer vision techniques to address the SLAM problem; this research followed the photogrammetry methods. As part of the SLAM process, a 3D point cloud is generated, which is a valuable asset not just to robotics but also for cultural heritage documentation, infrastructure inspection, urban design, search and rescue operations, surveying, *etc.* The inferred pose can aid pedestrians/robots in navigating through unfamiliar indoor labyrinths and aid location-based services [1].

The term SLAM was first coined in 1995, but its concept and probability basis dates back to the 1980s [2]. The theoretical concept of SLAM is a matured subject [3] and over the years many SLAM systems have been developed. These systems target specific environments (e.g., aerial [4], underwater [5], indoors [6], and outdoors [7]) because a generic system architecture and SLAM algorithm has yet to emerge. According to [8], assembling a mobile mapping and navigation system involves three main aspects: *System Design*, *Choice of Sensors*, and *Methods for Localization and Mapping*. In this paper, an innovative indoor 3D mapping system with novel solutions to all three aspects, called the Scannect, is presented. The objective is to improve the efficiency of data acquisition (when using 3D terrestrial laser scanners) in large indoor environments with lots of occlusions through sensor fusion. The main contributions of this paper are:

- A novel design for a "continuous stop-and-go" indoor mapping system by fusing a low-cost 3D terrestrial laser scanner with two Microsoft Kinect sensors and a micro-electro-mechanical systems (MEMS) inertial measurement unit (IMU);
- A solution to the 5-point monocular visual odometry (VO) problem using a tightly-coupled implicit iterative extended Kalman filter (IEKF) without introducing additional states; and
- A new point-to-plane iterative closest point (ICP) algorithm suitable for triangulation-based 3D cameras solved in a tightly coupled implicit IEKF framework.

Before describing the proposed system, recent scientific advancements in the three key topics are first highlighted. The advantages of the proposed indoor mapping system in the same three aspects relative to existing systems are then presented. The proposed mathematical model for separately updating the IEKF-SLAM using the RGB and depth information is explained. This is followed by results from real data captured with the Scannect.

1.1. System Design

For a lot of mobile robotic operations, efficient perception and pose estimation are critical. This motivated the design for most systems to function in full-kinematic mode, where the mapping operation is performed while the robot is moving [9]. More often than not, the robot's along-track movement extends the imaging sensor's coverage beyond its field of view [10]. However, because of imperfections in synchronization, pose estimation, control, and other sources, the resulting map accuracy and resolution can be rather poor. To quantify the mapping errors, a reference map is often captured in stop-and-go mode where the imaging sensor is only activated when the platform has halted its movements [11]. It is agreed for terrestrial applications that perception data captured while the platform is stationary can yield higher quality data because motion blur and any other robot motion induced errors are avoided [12]. Based on a recent survey done by [12] existing mobile mapping systems either solely operates in stop-and-go or in full-kinematic mode. The former approach is hindered by the slow data acquisition and the latter may lack accuracy. This survey further indicated a system that can operate in both modes has not yet been developed; specifically one that can capture higher density, stability and accuracy data when it is parked and lower quality but continuous data while it is moving.

1.2. Choice of Sensors

Indoor SLAM can be performed using many different sensors, for example it can be applied to Wi-Fi positioning [13]. However, due to the effectiveness and wide success of optical systems for mapping, vision-based SLAM is by far the most common scheme [14]. Popular sensors used for visual SLAM include monocular/stereo cameras [15,16], 2D/3D Light Detection and Ranging (LiDAR) systems [11,17], and 3D cameras [18], all of which have their own advantages and disadvantages.

Monocular cameras are passive sensors that can capture a 2D array of light intensity information instantaneously. Although they are sensitive to the ambient illumination and the information they encapsulate for SLAM is dependent on the scene's texture, being a bearing-only sensor [19] it is largely independent of the object's reflectance properties and can observe details as far as their pixel resolution allows. Stereo cameras can further perceive depth from a single exposure station based on triangulation, a process similar to how human vision operates [20]. LiDAR systems on the other hand are active sensors that acquire 2D scan lines by measuring distance (based on the time-of-flight principle) and bearing information point-by-point at high speed. The third dimension is obtained either by the platform's trajectory (for 2D scanners) or by a rotating head/mirror (for 3D scanners). They can observe dense geometric information over homogeneous surfaces under any illumination conditions. However, their observable range is hindered by the amount of energy they can emit and receive after the laser signal has been reflected and absorbed along its path. In addition, since 2D scan lines are acquired sequentially, the time it takes to sweep the laser over the object of interest is typically long, making them slower than cameras [10].

These complementary characteristics of LiDAR and imagery have been detailed and harvested for airborne and terrestrial mapping by many [21,22]. In recent years, 3D cameras have emerged integrating the benefits of time-of-flight (ToF) measurements with the fast acquisition and gridded structure of digital images at the hardware level. Despite the fact that many commercial 3D cameras are based upon the ToF principle, currently the most widely used 3D camera on the market is the structured-light RGB-D camera known as the Microsoft Kinect. The per pixel metric depth data enhances the information content of the RGB image significantly and can make tasks like obstacle avoidance, object tracking and recognition more robust and accurate [23]. While it appears that RGB-D cameras may have eliminated the need for LiDAR in robotic applications, LiDAR has retained its place in the robotics community [24], and among others, due to the limited range, resolution, accuracy, and field of view of modern RGB-D cameras. Instead, RGB-D cameras might be more suitable for filling in gaps in the LiDAR point cloud at close-range as it can be more efficient for covering occluded areas [25]. In addition, even though the cost of a Kinect is lower than most 3D ToF cameras, [26–28] have shown that the Kinect can produce more accurate results and are suitable for mapping; at close-range (less than 3.5 m) its accuracy can be similar to LiDAR and medium-resolution stereo cameras [26,27].

Existing indoor mapping systems based on the Kinect have so far been limited to a single system. In addition, they are often treated as substitutes for more expensive scanners, rather than being used as an aiding sensor to laser scanners. Furthermore, the integration of geometrically accurate LiDAR, semantic RGB, and efficient depth map for more informed mapping and pose estimation has not been thoroughly investigated, despite their known complementary characteristics.

1.3. Methods for Localization and Mapping

Indoor localization-only solutions exist, for example pedestrian navigation devices [29]. Often IMUs are used in conjunction with a spatial resection based on bearings, distances, or a combination of both from primitives [30]. The primitives can be points, higher-dimensional 2D/3D geometric features (e.g., lines, planes, cylinders, and spheres), or more complex surfaces (e.g., a sofa and statue of a lion). The main assumption is that the positions of the primitives are known, either from a computer-aided design (CAD) model, surveying, or other means.

On the contrary, mapping-only solutions are also common, and are typically based on bearings, distances, or both measurements from one or more locations. In this scenario, the location of the sensor needs to be known, and often in combination with orientation information [10].

Localization and mapping are highly correlated and are often described as the "chicken and egg" problem [31]. Naively performing mapping operations after localization without any accounts for their correlation can lead to poorer accuracy, or even divergence in the solution [3]. Therefore, state-of-the-art methods usually solve the two problems simultaneously using optimizers such as least-squares, Kalman filters (KF), information filters, and particle filters [32]. The measurement models used by these optimizers depend on the data source and the features being matched.

For 2D camera images, popular matching methods include area-based matching (e.g., normalized cross-correlation), point-based matching (e.g., blob detectors like scale-invariant feature transform (SIFT) or corner detectors like Harris), and line-based matching. Keypoint matching is usually more accurate than area-based methods [33], and these points can be tracked over consecutive frames (e.g., by the Kanade–Lucas–Tomasi (KLT) feature tracker) for efficiency and reliability at the risk of experiencing tracking drift. Usually a pin-hole camera model is used to relate pixel observations to their homologous point in the world coordinate system using the collinearity condition [20]. For cameras the matching in general can be formulated as 3D-to-3D, 3D-to-2D, or 2D-to-2D matching, with the accuracy and complexity generally increasing in the order presented [33].

To match point clouds from 3D cameras or laser scanners, point-based, feature-based (e.g., lines, planes, and tori), and freeform matching (e.g., ICP) are possible. Since dense point clouds are usually available and the nature of objects/shapes in the scene can be unpredictable, freeform matching is one of the most popular choices in robotics. It is very flexible but is less robust than matching signalized markers, and can be computationally intensive. The two original ICP implementations that are still widely used today are the point-to-point ICP by Besl and McKay [34] (conventionally solved using Horn's method [35]) and the point-to-plane ICP by Chen and Medioni [36] (formulated as a least-squares minimization problem). To improve the algorithm, over the years many variations of ICP have been proposed and tested. For more details, the reader can refer to taxonomy of ICP in [37]. As no ICP algorithm is superior in all situations, the designer is responsible for selecting and/or modifying the ICP algorithm to fit their needs.

Many different VO/SLAM algorithms based on the above concepts exist and a review of all of them is beyond the scope of this paper. Instead, a few popular state-of-the-art methods relevant to this project are explained with their pros and cons highlighted.

Typical VO operates like Parallel Tracking and Mapping (PTAM) [38]. A pair of 2D images is used to intersect points in the scene with the scale being arbitrary or introduced based on a priori information. Then consecutive images that are captured use these intersected 3D points to perform a resection to solve for their egomotion. Once this transformation is known, they are once again used in the intersection process to create more points in the scene. The end result is usually a sparse map of the environment that carries little value for most mapping operations compared to dense point clouds from LiDAR or 3D cameras. Furthermore, three drawbacks arise with PTAM: (1) the "catch 22 problem" where more reconstructed 3D points give a better camera tracking solution, but cause the state vector to grow rapidly; (2) the same 3D point needs to be observable in at least three images before it can be used for tracking the camera's motion; (3) an initial metric scale is difficult to obtain. The last two statements are less of a problem because through fusion with an IMU the scale can be estimated as quickly as 15 s [39]. Or if depth data are available it can be solved instantaneously from a single exposure and camera tracking can begin immediately following the first camera exposure [40].

Henry *et al.* [41] had a single PrimeSense RGB-D camera carried by a human in a forward-facing configuration for indoor mapping. They explained that the depth data ignores valuable cues in the RGB images and RGB images alone do not perform well in dark and sparsely textured areas. Therefore they presented 1-step and 2-step matching methods that exploit both pieces of information. In the 1-step case, a joint optimization was performed that minimized the point-to-point distances for the detected 3D keypoints and point-to-plane distances for the depth maps using ICP. To improve the speed of the algorithm, they suggested splitting this into two steps: running the keypoint matching first and only performing ICP after if the match was poor (e.g., insufficient keypoints were matched). This resulted in only minor compromise in terms of accuracy. In both cases, they performed RANSAC on the 3D keypoints for the initial alignment step. Compared to their previous work [42], they have replaced SIFT with the Features from Accelerated Segment Test (FAST) detector, minimized the reprojection errors of the matched 3D keypoints instead of Euclidean distances, and did a global optimization using sparse bundle adjustment (SBA) instead of the Tree-based netwORk Optimizer (TORO) [31]. In their indoor mapping experiment, they travelled over 71 m in an office space and their ICP-only solution showed 15 cm error, while their proposed RGB-D ICP method showed 10–11 cm error (*i.e.*, 0.15% error over the travelled distance). Their proposed RGB-D ICP scheme eliminated the need to weight the color space relative to the Euclidean space [43], but created the new problem of having to weight 3D keypoints against the point-to-plane ICP. In this paper they give the two different point matches the same weight in their joint-optimization. Moreover, they only mention the possibility of weighting the points in ICP based on distances, normal angle, *etc.* to improve the matching but the same weight was assigned to all points. The authors acknowledge and stress the importance of minimizing reprojection errors, but their point-to-plane ICP did not adopt this scheme, only the 3D keypoints used this cost function. They also mention that because they tightly-coupled depth and color information, whenever the depth sensor is out of range they do not have 3D keypoints for RANSAC matching even though they were detectable in 2D. This can be rather common with their single forward facing sensor configuration. For example, when looking down a hallway, the majority of the center pixels will likely be out of range.

The Kinect is capable of delivering 9.2 million points per second, a rate that far exceeds any terrestrial laser scanners. A popular method for handling this vast amount of data in real-time is the KinectFusion solution [44]. They parallelized the point-to-plane ICP algorithm to be executed on the graphics processing unit (GPU) for speed and fused the dense point clouds using the Truncated Signed Distance Function (TSDF) [45]. This novel implementation was considered for this project but at the time its measurement volume was restricted by the memory of the GPU. Even with a lower resolution for the TSDF, mapping over hundreds of meters in distances was not possible. This method can map in the dark because only the depth images were considered and all the semantic RGB data were ignored. Although it was designed for handheld sequences, slow and steady egomotion was assumed, as sudden jerks and insufficient overlap can cause the ICP to fail. In addition, a lack of geometry in the scene (a well-known problem with ICP) makes it difficult for the KinectFusion method to converge to the global minimum.

More recently [46] has lifted this limitation of measurement volume by downloading the old data onto the central processing unit (CPU) and using a moving TSDF to only map the most recent and immediate point clouds. In their later paper they even took advantage of the RGB information [47], and added loop-closure [48]. Keller *et al.* [49] improved upon KinectFusion by replacing the volumetric representation by a point-based representation of the scene. This eliminated many computational overheads and conversions between different scene representations, making it more scalable and faster. Each point in the depth map is projectively associated with the scene and reliable points are merged using a weighted average. Besides being able to map larger areas, the biggest improvement is the ability to handle a certain degree of dynamics in the scene. Even with these significant improvements, being a single Kinect system, its field of view (FOV) is limited to less than 60° and their ICP algorithms will fail when the scene lacks features (e.g., only a single homogenous wall is in view). Also, the timing error between the color and depth information was usually neglected [50].

Leutenegger *et al.* [51] tightly-coupled a stereo-camera with an IMU using keyframe-based nonlinear optimization instead of filtering techniques. Keypoints were identified using the Harris detector and descriptions were computed using BRISK. Unlike the Kinect, only a sparse 3D map was obtained and landmarks were removed in the marginalization step to reduce the computation load.

Unlike the above papers, another stream of research used Kalman filtering to solve the SLAM problem. Li *et al.* [52] demonstrated that the performance and convergence of SLAM is affected by landmark initialization uncertainty and linearization error. The former can be addressed by using the depth information provided by the Kinect to better initialize the 3D landmarks. The latter can be improved by replacing the extended Kalman filter (EKF) with the IEKF, which re-linearizes the measurement model by iterating an approximate maximum a posteriori estimate around the updated states rather than relying on the dynamics model. They showed using real data from their keypoint based SLAM that in some situations where EKF will fail, the IEKF would not. They also proposed a way to add and remove landmarks from the state vector to avoid memory overload. The IEKF can help reduce the effect of non-linearities in the measurement model; however just like the EKF they are still based on a measurement model where a single observation is expressed as a function of the unknown states.

Aghili *et al.* [53] realized the importance of initial alignment for the ICP and attempts to keep track of the pose when good measurement updates were unavailable. They solved this by fusing the ICP pose in a KF under a closed-loop configuration. In this case, the measurements were loosely-coupled and the EKF only provided an initial guess for the ICP method. With a similar motivation, Hervier *et al.* [54] fused the Kinect with a three axes gyroscope for improved mapping and localization accuracy. They used the ICP algorithm in the Visualization Toolkit (VTK) to first solve for the pose, and then fuse it in the KF in a loosely-coupled manner. In their current system, they mentioned it is difficult to evaluate movements parallel to a flat wall.

In the navigation domain, vision-assisted IMU is popular for localization in GNSS-denied environments, but the reconstructed scene is often viewed as a nuisance parameter. Kottas and Roumeliotis [55] presented results from monocular camera and IMU fusion that used both point and line features. Their proposed Observability-Constraint Multi-State Constraint Kalman Filter improved the consistency of the EKF results by ensuring rotational information about line, which is unobservable by vision, does not enter the filter. Using this method for a handheld sequence, they estimated sparse feature parameters and reported a localization error of 0.31 m over a 144 m trajectory (*i.e.*, 0.22% error of the distance travelled). Li *et al.* [56] fused the IMU and rolling-shutter camera onboard the Samsung Galaxy S2 for pose estimation. Using their modified Multi-State-Constraint Kalman Filter a sliding window of camera states was processed and an approximate positioning error of 0.8% over the travelled distance was reported.

All these SLAM solutions will fail in cases when a single homogenous plane is imaged. Often when the ICP method is used with a KF, they are loosely-coupled rather than tightly-coupled. Most of the work either ignores the RGB information or if included they require the depth and RGB information to be available simultaneously and assumed no synchronization errors.

In Grisetti *et al.* [31] they distinguished SLAM into the filtering approach and smoothing approach. Filtering approaches are incremental by nature and often use KF (also referred to as online SLAM), whereas smoothing approaches estimate the complete trajectory from all the measurements (a.k.a the full SLAM problem). This paper focuses on the forward filtering part of SLAM and proposes new ways to solve monocular VO and ICP in a tightly-coupled KF.

2. The Scannect Mobile Mapping System

The Scannect is an indoor mapping system that is unique in its *System Design*, *Choice of Sensors*, and *Methods for Localization and Mapping*. Each of these aspects of the Scannect is explained in the following sections.

2.1. Proposed System Design

The proposed system attempts to harvest the accuracy and stability of stop-and-go systems while maintaining a degree of speed from full-kinematic systems. High quality 3D scans are captured when the robot has "stopped", and lower quality data is captured while it is "going" for mapping and keeping track of the relative change in pose. For this reason, the system design can be seen as a hybrid between static and kinematic mapping, termed continuous stop-and-go mode in this paper. According to a recent review by [12], systems operating in continuous stop-and-go mode have not yet been released. If higher quality data are desired the robot can make more frequent stops, hence approaching the stop-and-go solution's quality or if speed and time is of the essence, it can make zero stops (approaching the full-kinematic solution). This trade-off between speed and accuracy does not affect other components of the Scannect, and can therefore be easily adapted to the project at hand.

Designed Robot Behavior

The robot's behavior during the mission is described in the following stages with the desirable trait of the system highlighted:

- Stage 1: The robot enters an unfamiliar environment without prior knowledge about the map or its current position.
    - The absolute position may never be known, but an absolute orientation based on the Earth's gravity and magnetic fields is determinable.
    - The system should "look" in every direction to its maximum range possible before moving around to establish a map at this arbitrary origin.
- Stage 2: Due to occlusions and limited perceptible range, the robot needs to move and explore the area concurrently.
    - When exploring new areas the map should be expanded while maintaining the localization solution based on the initial map.
- Stage 3: Occasionally when more detail is desired or the pose estimation is uncertain, the system can stop and look around again.
    - As the system is for autonomous robots with no assumptions about existing localization infrastructure (e.g., LED position systems), every movement based on the dead-reckoning principle will increase the position uncertainty. Typically it is more accurate to create a map using long-range static remote sensing techniques because they are typically less than the dead-reckoning errors.
    - Looking forward and backward over long ranges from the same position can possibly introduce loop-closure.
    - One of the Kinects and the IMU are rigidly mounted together and force centered on the mobile platform. Through a robotic arm, quadcopter or by other means the Kinect and IMU can be dismounted from the platform during "stop" mode for mapping, making it

more flexible/portable for occlusion filling. Afterwards it can return and dock at the same position and orientation and normal operations is resumed.

2.2. Proposed Choice of Sensors

The starting point of the Scannect is the FARO Focus^{3D} S terrestrial laser scanning (TLS) instrument. This is motivated by the accuracy, stability, speed, range, and field of view of this modern static 3D TLS instrument, (e.g., millimeter-level range accuracy and 360°/310° horizontal/vertical FOV, respectively). This sensor is very suitable for indoor mapping applications, especially with the recent upgrades (*i.e.*, digital compass and tilt sensors), and reductions in cost, size, and weight. Its success for indoor mapping can be seen in the papers published using this system for as-built surveys and cultural heritage documentation [57,58].

The Microsoft Kinect is a trinocular vision system capable of capturing dense RGB-D information at up to 30 Hz. The RGB information comes from an onboard VGA resolution camera while the depth information is determined using the coded-light principle: a pattern is emitted by the projector and simultaneously imaged by the infrared (IR) camera to perform photogrammetric intersection. More details about the Kinect can be found in many articles, e.g., [59,60].

In large open spaces (e.g., a gymnasium, shopping malls, and airports) TLS can be more efficient and effective at mapping than 3D cameras like the Kinect. However, TLS falls short when many occlusions exist in the scene because so many static scans are needed for complete coverage, which is very time-consuming. When setup on a mobile platform (e.g., trolley) the setup-move-setup time is reduced [12]. A selective FOV in theory can shorten the scan time when compared to a full 360° scan, however because of the rapid scan speed of modern scanners, the process of capturing a preview scan followed by the manual FOV selection and doing a dense scan can actually result in longer overall acquisition time. In cases where small occlusions needed to be filled, the Kinect is perhaps more efficient. Unlike 2D profile scanners which essentially have no overlap in kinematic mode, the Kinect provides a 3D point cloud within its FOV instantaneously at 30 Hz. Consequently, a high-degree of overlap exists between consecutive frames, which facilitate continuous point cloud stitching for higher quality mapping while maintaining the robot's pose. Because the Kinect, as well as other 3D cameras currently on the market, all have a relatively small FOV (the largest being 90°), a simple and cost-effective solution is to add extra Kinects [61].

The Scannect was equipped with two sideward facing Kinects (Figure 1); this architecture was chosen to maximize the overall FOV for indoor SLAM and to prevent possible interference between the two Kinects. If desired, additional Kinects are possible and the interference issue can be mitigated by introducing small vibrations to each Kinect [62]. Although a forward facing and skyward facing Kinect design has also been considered, it was not implemented because (1) a forward facing Kinect usually does not have sufficient range to reach the end of a long corridor to provide any useful information; (2) the considered situations in this paper assume a flat horizontal floor for the building, which is often the case, so a non-holonomic constraint was used for constraining the height of the robot instead since a sideward facing Kinect is more valuable for mapping.

Nonetheless, all the sensors discussed thus far require a direct line-of-sight to conjugate features in the overlapping regions. If the overlap between two visual datasets is small or lacking in features/texture, optical-based matching techniques may be inaccurate or even fail in some cases. A suitable aiding device is an inertial measurement unit, which operates well regardless of its environment. A MEMS-based IMU from Xsens, the MTi, was adopted for the task at hand. When the robot enters a dark corridor with walls too far for the Kinect, the integrated rotation rate will keep track of the orientation, and the rotated acceleration after double integration will keep track of the position of the platform.

Figure 1. Sensor configuration of the Scannect.

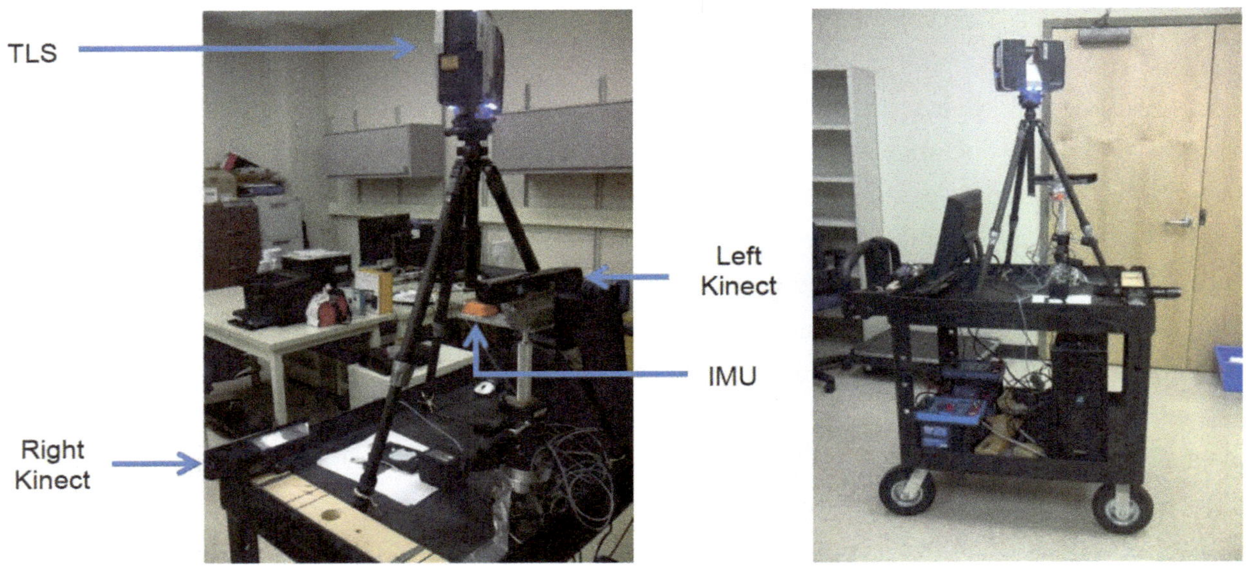

All data were logged and processed on a Windows 7 desktop computer equipped with an Intel® Core™ i7 processor (up to 3.07 GHz) and 12.0 GB of RAM. In this prototype, the processing was performed offline, which is not unreasonable as most stop-and-go mobile scanning systems process their data post-mission.

Overall, the Focus^{3D} S is envisioned for mapping large open areas in static mode. When the robot is moving the Kinects are used for localization and filling in the shadowed areas until the system stops to perform Zero velocity UPdaTes (ZUPT) and capture another laser scan. The IMU continuously assists the Kinect matching and bridges the gaps when vision-based localization fails due to lack of details in the scene. To the authors' best knowledge this is the first paper using more than one Kinect for an inside-out type system. Popular multi-Kinect systems are designed for body scanning or motion capture with an outside-in design [61]. Furthermore, the Kinect is usually used as the sole mapping sensor, instead of playing a more assistive role to a more accurate laser scanner, as in this paper.

System Calibration

Systematic errors are prevalent in all man-made instruments, for instance due to manufacturing flaws. Before integrating the four proposed sensors for SLAM, they were first individually calibrated for their intrinsic parameters. Afterwards, the relative rotation (boresight) and relative translation (leverarm) between all the sensors were determined through an extrinsic calibration process. Only the intrinsic calibrations of the optical sensors are considered in this paper. The calibration procedure for MEMS-based IMUs has reached a level of maturity that is widely accepted [63]. The commonly adopted methods solve for biases, scale factor errors, and axes non-othogonalities for the triad of accelerometers and gyroscopes, and misalignments between them. For the accelerometer calibration, the sensor can be placed in various orientations and capturing a sequence of static data. These measurements are related to the local gravity vector, which is used as the reference signal. For the gyroscopes, a rotation rate table can be used to establish the reference signal. Since such a calibration requires specialized equipment that is not easily accessible, the manufacturer's calibration was relied upon. Furthermore, MEMS IMUs are known to have a poor bias stability, so the residual systematic errors need to be modeled stochastically in every operation.

Microsoft Kinect RGB-D Camera Calibration

The Kinect is originally intended for gaming applications, where metric measurements are not critical. In order to integrate the Kinect with a laser scanner for mapping, the sensor needs to be individually calibrated to ensure the residual systematic errors from the manufacturer's calibration are minimized. Distortions in the Kinect's depth map have previously been reported and various calibration approaches have been proposed [59,64]. As the Kinect is a relatively new sensor, a commonly accepted calibration procedure has not yet been established, which is in contrast to MEMS IMU calibration. For convenience, a user-self calibration approach that requires only a planar checkerboard pattern for performing a total system calibration was selected.

The Kinect consists of three optical sensors, all of which can be modeled using the pin-hole camera model. The well-established bundle adjustment with self-calibration method was modified in [65] to solve for all the intrinsic and extrinsic parameters of all the optical sensors in the Kinect simultaneously. This method was adopted in this paper and was responsible for independently calibrating the two Kinects installed onboard the Scannect. The results from the calibration are published in [65].

FARO Focus^{3D} S 120 3D Terrestrial Laser Scanner Calibration

The Focus^{3D} S is the lowest-cost sensor in its category of laser scanners. Compared to more expensive scanners within the same class, it has been found to exhibit more significant systematic distortions, in particular angular errors. The multi-station self-calibration approach is a popular and effective means for reducing the systematic errors internal to TLS instruments without the need of specialized tools [66,67]. Analogous to camera calibration, by observing the same targets from different stations a least-squares adjustment can be performed for solving the biases in its distance and bearing measurements and various axes misalignments, eccentricities, and wobbling. Recently, [68] showed that signalized targets can be replaced by planar features to achieve similar calibration results with reduced manual labor. The Focus^{3D} S on the Scannect was calibrated using this plane-based self-calibration method, and the calibration results can be found in [69].

Boresight and Leverarm Calibration between the Kinect and Laser Scanner

Both the Kinect and laser scanner are visual sensors, therefore target fields such as a planar checkerboard pattern is visible to both. These common targets can be independently extracted in their corresponding point clouds and then related through a 3D rigid-body transformation [70]. This approached was adopted to simultaneously solve for their boresight and leverarm parameters. Since the Kinects point away from each other, they were related through the intermediate laser scanner.

Boresight and Leverarm Calibration between the Kinect and IMU

The problem of solving for the 3D rigid body transformation parameters between an IMU and camera is often encountered in multi-sensor mapping systems [10]. Some common methods include simple surveying techniques via a total station, solving it as a state estimation problem in an EKF, or formulating it as an offline global optimization problem. The first method is most suitable for higher-end IMUs and metric cameras with clear markers on the exterior casing, for instance in airborne mapping systems. The locations of the Kinect's perspective centres are unknown and CAD models are not published by Microsoft. Therefore, the other two algebraic calibration methods are perhaps more suitable.

Lobo and Dias [71] proposed a popular two-step calibration procedure that first estimates the boresight parameters and then uses a turntable that spins about the IMU's origin to solve for the leverarm parameters. Not only does this require specialized tools and careful placement of the system on the turntable, correlations between the rotational and translational parameters are ignored in this method. Hol *et al.* [72] formulated the calibration as a gray-box model where the navigation states were solved in the EKF using their current best estimate of the unknown parameters (boresight, leverarm, gyro bias, accelerometer bias, and gravity vector). A post-filtering adjustment was then used to improve their estimate of the unknown parameters simultaneously by minimizing the differences between the predicted and measured target positions in image space. This online filtering and offline adjustment is then repeated until convergence. This type of calibration is preferred in this project because it is more user-friendly and can account for the correlations between the boresight and leverarm parameters. Based on the approach in [72], the measurement model shown in Equation (1) was used for calibration in this project. Two subtle differences exist in the implementation presented herein: first, the slowly time-varying biases are solved in the IEKF like [73]; second, the gravity vector in the navigation frame has been replaced by the rotation angles between the navigation frame and world frame of the checkerboard. The rotation matrix notation was used here and in other equations for convenience in representation; in the actual adjustment, quaternions were used to give the system full rotational freedom without encountering gimbal lock.

$$p_i^{RGB}(t) = \mu_i^{RGB}(t) R_{IR}^{RGB}\left\{R_s^{IR}\left\{R_n^s(t)\left\{R_W^n O_i^W - T^n(t)\right\} - \Delta b^s\right\} - \Delta b^{IR}\right\} \tag{1}$$

where $p_i^{RGB}(t)$ is the observation vector for point i expressed in the RGB image space at time t; $\mu_i^{RGB}(t)$ is the scale factor for point i at time t (which can be eliminated by dividing the image measurements model by the principal distance model [20]); R_{IR}^{RGB} is the relative rotation between the IR and RGB cameras; R_s^{IR} is the relative rotation between the IMU sensor and IR camera; $R_n^s(t)$ is the orientation of the IMU sensor relative to the navigation frame at time t; R_W^n is the relative rotation between the world frame defined by the checkerboard and the navigation frame; O_i^W is the 3D coordinates of point i in the world frame; $T^n(t)$ is the position of the IR camera in the navigation frame at time t; Δb^s is the leverarm between the IMU sensor and IR camera; Δb^{IR} is the relative translation between the IR and RGB cameras.

2.3. Proposed Methods for Localization and Mapping

Figure 2 shows the interaction between the sensors onboard the Scannect. The different colors separate the various components of the Scannect (e.g., cyan highlights the egomotion estimation using RGB images and purple highlights the 3D image processing). The IMU observations after correction (Equation (2)) were integrated to provide changes in orientation and position (Equation (3)), which were useful for initializing the 2D/3D matching and keep track of the states between measurement updates. The accelerometer and gyroscope biases were solved in the KF as a first order Gauss-Markov process (Equation (4)). The Gauss-Markov parameters were determined from autocorrelation analysis [74]. The TLS point clouds were matched using the conventional point-to-plane ICP [36] with the first point cloud oriented relative to magnetic north in the local level frame. The proposed RGB-D localization and mapping method is based on a novel ICP implementation suitable for triangulation-based 3D cameras such as the Kinect. The new cloud was always registered with the global scene to partially address the loop-closure problem. The localization was separated into depth-only and RGB-only for two reasons: (1) the depth and RGB streams are not perfectly synchronized; and (2) the depth measurements have a limited range—by processing the RGB data separately, localization of the robot is still possible when depth data are unavailable. Both depth-only and RGB-only processing used a tightly-coupled implicit iterative extended Kalman filter to estimate the states.

Figure 2. The overall workflow of performing Simultaneous Localization and Mapping (SLAM) using the Scannect.

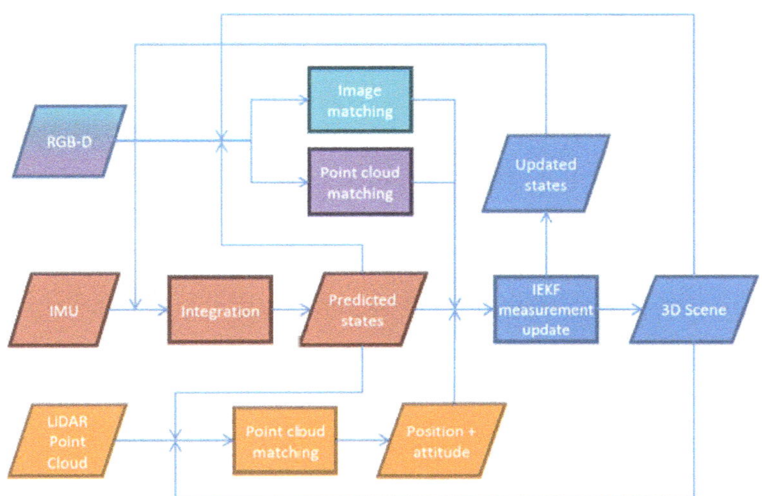

$$\begin{aligned} \ddot{x}^n(t) &= R_s^n(t)\left[f^s(t) - \delta^a(t)\right] - 2\omega_n^{e\to i} \times \dot{x}^n(t) + g^n \\ \omega_s^{s\to n}(t) &= \omega_s^{s\to i}(t) - R_n^s(t)\omega_n^{e\to i} - \delta^\omega(t) \end{aligned} \quad (2)$$

$$\begin{aligned} x^n(t+1) &= x^n(t) + \dot{x}^n(t)\Delta t + \ddot{x}^n(t)\,{}^{\Delta t^2}\!/_2 \\ \dot{x}^n(t+1) &= \dot{x}^n(t) + \ddot{x}^n(t)\Delta t \\ q^{s\to n}(t+1) &= q^{s\to n}(t) \otimes \exp\left[\omega_s^{s\to n}(t)\,{}^{\Delta t}\!/_2\right] \end{aligned} \quad (3)$$

$$\delta^a(t+1) = e^{-\Delta t/\tau^a}\delta^a(t) + \eta^a(t)$$
$$\delta^\omega(t+1) = e^{-\Delta t/\tau^\omega}\delta^\omega(t) + \eta^\omega(t) \tag{4}$$

where $x^n, \dot{x}^n, and\ \ddot{x}^n$ are the position, velocity, and acceleration of the IMU sensor in navigation frame, respectively; $f^s(t)$ and $\omega_s^{s\to i}(t)$ are the accelerometer and gyroscope measurements at time t, respectively; $\varepsilon^a(t)$ and $\varepsilon^\omega(t)$ are their corresponding measurement noises; $\omega_n^{e\to i}$ is the rotation rate of the earth as seen in the navigation frame; g^n is the local gravity in the navigation frame; $q^{s\to n}$ is the rotation from IMU sensor frame to navigation frame expressed using quaternions; $\delta^a(t)$ and $\delta^\omega(t)$ are the slowly time-varying accelerometer and gyroscope biases, respectively, which are modelled as a first order Gauss-Markov process; $\eta^a(t)$ and $\eta^\omega(t)$ are the corresponding Gauss-Markov process driving noises; and Δt is the time interval between the current and previous measurement.

2.3.1. Tightly-Coupled Implicit Iterative Extended Kalman Filter

The Kalman filter has been a popular choice for sensor fusion and navigation for decades. The Kalman filter assumes a linear relationship between the measurements and the states. Non-linearities can be approximated by using a first order Taylor series expansion. If the non-linearities are severe, higher order terms of the Taylor series can be included, and/or an iterative approach can be taken. All the filtering in this paper was done following the latter approach where an IEKF is used for updating the states [75]. A total of 16 states were used by the system (namely 3 for position, 3 for velocity, 4 for attitude in the quaternion, 3 for gyro bias, and 3 for accelerometer bias) and the size of this vector remains constant throughout the entire large-scale mapping project. This is desired and done to reduce the likelihood of divergence with a large state vector, and the reduction of system performance over time as more states are added [75]. In addition, many geometrical formulas used in vision actually violate the parametric model assumed by the KF (*i.e.*, a single observation is a function of the unknown states, as shown in Equation (5)); instead they are implicitly defined (*i.e.*, the observations and unknown states are inseparable in the measurement update model, as shown in Equation (6)). The extension of the conventional IEKF for implicit math models are given in [76], and was used in all the matching processes described in this paper.

$$y = h(X) \tag{5}$$

$$h(X,Y) = 0 \tag{6}$$

where y represents a single observation; Y represents multiple observations; X represents the unknown states; and h denotes a functional model relating the observation(s) and unknowns.

For outlier removal, the normalized residuals were used instead of testing the innovations [77]. This implicit IEKF formulation permits the Kinect measurements to update the state vector in a tightly-coupled manner, which allows for the system to be updated even if the matching is underdetermined (e.g., less than the minimum number of matches is found, or their distribution is collinear).

2.3.2. Dense 3D Point Cloud Matching for the Kinect

Computer vision techniques for egomotion estimation and 3D reconstruction are very suitable for robotics SLAM applications because of their balance between speed, accuracy, and reliability. Often in freeform scan-matching, the ICP method is used. The ICP matching is usually first solved independently to reconstruct the scene, and then the relative translation and rotation between the point clouds are fused using a Kalman filter to update the pose. When the point clouds being matched are rich in geometric features, this method can provide high quality pose estimates. Typical indoor urban environments present additional challenges such as a single flat wall and sparse/poorly-distributed point clouds due to sensor saturation (the Kinect has difficulties producing a point cloud under direct sunlight, even through windows) and scene's reflectivity (e.g., mirrors and glass). For the former case, point-to-point ICP can converge to an infinite number of solutions based on the initial alignment, and the point-to-plane ICP will have a singular matrix that would not invert due to lack of geometric constraints. If the wall exhibits a lot of texture, RGB information can be combined with the standard geometry-only ICP in a joint optimization to obtain a solution [43], however if the wall is homogenous (e.g., white walls without textures are common) this approach is not applicable. Furthermore, based on the OpenNI programmer's guide [78] the depth and color streams of the Kinect are not perfectly synchronized. In some extreme cases, data captured using Kinect Studio from the Microsoft Kinect SDK even revealed significant lag between the two streams. Figure 3 shows the temporal availability of color and depth information while a Kinect was logging both simultaneously. The red and blue lines indicate when a RGB image and depth map was captured, respectively; lag up to 0.26 s can be observed.

Figure 3. Recorded color and depth data using Kinect Studio with the red and blue lines showing the time when an image is acquired.

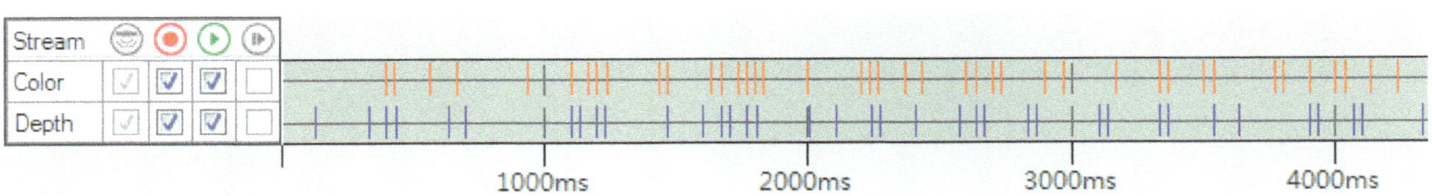

A general overview of the depth matching methodology is summarized graphically in Figure 4. A calibrated point cloud augmented with RGB and normals was generated in near real-time using OpenMP, Point Cloud Library (PCL), and [65]. Despite the synchronization errors, the out-of-sync color information was included in the initial alignment step to improve the distinctiveness of the selected keypoints. The stable keypoints were then extracted from this new point cloud and matched to a previous keyframe. The estimated pose after outlier removal (using the uniqueness constraint, Signature of Histograms of OrienTations COLOR (SHOTColor) descriptor, IMU predicted image point locations, and RANSAC) was used as the initial alignment for the tightly-coupled ICP. During scan-matching, the input point cloud (without considering the RGB information) was iteratively matched with the reconstructed scene (*i.e.*, the map) while incorporating the a priori information from the IMU. The synchronization problem was mitigated by using the measurement model described in the section *Cost Function*, which updates the KF using solely the geometry information. After a laser scan, when mobile mode was resumed, this approach becomes comparable to a localization only solution in a known environment until reaching occluded areas. The output of this method includes the 16 updated states and an extended map of the environment. The state parameters were estimated in the KF, while the map was obtained from averaging new points at the current pose with the scene using a voxel grid. The adopted model-to-scene scan-matching approach estimates the map by allowing the scene to be updated through a moving average instead of augmenting the state vector with the map, which is computationally demanding for the KF since every Kinect exposure contains up to 307,200 points.

Figure 4. SLAM using the Kinect's depth information.

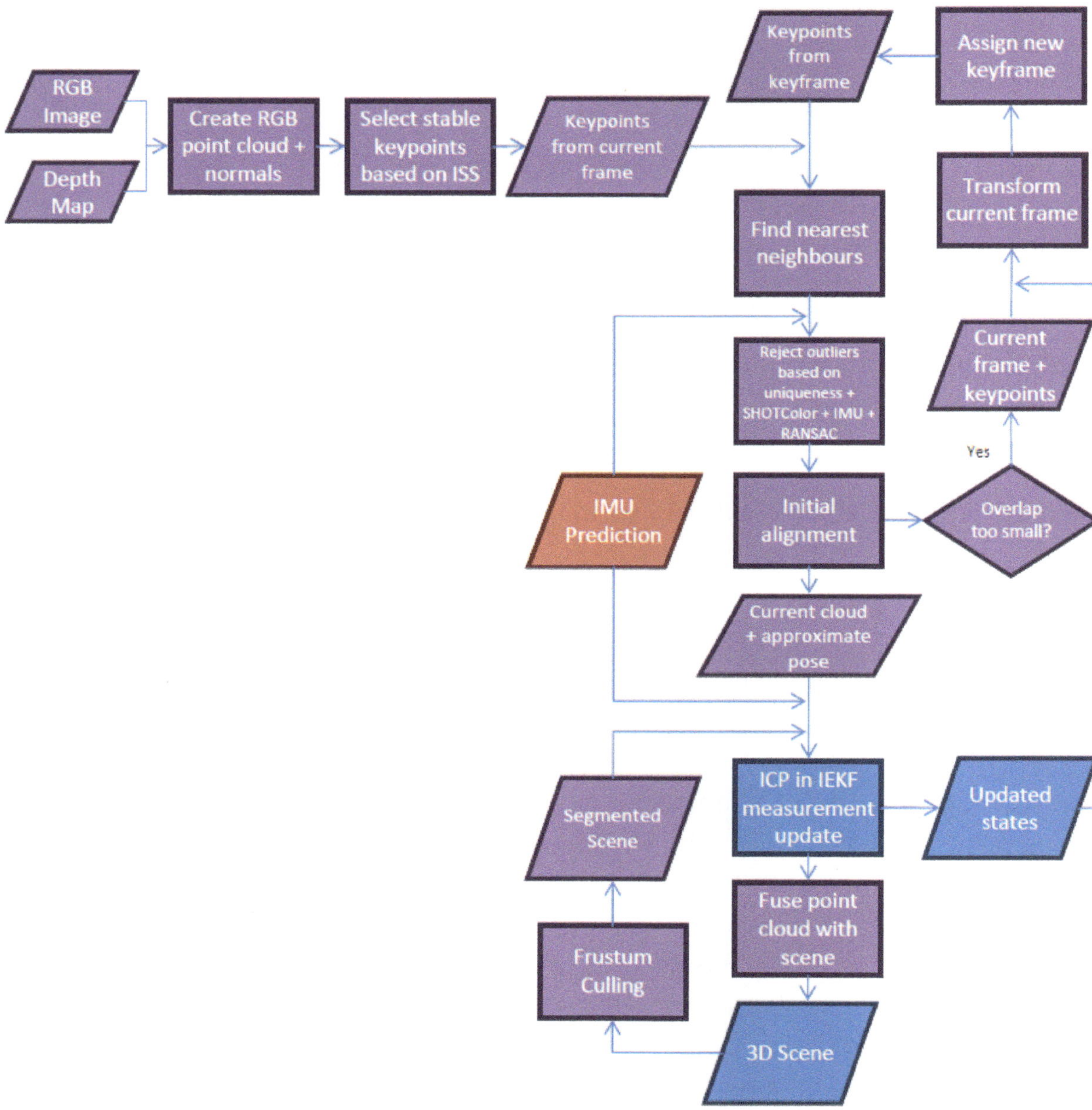

The central concept of the proposed depth processing pipeline is the ICP algorithm. Numerous variations of ICP can be found in literature, each with their own advantages and disadvantages. Rusinkiewicz and Levoy [37] summarized five key components for tuning the ICP algorithm: the sampling method, searching, cost function, outlier rejection, and weighting. Following this guideline, a new point-to-plane ICP solved in a tightly-coupled IEKF is proposed to account for the solo homogenous flat wall situation described above. Through the additional information provided by the dynamics model, not only does the initial alignment become trivial, it makes the single plane case solvable. The following sections explain the proposed ICP algorithm in relation to the five key components, initial alignment, and the loop closure problem.

Sampling

A single Kinect depth map is 640 by 480 pixels, which can result in a point cloud with a maximum size of 307,200 points. This is a lot of data, especially considering the Kinect simultaneously captures RGB images at 30 Hz. Since the corresponding points between the IR camera and projector are determined using a 9 × 9 kernel [59], depth values of adjacent pixels are highly correlated and do not carry as much information as pixels that are further away. Therefore, to reduce the computation time, the point clouds were downsampled. For this, the normal downsampling technique suggested by [37] was used, because it showed better performance than uniform and random downsampling.

Searching

The popular Fast Library for Approximate Nearest Neighbors (FLANN) [79], which is a fast KD-tree, was used for speeding up the query process, which is one of the most computationally intensive steps of the ICP algorithm. Instead of just finding a single nearest neighbor in the target cloud for every point in the source cloud, the k-nearest points in the target cloud based on the Euclidean distance to the query point were selected, and the point with the minimum orthogonal distance was chosen as the corresponding point. This was then repeated with every point in the target cloud as query points to find the correspondences. Only correspondences that pass this uniqueness constraint were considered as matches.

Instead of building a KD-tree for a growing scene and querying a large amount of points, two steps were taken to reduce the memory and the processing load. First, every new ICP-registered point cloud was fused with the scene model using a voxel grid representation, where points within the same voxel were reduced to their centroid. The second step involves a fast segmentation of the scene to extract only regions that were overlapping with the new incoming point cloud using frustum culling [80] based on the FOV of the Kinect and IMU-predicted pose. This essentially sets a threshold on the maximum number of points to be queried, thus prevented certain iterations from running significantly longer than others.

Cost Function

The point-to-plane cost function minimizes the orthogonal distances instead of the Euclidean distances as in the point-to-point ICP; this is known to perform better in terms of accuracy and convergence speed for most cases [37]. More importantly, it is preferred in this application because it would not “anchor” the along-track translation of the robot when it is travelling down a corridor. The point-to-plane ICP will only minimize the 1D orthogonal distance, therefore for a side-looking Kinect it will not prevent it from sliding along the wall as desired. When solving it in a KF framework, information from the dynamics model (*i.e.*, IMU) can help push the solution forward in the along-track direction while the ICP corrects the across-track position and the two orientations that are not parallel to the normal of that wall. Furthermore, the proposed tightly-coupled point-to-plane ICP will not be singular even if only one plane is within view.

As for all camera-based models, minimization of the reprojection errors yields higher quality results than simply minimizing the Euclidean distances because it better describes the physical measurement principle. Following the derivations from [65], for a rigidly attached stereo camera, its measurement model can be written as Equation (7) (for the master camera) and (9) (for the slave camera), where the projector is treated as a reverse camera. The projector's observation vector $p_{i,k}^{PRO}(t)$ is derived by backprojecting the 3D points into the projector's image space using its extrinsic and intrinsic parameters [65]. The inherent scale ambiguity for both the IR camera ($\mu_{i,k}^{IR}$) and the projector ($\mu_{i,k}^{PRO}$), which are traditionally eliminated from the model by dividing the first and second components by the third component of Equations (7) and (8) in parametric form [20], are solved through spatial intersection with the point-to-plane constraint (realized by substituting Equations (7) and (8) into Equation (9), respectively). The depth camera measurement update model implemented in the proposed KF framework is then obtained by equating Equations (7) and (8). It is worth noting that the functional relationship between the observations (*i.e.*, image coordinate measures for the IR camera and projector) and unknown states (*i.e.*, the pose) are implicitly defined. The math model presented here minimizes the reprojection errors of the Kinect's IR camera and projector pair implicitly to reduce the orthogonal distance of every point in the source point cloud relative to the target point cloud.

$$O_i^n = R_s^n(t)\left\{R_{IR}^s R_k^{IR}\left\{\mu_{i,k}^{IR}(t)p_{i,k}^{IR}(t) - \Delta b^k\right\} + \Delta b^s\right\} + T^n(t) \tag{7}$$

$$O_i^n = R_s^n(t)\left\{R_{IR}^s R_k^{IR}\left\{\left[R_{PRO}^{IR}\right]_k \mu_{i,k}^{PRO}(t)p_{i,k}^{PRO}(t) + \Delta c^k - \Delta b^k\right\} + \Delta b^s\right\} + T^n(t) \tag{8}$$

$$\begin{bmatrix} a_i & b_i & c_i \end{bmatrix} \bullet \left[O_i^n\right] - d_i = 0 \tag{9}$$

where, R_k^{IR} is the relative rotation from the IR camera of Kinect k to the IR camera of the reference Kinect; $\mu_{i,k}^{IR}(t)$ and $\mu_{i,k}^{PRO}(t)$ are the scale factors for point i observed by the IR camera and projector of Kinect k at time t, respectively; $p_{i,k}^{IR}(t)$ and $p_{i,k}^{PRO}(t)$ are the image measurement vectors of point i observed by the IR camera and projector of Kinect k at time t, respectively; $[R_{PRO}^{IR}]_k$ is the relative rotation between the IR camera and projector of Kinect k; Δb^k is the relative translation between the IR camera of Kinect k and the reference Kinect; Δc^k is the relative translation between the IR camera and projector of Kinect k; a_i, b_i, c_i, and d_i are the best-fit plane parameters of point i in the navigation frame.

Outlier Rejection

The correspondence matching between two point clouds can only be seen as approximate because exact point matches cannot be assumed in dense point clouds captured by the Kinect or laser scanner. This causes the ICP algorithm to be highly sensitive to outliers in the approximated correspondences. To minimize the possibility of outliers, several precautions were taken, resulting in the following three outlier removal steps.

Rejection When Estimating Correspondences

One of the benefits of including an IMU is that it can provide good initial alignment for the ICP. Using this prediction and a fixed radius around each query point, correspondences established during the searching step above that were outside of each sphere can be removed. The remaining correspondences were further compared using their normal vector determined in their local neighborhood using orthogonal regression. Only points with differences in spatial angle less than a threshold were retained.

Rejection Using RANSAC

When more than 3 non-collinear points were matched between the source and target clouds, RANSAC based on the point-to-point ICP from Besl and Mckay [34] can be used for finding the inliers efficiently. This approach is similar to the RANSAC approach explained in [41].

Rejection in the Kalman Filtering

In case outliers still exist in the correspondences, standard outlier detection method based on the normalized residuals was adopted [77].

Weighting

Assigning weights that are inversely proportional to the depth measurements when registering point clouds coming from the Kinect has shown to have significant improvements in terms of accuracy [50]. For the Scannect this was achieved based on the pin-hole camera model and collinearity equations with assumed additive Gaussian noise for the pixel measurements, which have been shown to be capable of describing the distance measurement uncertainty of the Kinect over its full range [28]. This weighting scheme simplifies the observation variance-covariance matrix to a diagonal matrix while properly describing uncertainties in the point cloud due to the baseline-to-range ratio and parallax angle over the Kinect's full frame. In contrast to [50], which would assign the same weight to all the points when registering coplanar points between two point clouds while the Kinect is oriented orthogonal to the wall, the adopted method will assign slightly different weights based on the variation of the parallax angles in these coplanar points. The end result is similar to the empirical weighting scheme adopted in [50], but with the underlying physical measurement principles of the Kinect expressed mathematically.

Initial Alignment

Generally, consecutive Kinect depth maps are captured merely fractions of a second after another. This plus transporting the Kinect slowly and smoothly during the SLAM process results in the initial transformation of the ICP to be highly predictable by a suitable motion model, and sometimes can even be assumed to be the same as the previous pose. Nevertheless, the IMU predictions are usually sufficient to get the ICP to converge to the correct solution because of the relatively short time update. However, because the ICP tends to converge to a local minimum as oppose to the global minimum, whenever possible the initial alignment is derived via other means to keep the ICP updates as independent from the IMU predictions as possible. This reduces the chance of a poor IMU prediction causing the ICP to converge to a local minimum and then using this distorted scene for camera localization.

This alternative method for deriving the initial alignment is based on 3D keypoint matching [41]. Similar to the conventional 2D keypoint matching technique [81], 3D keypoint matching begins with keypoint detection, every keypoint is then associated with a description, and points with the most similar description are matched. Most of the papers directly solve for the egomotion in EKF using this keypoint method, indicating its fidelity. However, in this paper they are only treated as initial approximations for ICP because a reliable implementation typically requires texture information, which as mentioned, RGB images are not triggered simultaneously as the depth images in the Kinect.

The Implicit Shape Signature (ISS) [82] was used for extracting keypoints with a unique and stable underlying surface. Many existing 3D keypoint detectors are just extensions of 2D image edge/corner detectors (e.g., Harris). While edges/corners are good for 2D images, in 3D their depth measurements are usually less reliable due to for example the mixed pixels effect. Therefore, the ISS, which was originally designed for 3D point clouds was favored. Also, it was chosen because [83] showed it has better overall performance than other 3D keypoint detectors that are publically available in PCL. To have a strong description for the detected keypoints, SHOTColor (a.k.a. Color Signature of Histograms of OrienTations (CSHOT)) in PCL was used [84]. This is one of the few descriptors that incorporates both shape and color information into their description, making its signature the most unique (*i.e.*, highest dimension) in PCL. RANSAC was then used to select a group of inliers and estimate the initial alignment between the two clouds to be used in ICP. Since RANSAC is an expensive process, the IMU predictions were used to eliminate obvious false positives by setting a radius threshold around each keypoint.

Loop-Closure

The proposed method presented here focused only on the filtering solution, also known as the SLAM frontend. For a full SLAM solution, the global relaxation and optimization is crucial. To partially address the issue of loop closure in this paper, the ICP is carried out in a model-to-scene fashion like KinectFusion [44]. This way, every new cloud coming from the sensors was matched to the global scene. This approach can be even more effective when performing on laser scanner data because of their wide FOV and long-ranges.

2.3.3. RGB Visual Odometry

As mentioned, RGB data was processed without consideration of the depth information. This reduces errors in egomotion estimation introduced through time synchronization errors [50] and more importantly keeps track of the pose even when depth data is unavailable and/or unreliable (e.g., imaging glass) [40]. The basic assumption is that the scene is static and any changes in the appearance of the RGB image must be caused by the camera's movement. For this task of identifying changes in consecutive images, the popular SIFT keypoints are detected along with descriptions computed by SURF using OpenCV. A blob detector is preferred over edge/corner detectors because such geometric information is already encapsulated in the depth data processing pipeline [85]. Correspondences were established by searching for points in the two images with the most similar description. These potential matches were then filtered based on uniqueness constraint, the IMU predicted keypoint positions in the new image, 5-point RANSAC in OpenCV, and later via outlier rejection at the measurement update stage. When the pixel differences between conjugate image points are smaller than a threshold ZUPT is performed, otherwise the camera measurement update is executed. The proposed VO workflow is summarized in Figure 5.

Figure 5. Visual odometry using monocular vision.

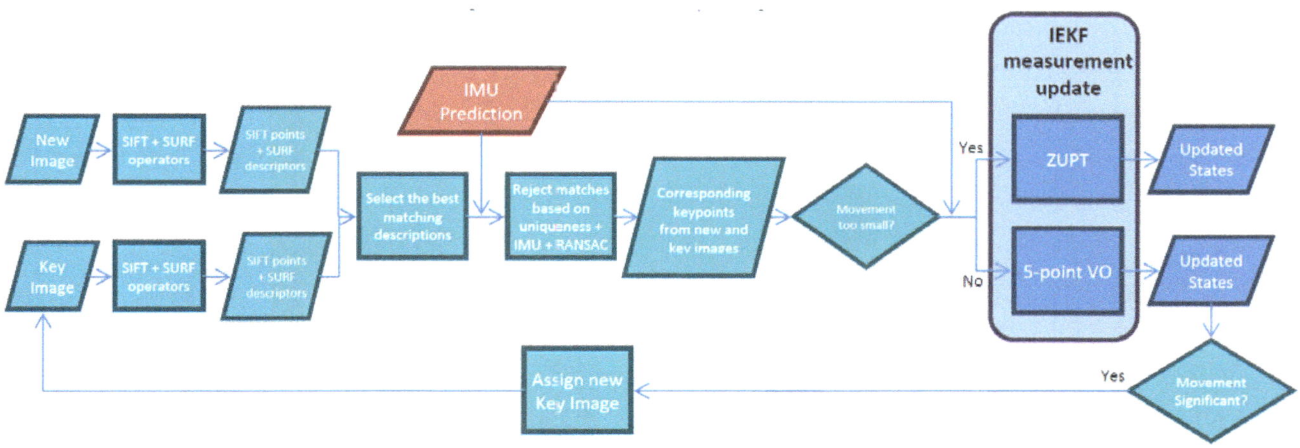

Monocular camera trajectory estimation in general can be solved as a SLAM problem or using VO [33]. The former achieves better global consistency at the expense of higher complexity and being more computationally demanding. While the latter focuses on local consistency and has shown great success for example in the Mars Rover exploration. VO is preferred in this project because the mapping and localization are mainly achieved using ICP, while the RGB data is just an aid to improve localization when depth is not available or when traversing over highly texturized areas of the map. The proposed VO algorithm differs from PTAM in the sense that it implicitly solves for the 3D points from a pair of images. This simultaneous photogrammetric intersection and resection is mathematically realized through solving the set of linear equations in Equation (10) and obtaining an implicit functional model. This was the main motivation to choose the VO route because it is possible to formulate a 2D-to-2D matching VO so as to not increase the size of the state vector while incorporating all the camera geometry information, correlation between the pose and reconstructed points, and minimizing the reprojection errors.

Its advantages over PTAM include egomotion estimation by matching across two images instead of thrree images, no need for a recovery routine in case camera tracking fails, and it is more scalable because of the fixed state vector dimension. A noticeable disadvantage arises from the fact that reconstructed points are not memorized; in other words they are immediately forgotten in the KF and later images cannot resect from continuously improving 3D object points, but only from selected key frames.

$$\begin{aligned}&R_s^n(t+1)\left\{R_{IR}^s R_k^{IR}\left\{\left[R_{RGB}^{IR}\right]_k \mu_i^{RGB}(t+1) p_i^{RGB}(t+1)+\Delta d^k-\Delta b^k\right\}+\Delta b^s\right\}+T^n(t+1)\\&-\left\{R_s^n(t)\left\{R_{IR}^s R_k^{IR}\left\{\left[R_{RGB}^{IR}\right]_k \mu_i^{RGB}(t) p_i^{RGB}(t)+\Delta d^k-\Delta b^k\right\}+\Delta b^s\right\}+T^n(t)\right\}=0\end{aligned} \tag{10}$$

where $[R_{RGB}^{IR}]_k$ is the relative rotation between the RGB and IR camera of Kinect k; Δd^k is the relative translation between the RGB and IR camera of Kinect k.

3. Results and Discussion

3.1. Boresight and Leverarm Calibration between the Kinect and Laser Scanner

Once all the optical systems were independently calibrated their measurements need to be related in 3D space before they can be tightly-coupled and fused in a KF. A standard checkerboard pattern was placed in 17 different static positions and orientations in front of one of the Kinects, where point clouds with the RGB texture overlaid were captured. At every setup, a laser scan of the checkerboard was also captured. The corners were extracted in 3D using a least-square cross-fitting [86]. The necessary boresight and leverarm parameters were then computed using the 3D rigid-body transformation equations solved in a least-squares adjustment. The same procedure was then repeated between the other Kinect and the Focus^{3D} S. Although the two sets of boresight and leverarm parameters were defined with respect to the Focus^{3D} S, for the convention adopted, they were redefined to use the left Kinect (mounted with the IMU) as the reference coordinate system. To assess the fit between the point clouds, 10 planes at various orientations and positions were extracted from both Kinects and compared with the Focus^{3D} S as the reference. The RMSE computed using the orthogonal distances between the Kinect point clouds and the plane defined by the Focus^{3D} S is 4.5 mm and 6.2 mm for the left and right Kinects, respectively. These estimates were higher than anticipated based on the expected accuracy from [65,69] but can probably be attributed to the instability of the camera mounts relative to the trolley in this prototype.

3.2. Boresight and Leverarm Calibration between the Kinect and IMU

As all the optical sensors were already calibrated both internally and externally relative to each other, only the boresight and leveram parameters between the IMU and one of the optical sensors, in this case the left Kinect was necessary to complete the system calibration. A checkerboard pattern was placed horizontally on a leveled surface. The rigidly-attached Kinect and IMU were then moved in front of it to ensure sufficient excitations about all three axes. To remove any range restrictions, only a video sequence from the RGB camera was captured at 10 Hz along with the IMU data at 100 Hz. A photogrammetric resection was performed in a tightly-coupled IEKF to update the IMU predictions online. After filtering, a global adjustment was used to estimate the boresight and leverarm parameters simultaneously offline. The errors between the IMU predicted corner locations and the measured locations in the RGB images before and after calibration is given in Table 1.

Table 1. Errors between the predicted and measured target positions in image space before and after calibration.

	Before Calibration	After Calibration	Improvements
RMSE x	29.1 pix	0.8 pix	97%
RMSE y	23.1 pix	1.1 pix	95%

3.3. Point-to-Plane ICP for the Kinect

To compare the proposed ICP algorithm to the widely accepted point-to-plane ICP from Chen and Medioni, the Kinect was mounted on a translation stage while imaging a complex scene between 0.9 m and 2.9 m away (Figure 6). A known translation of 5 cm was then introduced in the depth direction of the Kinect with a standard deviation of 0.05 mm and then a second point cloud was captured. The two point clouds were then registered using the two ICP algorithms. The RMSE of the fit between the two point clouds computed using the point-to-plane distances metric was 7 mm using either algorithm. Also, based on a qualitative analysis of the registered point clouds, no apparent differences between the two methods were identifiable. However, there is a slight difference between the recovered movements of the Kinect. The method by Chen and Medioni [36] underestimated the translation by 1.1 mm and the proposed method underestimated the translation by 0.1 mm. This small improvement comes from the weights assigned to correspondences (based on the triangulation geometry) and a cost function which minimizes the reprojection error of the IR camera and projector pair. Differences of a millimeter may seem insignificant at first, but in the absence of additional information the errors can accumulate rather quickly in a Kinect-only dead-reckoning solution, especially when the Kinect is capturing with a frequency as high as 30 Hz.

Figure 6. Calibrated RGB-D data from Kinect of the scene used for testing iterative closest point (ICP).

To test the suitability of the devised ICP algorithm for mapping applications when assisted by an Xsens MTi-30 IMU, a sequence of handheld data was captured. This illustrates the mapping abilities when this module is removed from the main trolley platform for exploring inaccessible spaces like stairs before returning to its forced centered position. The Kinect and the MTi-30 were rigidly mounted together with the Kinect logging at 10 Hz and the MTi-30 logging at 100 Hz. A person carried the Kinect and IMU by hand, ascended a flight of stairs, turned around and then descended the same flight of stairs. A total of 315 RGB-D datasets were captured and processed. The final 3D point cloud of the staircase is shown in Figure 7 where the colors are assigned based on the local normals. Considering that the distances travelled was short, the model-to-scene registration scheme was capable of partially handle the loop-closure problem and no apparent misfits can be identified near the bottom of the staircase. For a quantitative accuracy assessment, 20 horizontal and vertical planes were extracted and the overall RMSE from this check plane analysis was 7.0 mm.

Figure 7. Stairs reconstructed using the Kinect and inertial measurement unit (IMU) in a human carried walking sequence.

3.4. 2D-to-2D Visual Odometry for Monocular Vision

A checkerboard pattern with known dimensions was imaged with the Kinect's RGB camera. The Kinect was handheld and waved in front of the checkerboard for approximately 80 s. The reference solution was generated using photogrammetric resection solved in a least-squares adjustment frame-by-frame. This represents the best case scenario in camera navigation where known control points are always visible. VO was performed by solving for the camera's pose using the mathematical model presented in Equation (10) with the first image held as the reference keyframe the entire time. The camera orientations converted from quaternion to Euler angles are shown in Figure 8. It can be observed that the VO solution follows the reference solution quite closely over certain sections of the plot (e.g., between 350 and 362 s). But in other sections a noticeable difference in their estimation can be observed, with errors reaching as high as 80°. The main cause for such a discrepancy is the intersection geometry. When the pair of images were captured near proximity of each other, the parallax angle is small, this results in a less accurate scene reconstruction, which translates into poor egomotion estimation. In fact, the proposed math model has a singularity when the light rays in the pair of images are all parallel with each other. However, this can easily be identified by checking the disparity of conjugate points. The case when the disparities are all below a threshold actually indicates the camera has not moved significantly, suggesting the possibility of the camera being static, therefore ZUPT was performed. For a MEMS-based IMU, integration of the gyroscope readings alone over a period of approximately 80 s would have introduced a noticeable rotation drift. But as witnessed in Figure 8, the fusion of VO with IMU prevented the estimated orientation from drifting. At the same time the IMU smoothed out the vision-based orientation estimates and provided a better orientation estimate when the uncertainty in VO is high, resulting in a final solution that follows the reference solution quite closely over the full trial. This suggests that VO and gyroscope combined can provide a better orientation estimate than either method alone.

Figure 8. Comparison between various ways of estimating monocular camera rotations. α, β, and γ represents the rotation about the x, y, and z axis, respectively. Res represents the photogrammetric resection solution.

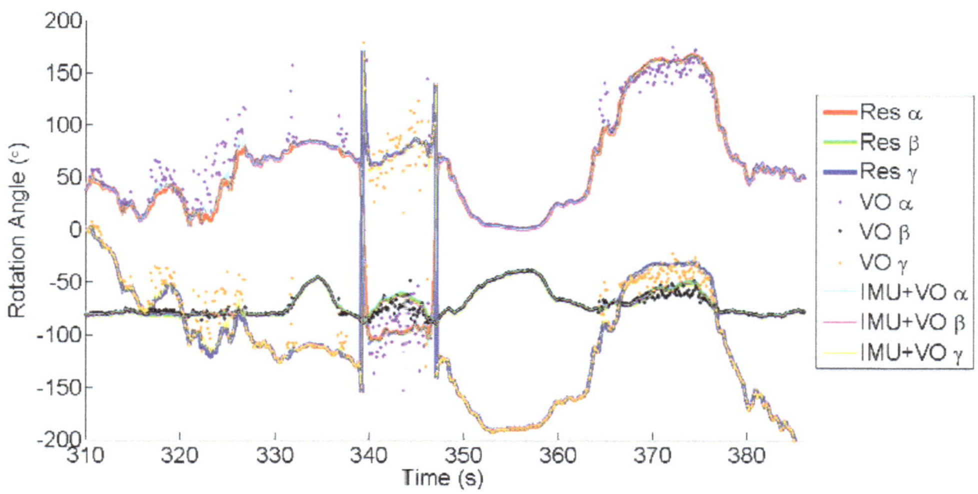

3.5. Scannect Testing at the University of Calgary

The full Scannect system was tested by mapping a floor in the Math Science building located at the University of Calgary, Canada (Figure 9). The hallways are mainly composed of doors (some are recessed), posters, bulletin boards, and white homogenous flat walls. The Scannect travelled 120 m in continuous mode and only stopped at the four corners to capture a 7 minutes scan with the Focus^{3D} S. While in continuous mode, both Kinects were capturing data at 10 Hz while the IMU was logging at 200 Hz. Because the floor was relatively flat, a height constraint was applied. Also, ZUPTS were applied every time sufficiently small movements were detected by the RGB VO and when the Focus^{3D} S was scanning. The total time it took to map the floor in the field was 35 minutes, significantly less than a typical laser scanner only project over such a large area. The processing time for both Kinects with the IMU was about two hours, however it should be stressed that the C++ code implemented has yet to be optimized. The map generated by the Kinect with RGB texture information added for visual appearance is shown in Figure 10. Although only the points measured by the Kinects are shown, these results were generated with the assistance of the Focus^{3D} S. At the beginning a 360° scan was first captured by the scanner and added to the scene. Then the Kinects used this as the scene in the dense 3D matching step. Meanwhile it filled in some occlusions and expands the map until it arrived at the next corner where another laser scan was captured and matched to the scene through the point-to-plane ICP with the IMU predicted pose as the initial alignment. This process was repeated until the floor had been explored (completely mapped).

Throughout the floor, 30 checkpoints were also captured by the Kinects. A top view of the Kinects' map along with the horizontal distribution of these signalized targets is shown in Figure 11. Due to the limited vertical FOV of the Kinects, all the checkpoints were approximately at the same height. The RMSE in the X, Y, and Z direction was 10.0 cm, 10.8 cm, and 9.0 cm, respectively. This is approximately 0.14% error over the travelled distance. These error estimates only indicates the forward filtering solution and is already similar if not better than the existing methods described in Section 1.3 (*i.e.*, [42,56,57]). After backward smoothing or SBA, the results are expected to improve [41]. Note that the laser scanner only solution shown in Figure 12 (with the green stars indicating the static scan locations) missed a lot of the doorways (in particular the top corridor between targets A26 and A18) but the general shape of the floor is prevalent. This general shape from the laser scans aided the Kinects' localization and provided some degree of loop-closure. From Figure 11 it can be perceived that the details such as doorways that were missed by the laser scanner were well localized in the same coordinate system by the Kinect. Some misalignments in the scanner data can be observed (e.g., the top corridor), this is likely a result of errors in the initial alignment provided by the Kinect and IMU filter and the accumulated mapping errors.

Figure 9. Image of one of the hallways in the Math Science building at the University of Calgary.

Figure 10. Oblique view of the Kinect point clouds from the 3D reconstructed scene.

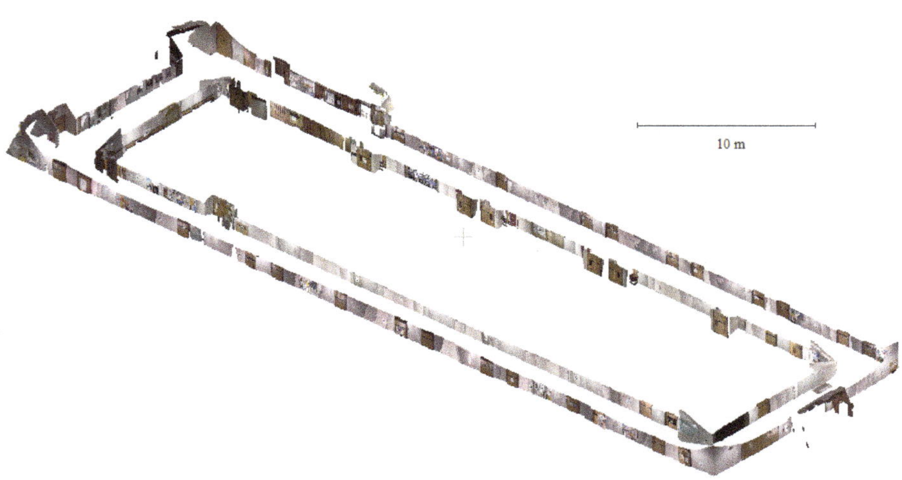

Figure 11. Top view of the *Kinect* point clouds from the 3D reconstructed scene. The magenta labels indicate the location of the check points.

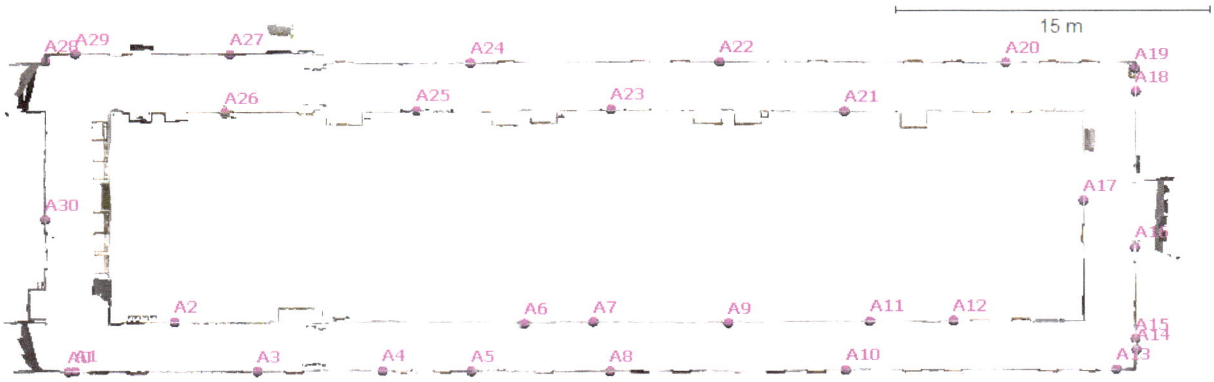

Figure 12. Top view of the *laser scanner* point clouds from the 3D reconstructed scene with the ceiling and floor removed. The green stars indicate the scan stations.

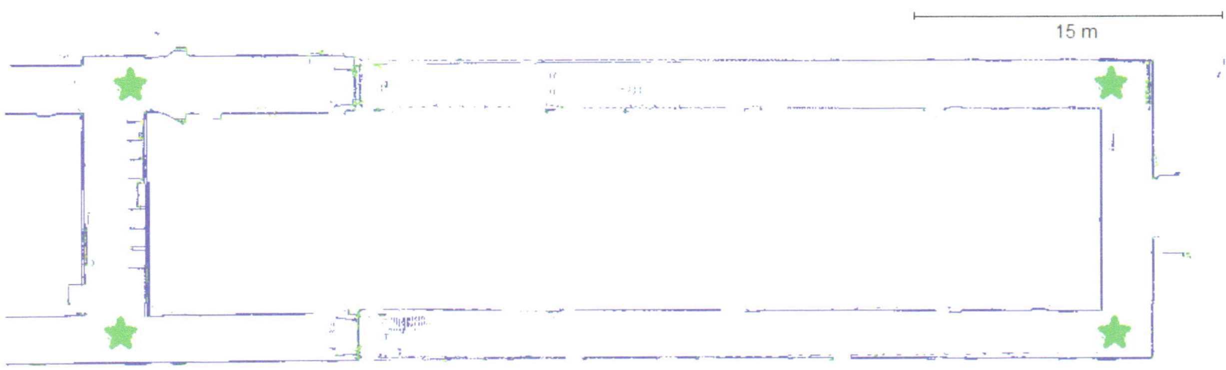

4. Conclusions and Future Work

SLAM is a theoretically matured subject with many practical implementation challenges. A few of these challenges were addressed in this paper: the implicit iterative extended Kalman filter was responsible for fusing measurement updates from the Kinect's RGB and depth stream separately to account for their timing errors and non-linearity of implicit photogrammetry equations; using only the RGB images, a 5-point visual odometry was performed in a tightly-coupled manner while eliminating the need to add or remove states in the filter; a new tightly-coupled ICP method for stereo-vision systems was used for processing the depth maps. The presented Scannect system is the first multi-Kinect mobile mapping system and the first to adopt the continuous stop-and-go design. This system can map an entire building floor efficiently with a Mean Radial Spherical Error of 17 cm. It was capable of navigating in areas with as little features as a single white wall.

In the future, more Kinects will be added to further increase the field of view of the Scannect. For instance, to make the Scannect more flexible and account for sloped floors, a skyward facing Kinect can be added without any modifications to the algorithms and processing pipeline presented in this paper. Newer 3D cameras with a wider field of view (e.g., PMD CamBoard nano) or higher accuracy (e.g., Kinect 2) will be considered. The LiDAR system will be tightly-coupled to allow weighting in the spherical coordinate system. Adding other autonomous non-vision-based sensors such as wheel odometers, magnetometers, and barometers will be beneficial for localization. Other non-holonomic constraints such as leveling updates can also be included if applicable. A more accurate scene representation such as surfels will be adopted [87]. To better handle loop-closures, methods like sparse bundle adjustment will be implemented to improve the global consistency. This was shown to be more accurate than TORO (a graph-based method) when using the Kinect [41]. For real-time online processing and visualization, general-purpose computing on graphics processing units (GPGPU) will be investigated to expedite the filtering process.

Acknowledgments

This research is funded by the Natural Science and Engineering Research Council (NSERC) of Canada, Alberta Innovates, the Canada Foundation for Innovation, and the Killam Trust.

Author Contributions

Jacky C.K. Chow prepared the manuscript; acquired the datasets using the laser scanner, Kinects, and IMU; developed software for manipulating, processing, and integrating data from the sensors; formulated the modified ICP algorithm and 5-point VO algorithm; deciding on the choice of sensors; coming up with the hybrid system design; and testing the Scannect and analyzing its results. Derek D. Lichti provided valuable input to the manuscript. Jeroen D. Hol, Giovanni Bellusci, Henk Luinge assisted in the IMU processing, had valuable insights for the project, and contributed to the writing of this manuscript.

Conflicts of Interest

The authors declare no conflict of interest.

References

1. Gadeke, T.; Schmid, J.; Zahnlecker, M.; Stork, W.; Muller-Glaser, K. Smartphone pedestrian navigation by foot-IMU sensor fusion. In Proceedings of the 2nd International Conference on Ubiquitous Positioning, Indoor Navigation, and Location Based Service, Helsinki, Finland, 3–4 October 2012; pp. 1–8.
2. Durrant-Whyte, H.; Bailey, T. Simultaneous localization and mapping: Part I. *IEEE Robot. Autom. Mag.* **2006**, *13*, 99–110.
3. Bailey, T.; Durrant-Whyte, H. Simultaneous localisation and mapping (SLAM): Part II state of the art. *IEEE Robot. Autom. Mag.* **2006**, *13*, 108–117.
4. Huang, A.; Bachrach, A.; Henry, P.; Krainin, M.; Maturana, D.; Fox, D.; Roy, N. Visual odometry and mapping for autonomous flight using an RGB-D camera. In Proceedings of the 15th International Symposium on Robotics Research, Flagstaff, AZ, USA, 28 August–1 September 2011.
5. He, B.; Yang, K.; Zhao, S.; Wang, Y. Underwater simultaneous localization and mapping based on EKF and point features. In Proceedings of the International Conference on Mechatronics and Automation, Changchun, China, 9–12 August 2009; pp. 4845–4850.

6. Kuo, B.; Chang, H.; Chen, Y.; Huang, S. A light-and-fast SLAM algorithm for robots in indoor environments using line segment map. *J. Robot.* **2011**, doi:10.1155/2011/257852.
7. Newman, P.; Cole, D.; Ho, K. Outdoor SLAM using visual appearance and laser ranging. In Proceedings of the IEEE International Conference on Robotics and Automation, Orlando, FL, USA, 15–19 May 2006; pp. 1180–1187.
8. Georgiev, A.; Allen, P. Localization methods for a mobile robot in urban environments. *IEEE Trans. Robot.* **2004**, *20*, 851–864.
9. Kümmerle, R.; Steder, B.; Dornhege, C.; Kleiner, A.; Grisetti, G.; Burgard, W. Large scale graph-based SLAM using aerial images as prior information. *J. Auton. Robots* **2011**, *30*, 25–39.
10. Vosselman, G.; Maas, H.-G. *Airborne and Terrestrial Laser Scanning*; Whittles Publishing: Caithness, UK, 2010.
11. Nüchter, A. Parallelization of scan matching for robotic 3D mapping. In Proceedings of the 3rd European Conference on Mobile Robots, Freiburg, Germany, 19–21 September 2007.
12. Lin, Y.; Hyyppä, J.; Kukko, A. Stop-and-go mode: Sensor manipulation as essential as sensor development in terrestrial laser scanning. *Sensors* **2013**, *13*, 8140–8154.
13. Ferris, B.; Fox, D.; Lawrence, N. WiFi-SLAM using Gaussian process latent variable models. In Proceedings of the 20th International Joint Conference on Artificial Intelligence, Hyderabad, India, 6–12 January 2007; pp. 2480–2485.
14. Des Bouvrie, B. Improving rgbd indoor mapping with imu data. Master's Thesis, Delft University of Technology, Delft, The Netherlands, 2011.
15. Strasdat, H.; Montiel, J.; Davison, A. Real-time monocular SLAM: Why filter? In Proceedings of the IEEE International Conference on Robotics and Automation, Anchorage, AK, 3–7 May 2010; pp. 2657–2664.
16. Tomono, M. Robust 3D SLAM with a stereo camera based on an edge-point ICP algorithm. In Proceedings of the IEEE International Conference on Robotics and Automation, Kobe, Japan, 12–17 May 2009; pp. 4306–4311.
17. Borrmann, D.; Elseberg, J.; Lingemann, K.; Nüchter, A.; Hertzberg, J. Globally consistent 3D mapping with scan matching. *Robot. Auton. Syst.* **2008**, *56*, 130–142.
18. May, S.; Fuchs, S.; Droeschel, D.; Holz, D.; Nüchter, A. Robust 3D-mapping with time-of-flight cameras. In Proceedings of the IEEE/RSJ International Conference on Intelligent Robots and Systems, St. Louis, MO, USA, 10–15 October 2009; pp. 1673–1678.
19. Siegwart, R.; Nourbakhsh, I.; Scaramuzza, D. *Introduction to Autonomous Mobile Robots*, 2nd ed.; The MIT Press: Cambridge, MA, USA, 2011.
20. Luhmann, T.; Robson, S.; Kyle, S.; Harley, I. *Close Range Photogrammetry: Principles, Techniques and Applications*; Whittles Publishing: Caithness, UK, 2006.
21. Mishra, R.; Zhang, Y. A review of optical imagery and airborne LiDAR data registration methods. *Open Remote Sens. J.* **2012**, *5*, 54–63.
22. Kurz, T.; Buckley, S.; Schneider, D.; Sima, A.; Howell, J. Ground-based hyperspectral and lidar scanning: A complementary method for geoscience research. In Proceedings of the International Association for Mathematical Geosciences, Salzburg, Austria, 5–9 September 2011.
23. Han, J.; Shao, L.; Xu, D.; Shotton, J. Enhanced computer vision with Microsoft Kinect sensor: A review. *IEEE Trans. Cybern.* **2013**, *43*, 1318–1334.

24. Zug, S.; Penzlin, F.; Dietrich, A.; Nguyen, T.; Albert, S. Are laser scanners replaceable by Kinect sensors in robotic applications? In Proceedings of the IEEE International Symposium on Robotic and Sensors Environments (ROSE), Magdeburg, Germany, 16–18 November 2012; pp. 144–149.
25. Meister, S.; Kohli, P.; Izadi, S.; Hämmerle, M.; Rother, C.; Kondermann, D. When can we use KinectFusion for ground truth acquisition? In Proceedings of the IEEE/RSJ International Conference on Intelligent Robots and Systems, Workshop on Color-Depth Camera Fusion in Robotics, Vilamoura, Algarve, Portugal, 7–11 October 2012.
26. Smisek, J.; Jancosek, M.; Pajdla, T. 3D with kinect. In Proceedings of the Consumer Depth Cameras for Computer Vision, Barcelona, Spain, 12 November 2011.
27. Stoyanov, T.; Louloudi, A.; Andreasson, H.; Lilienthal, A. Comparative evaluation of range sensor accuracy for indoor environments. In Proceedings of the European Conference on Mobile Robots, Örebro, Sweden, 7–9 September 2011; pp. 19–24.
28. Chow, J.; Ang, K.; Lichti, D.; Teskey, W. Performance analysis of a low-cost triangulation-based 3D camera: Microsoft Kinect system. In Proceedings of the International Society of Photogrammetry and Remote Sensing Congress, Melbourne, Australia, 25 August–1 September 2012.
29. Renaudin, V.; Yalak, O.; Tomé, P.; Merminod, B. Indoor navigation of emergency agents. *Eur. J. Navig.* **2007**, *5*, 36–45.
30. Ghilani, C.; Wolf, P. *Elementary Surveying: An Introduction to Geomatics*, 12th ed.; Prentice Hall: Upper Saddle River, NJ, USA, 2008.
31. Grisetti, G.; Stachniss, C.; Burgard, W. Improving grid-based SLAM with with Rao-Blackwellized particle filters by adaptive proposals and selective resampling. In Proceedings of the IEEE International Conference on Robotics and Automation, Barcelona, Spain, 18–22 April 2005; pp. 2432–2437.
32. Gustafsson, F. *Statistical Sensor Fusion*, 2nd ed.; Studentlitteratur AB: Lund, Sweden, 2010.
33. Scaramuzza, D.; Fraundorfer, F. Visual odometry part I: The first 30 years and fundamentals. *IEEE Robot. Autom. Mag.* **2011**, *18*, 80–92.
34. Besl, P.; McKay, N. A method for registration of 3D shapes. *IEEE Trans. Pattern Anal. Mach. Intell.* **1992**, *14*, 239–256.
35. Horn, B. Closed-form solution of absolute orientation using unit quaternions. *J. Opt. Soc. Am.* **1987**, *4*, 629–642.
36. Chen, Y.; Medioni, G. Object modelling by registration of multiple range images. *Image Vis. Comput.* **1992**, *10*, 145–155.
37. Rusinkiewicz, S.; Levoy, M. Efficient variants of the ICP algorithm. In Proceedings of the 3rd International Conference on 3-D Digital Imagining and Modeling, Quebec, QC, Canada, 28 May–1 June 2001; pp. 145–152.
38. Klein, G.; Murray, D. Parallel tracking and mapping for small AR workspaces. In Proceedings of the 6th IEEE and ACM International Symposium on Mixed and Augmented Reality, Nara, Japan, 13–16 November 2007; pp. 225–234,
39. Nützi, G.; Weiss, S.; Scaramuzza, D.; Siegwart, R. Fusion of IMU and vision for absolute scale estimation in monocular SLAM. *J. Intell. Robot. Syst.* **2011**, *61*, 287–299.

40. Scherer, S.; Dube, D.; Zell, A. Using depth in visual simultanesous localisation and mapping. In Proceedings of the IEEE International Conference on Robotics and Automation, St. Paul, MN, USA, 14–18 May 2012.
41. Henry, P.; Krainin, M.; Herbst, E.; Ren, X.; Fox, D. RGB-D mapping: Using Kinect-style depth cameras for dense 3D modeling of indoor environments. *Int. J. Robot. Res.* **2012**, *31*, 647–663.
42. Henry, P.; Krainin, M.; Herbst, E.; Ren, X.; Fox, D. RGB-D mapping: Using depth cameras for dense 3d modeling of indoor environments. In Proceedings of the RGB-D: Advanced Reasoning with Depth Cameras in Conjunction with Robotics Science and Systems, Zaragoza, Spain, 27 June 2010.
43. Johnson, A.; Kang, S. Registration and integration of textured 3D data. *Image Vis. Comput.* **1999**, *17*, 135–147.
44. Izadi, S.; Kim, D.; Hilliges, O.; Molyneaux, D.; Newcombe, R.; Kohli, P.; Shotton, J.; Hodges, S.; Freeman, D.; Davison, A.; et al. KinectFusion: Real-time 3D reconstruction and interaction using a moving depth camera. In Proceedings of the 24th ACM Symposium on User Interface Software and Technology, Santa Barbara, CA, USA, 16–19 October 2011; pp. 559–568.
45. Curless, B.; Levoy, M. A volumetric method for building complex odels from range images. In Proceedings of the 23rd Annual Conference on Computer Graphics and Interactive Techniques—SIGGRAPH, New York, NY, USA, 4–9 August 1996; pp. 303–312.
46. Whelan, T.; McDonald, J.; Kaess, M.; Fallon, M.; Johannsson, H.; Leonard, J. Kintinuous: Spatially extended KinectFusion. In Proceedings of the 3rd RSS Workshop on RGB-D: Advanced Reasoning with Depth Cameras, Sydney, Australia, 9–10 July 2012.
47. Whelan, T.; Johannsson, H.; Kaess, M.; Leonard, J.; McDonald, J. Robust real-time visual odometry for dense RGB-D mapping. In Proceedings of the IEEE International Conference on Robotics and Automation, Karlsruhe, Germany, 6–10 May 2013; pp. 5724–5731.
48. Whelan, T.; Kaess, M.; Leonard, J.; McDonald, J. Deformation-based loop closure for large scale dense RGB-D SLAM. In Proceedings of the IEEE/RSJ International Conference on Intelligent Robots and Systems, Tokyo, Japan, 3–8 November 2013.
49. Keller, M.; Lefloch, D.; Lambers, M.; Izadi, S.; Weyrich, T.; Kolb, A. Real-time 3D reconstruction in dynamic scenes using point-based fusion. In Proceedings of the International Conference on 3D Vision, Seattle, WA, USA, 29 June–1 July 2013; pp. 1–8.
50. Khoshelham, K.; Dos Santos, D.; Vosselman, G. Generation and weighting of 3D point correspondences for improved registration of RGB-D data. In Proceedings of the ISPRS Annals of the Photogrammetry and Remote Sensing and Spatial Information Sciences, Volume II-5/W2, Antalya, Turkey, 11–13 November 2013.
51. Leutenegger, S.; Furgale, P.; Rabaud, V.; Chli, M.; Konolige, K.; Siegwart, R. Keyframe-based visual-inertial SLAM using nonlinear optimization. In Proceedings of the Robotics: Science and Systems, Berlin, Germany, 24–28 June 2013.
52. Li, L.; Cheng, E.; Burnett, I. An iterated extended Kalman filter for 3D mapping via Kinect camera. In Proceedings of the IEEE International Conference on Acoustics, Speech and Signal Processing, Vancouver, Canada, 26–31 May 2013; pp. 1773–1777.

53. Aghili, F.; Kuryllo, M.; Okouneva, G.; McTavish, D. Robust pose estimation of moving objects using laser camera data for autonomous rendezvous and docking. In Proceedings of the International Society of Photogrammetry and Remote Sensing Archives Volume XXXVIII-3/W8, Paris, France, 1–3 September 2009; pp. 253–258.
54. Hervier, T.; Bonnabel, S.; Goulette, F. Accurate 3D maps from depth images and motion sensors via nonlinear Kalman filtering. In Proceedings of the IEEE/RSJ International Conference on Intelligent Robots and Systems, Vilamoura, Portugal, 7–12 October 2012; pp. 5291–5297.
55. Kottas, D.; Roumeliotis, S. Exploiting urban scenes for vision-aided inertial navigation. In Proceedings of the Robotics: Science and Systems, Berlin, Germany, 24–28 June 2013.
56. Li, M.; Kim, B.; Mourikis, A. Real-time motion tracking on a cellphone using inertial sensing and a rolling shutter camera. In Proceedings of the IEEE International Conference on Robotics and Automation, Karlsruhe, Germany, 6–10 May 2013.
57. Burens, A.; Grussenmeyer, P.; Guillemin, S.; Carozza, L.; Lévêque, F.; Mathé, V. Methodological developments in 3D scanning and modelling of archaeological French heritage site: The bronze age painted cave of Les Fraux, Dordogne (France). In Proceedings of the International Archives of the Photogrammetry, Remote Sensing and Spatial Information Sciences, Volume XL-5/W2, Strasbourg, France, 2–6 September 2013; pp. 131–135.
58. Chiabrando, F.; Spanò, A. Point clouds generation using TLS and dense-matching techniques. A test on approachable accuracies of different tools. In Proceedings of the ISPRS Annals of the Photogrammetry, Remote Sensing and Spatial Information Sciences, Volume II-5/W1, Strasbourg, France, 2–6 September 2013; pp. 67–72.
59. Khoshelham, K.; Oude Elberink, S. Accuracy and resolution of kinect depth data for indoor mapping applications. *Sensors* **2012**, *12*, 1437–1454.
60. Dal Mutto, C.; Zanuttigh, P.; Cortelazzo, G. *Time-of-Flight Cameras and Microsoft Kinect*; Springer: New York, NY, USA, 2013.
61. Kainz, B.; Hauswiesner, S.; Reitmayr, G.; Steinberger, M.; Grasset, R.; Gruber, L.; Veas, E.; Kalkofen, D.; Seichter, H.; Schmalstieg, D. OmniKinect: Real-time dense volumetric data acquisition and applications. In Proceedings of the 18th ACM Symposium on Virtual Reality Software and Technology, Toronto, ON, Canada, 10–12 December 2012; pp. 25–32.
62. Butler, A.; Izadi, S.; Hilliges, O.; Molyneaux, D.; Hodges, S.; Kim, D. Shake and sense: Reducing interference for overlapping structured light depth cameras. In Proceedings of the SIGCHI Conference on Human Factors in Computing Systems, Austin, TX, USA, 5–10 May 2012; pp. 1933–1936.
63. Aggarwal, P.; Syed, Z.; Noureldin, A.; El-Sheimy, N. *MEMS-Based Integrated Navigation*; Artech House: Norwood, MA, USA, 2010.
64. Herrera, D.; Kannala, J.; Heikkilä, J. Joint depth and color camera calibration with distortion correction. *IEEE Trans. Pattern Anal. Mach. Intell.* **2012**, *34*, 2058–2064.
65. Chow, J.; Lichti, D. Photogrammetric bundle adjustment with self-calibration of the PrimeSense 3D camera technology: Microsoft Kinect. *IEEE Access* **2013**, *1*, 465–474.
66. Lichti, D. Error modelling, calibration and analysis of an AM-CW terrestrial laser scanner system. *ISPRS J. Photogramm. Remote Sens.* **2007**, *61*, 307–324.

67. Reshetyuk, Y. A unified approach to self-calibration of terrestrial laser scanners. *ISPRS J. Photogramm. Remote Sens.* **2010**, *65*, 445–456.
68. Chow, J.; Lichti, D.; Glennie, C.; Hartzell, P. Improvements to and comparison of static terrestrial LiDAR self-calibration methods. *Sensors* **2013**, *13*, 7224–7249.
69. Chow, J.; Lichti, D.; Teskey, W. Accuracy assessment of the Faro Focus3D and Leica HDS6100 panoramic type terrestrial laser scanner through point-based and plane-based user self-calibration. In Proceedings of the FIG Working Week 2012: Knowing to Manage the Territory, Protect the Environment, Evaluate the Cultural Heritage, Rome, Italy, 6–10 May 2012.
70. Al-Manasir, K.; Fraser, C. Registration of terrestrial laser scanner data using imagery. *Photogramm. Rec.* **2006**, *21*, 255–268.
71. Lobo, J.; Dias, J. Relative pose calibration vetween visual and inertial sensors. *Int. J. Robot. Res.* **2007**, *26*, 561–575.
72. Hol, J.; Schön, T.; Gustafsson, F. Modeling and calibration of inertial and vision sensors. *Int. J. Robot. Res.* **2009**, *29*, 231–244.
73. Mirzaei, F.; Roumeliotis, S. A Kalman filter-based algorithm for IMU-camera calibration: Observability analysis and performance evaluation. *IEEE Trans. Robot.* **2008**, *24*, 1143–1156.
74. Titterton, D.; Weston, J. *Strapdown Inertial Navigation Technology*; IEE Radar, Sonar, Navigation and Avionics Series; Peter Peregrinus Ltd.: Stevenage, UK, 1997.
75. Gelb, A. *Applied Optimal Estimation*; The MIT Press: Cambridge, MA, USA, 1974.
76. Steffen, R.; Beder, C. Recursive estimation with implicit constraints. In Proceedings of the 29th DAGM Conference on Pattern Recognition, Heidelberg, Germany, 12–14 September 2007; pp. 194–203.
77. Steffen, R. A robust iterative Kalman filter based on implicit measurement equations. *Photogramm. Fernerkund. Geoinf.* **2013**, *2013*, 323–332.
78. OpenNI Programmer's Guide. OpenNI 2.0 API. Available online: http://www.openni.org/openni-programmers-guide/ (accessed on 16 December 2013).
79. Muja, M.; Lowe, D. Fast approximate nearest neighbors with automatic algorithm configuration. In Proceedings of the International Conference on Computer Vision Theory and Applications (VISAPP'09), Lisboa, Portugal, 5–9 February 2009; pp. 331–340.
80. Luck, J.; Little, C.; Hoff, W. Registration of range data using a hybrid simulated annealing and iterative closest point algorithm. In Proceedings of the IEEE International Conference on Robotics and Automation, San Francisco, CA, USA, 24–28 April 2000; pp. 3739–3733.
81. Bradski, G.; Kaehler, A. *Learning OpenCV*; O'Reilly Media: Sabastopol, CA, USA, 2008.
82. Zhong, Y. Intrinsic shape signatures: A shape descriptor for 3D object recognition. In Proceedings of the IEEE 12th International Conference on Computer Vision, Kyoto, Japan, 29 September–2 October 2009; pp. 689–696.
83. Filipe, S.; Alexandre, L. A comparative Evaluation of 3D keypoint detectors. In Proceedings of the 9th Conference on Telecommunications, Castelo Branco, Portugal, 8–10 May 2013; pp. 145–148.
84. Tombari, F.; Salti, S.; Di Stefano, L. A combined texture-shape descriptor for enhanced 3D feature matching. In Proceedings of the 18th IEEE International Conference on Image Processing, Brussels, Belgium, 11–14 September 2011. pp. 809–812.

85. Fraundorfer, F.; Scaramuzza, D. Visual odometry part II: Matching, robustness, optimization, and applications. *IEEE Robot. Autom. Mag.* **2012**, *19*, 78–90.
86. Chow, J.; Ebeling, A.; Teskey, W. Point-based and plane-based deformation monitoring of indoor environments using terrestrial laser scanners. *J. Appl. Geod.* **2012**, *6*, 193–202.
87. Pfister, H.; Zwicker, M.; van Baar, J.; Gross, M. Surfels: Surface elements as rendering primitives. In Proceedings of the 27th Annual Conference on Computer Graphics and Interactive Techniques, New Orleans, LA, USA, 23–28 July 2000; pp. 335–342.

7

Neural Networks Integrated Circuit for Biomimetics MEMS Microrobot

Ken Saito [1,*], Kazuaki Maezumi [2], Yuka Naito [2], Tomohiro Hidaka [2], Kei Iwata [2], Yuki Okane [2], Hirozumi Oku [2], Minami Takato [2] and Fumio Uchikoba [1]

[1] Department of Precision Machinery Engineering, College of Science and Technology, Nihon University, 7-24-1 Narashinodai, Funabashi-shi, Chiba 274-8501, Japan; E-Mail: uchikoba@eme.cst.nihon-u.ac.jp

[2] Precision Machinery Engineering, Graduate School of Science and Technology, Nihon University, 1-8-14 Kanda Surugadai, Chiyoda-ku, Tokyo 101-8308, Japan; E-Mails: cska13053@g.nihon-u.ac.jp (K.M.); csyk13045@g.nihon-u.ac.jp (Y.N.); csto10130@g.nihon-u.ac.jp (T.H.); cski10025@g.nihon-u.ac.jp (K.I.); csyu10037@g.nihon-u.ac.jp (Y.O.); cshr10043@g.nihon-u.ac.jp (H.O.); takato@eme.cst.nihon-u.ac.jp (M.T.)

* Author to whom correspondence should be addressed; E-Mail: kensaito@eme.cst.nihon-u.ac.jp

---

Abstract: In this paper, we will propose the neural networks integrated circuit (NNIC) which is the driving waveform generator of the 4.0, 2.7, 2.5 mm, width, length, height in size biomimetics microelectromechanical systems (MEMS) microrobot. The microrobot was made from silicon wafer fabricated by micro fabrication technology. The mechanical system of the robot was equipped with small size rotary type actuators, link mechanisms and six legs to realize the ant-like switching behavior. The NNIC generates the driving waveform using synchronization phenomena such as biological neural networks. The driving waveform can operate the actuators of the MEMS microrobot directly. Therefore, the NNIC bare chip realizes the robot control without using any software programs or A/D converters. The microrobot performed forward and backward locomotion, and also changes direction by inputting an external single trigger pulse. The locomotion speed of the microrobot was 26.4 mm/min when the step width was 0.88 mm. The power consumption of the system was 250 mWh when the room temperature was 298 K.

Keywords: biomimetics; MEMS; microrobot; integrated circuit; neural networks

1. Introduction

Many studies have been done on microrobots for several applications such as precise manipulation, medical field, and so on [1–5]. However, further miniaturizations and higher functionalization on the microrobot system are required to play an important role in these fields. Although the miniaturization of the robot has conventionally been progressed by mechanical machining and assembling, some difficulty has appeared in order to achieve further miniaturizations. In particular, frame parts, actuators, motion controllers, power sources and sensors [6]. Instead of the conventional mechanical machining, micro fabrication technology based on the IC production lines has been studied for making the simple components of the microrobot [7–9]. In addition, the development of the actuator is an important subject. The type of the micro actuator by micro fabrication technology is categorized into two groups. For example: one group uses the field forces and the other group uses the property of the material itself [10–14]. In particular, shape memory alloy actuator shows a large displacement, such as 50%, of the total length in millimeter size. However, micro actuators using field forces or piezoelectric elements to the microrobot had a weakness for moving on the uneven surface. Therefore, a microrobot which could locomote by step pattern was desired.

Programmed control by digital systems based on a microcontroller has been the dominant system among the robot control. On the other hand, insects realize the autonomous operation using excellent structure and active neural networks control by compact advanced systems. Therefore, some advanced studies of artificial neural networks have been focused on for applying to the robot systems. A lot of studies have reported both on software models and hardware models [15–17]. However, using the mathematical neuron models in large scale neural network is difficult to process in continuous time because the computer simulation is limited by the computer performance, such as the processing speed and memory capacity. In contrast, using the hardware neuron model is advantageous because even if a circuit scale becomes large, the nonlinear operation can perform at high speed and process in continuous time. Therefore, the construction of a hardware model that can generate oscillatory patterns was desired. For this reason, we are studying about millimeter size microrobot system which can control the locomotion by active hardware neural networks.

Previously, we constructed the 4.0, 2.7, 2.5 mm, width, length, and height size microrobot fabricating the silicon wafer by micro fabrication technology, reported the rotary-type actuator composed of heat stimulated artificial muscle wires in the robot, and also driving waveform of the robot was generated by using packaged neural networks integrated circuit (NNIC) which was externally connected [18].

In this paper, we will propose the NNIC bare chip with peripheral circuit which can mount on the top of the microrobot. The NNIC bare chip with peripheral circuit is a microelectro system. The microrobot fabricated by micro fabrication technology is a micromechanical system. Therefore, the whole system is a microelectromechanical systems (MEMS) microrobot.

2. Biomimetics MEMS Microrobot

We constructed the miniaturized robot by micro fabrication technology. In this chapter, the basic components of the fabricated MEMS microrobot were shown. The number of the legs of the MEMS microrobot was six. The structure and the step pattern of the robot was emulated those of an ant. The microrobot consisted of frame parts, rotary type actuators and link mechanisms.

2.1. Mechanical Components

Figure 1 shows the mechanical parts of the MEMS microrobot. The fabrication of the silicon wafer was done by the micro fabrication technology. The shapes were machined by photolithography based inductively coupled plasma (ICP) dry etching [19]. We used the 100, 200, 385, 500 μm thickness silicon wafer depending on each mechanical part.

Figure 1. Mechanical parts of the microelectromechanical systems (MEMS) microrobot.

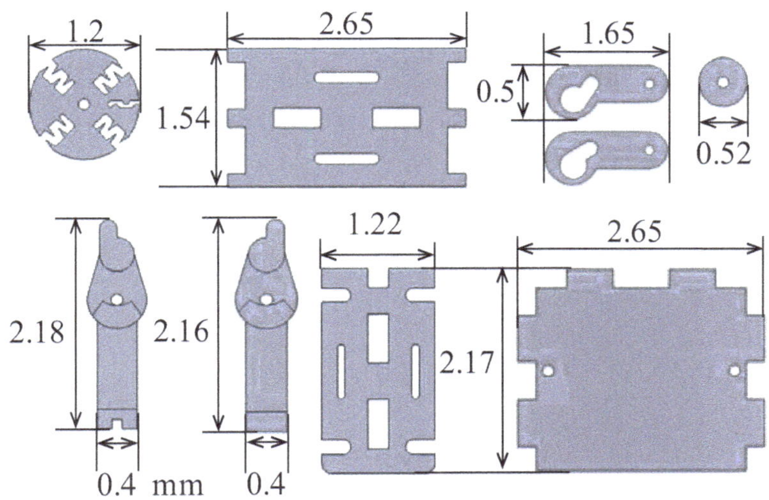

Figure 2 shows the developed figure of the MEMS microrobot. The frame components, the rotary type actuators and the link mechanism were made from silicon wafer. The frame parts consisted of a top frame, a rear frame, a front frame and a center frame. The rotary type actuators consisted of a rotor, GND wire, shaft and four pieces of helical artificial muscle wires. The rotary type actuator generated the locomotion of the robot by supplying the electrical current to the helical artificial muscle wires. The wire shrank at high temperature and extended at low temperature. In this study, the wire was heated by electrical current flowing, and cooled by stopping the flowing. The rotational movement of the each actuator was obtained by changing the flowing sequence. The link mechanisms consisted of a link bars, shafts and legs. The front leg and the rear leg were connected to the middle leg by link bars, respectively. The middle leg is connected to the rotor by the shaft. Therefore, the rotational phase was the same as the rotor. In contrast, the other two legs are connected by the link bar that generates 90° phase shift. Also, a backward step was obtained by the counter rotation of the actuator.

Figure 2. Developed figure of the MEMS microrobot.

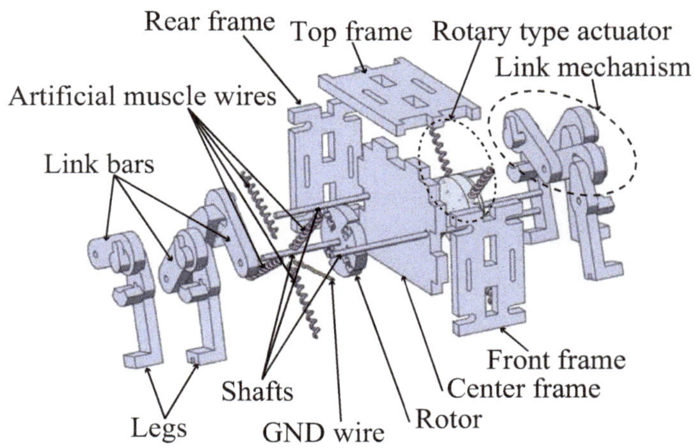

Figure 3 shows the picture of the rotary type actuator. The one side of the artificial muscle wires connected to the copper wires and the other side connected to the rotor by using solder paste. The GND wire was connected to the rotor directly. The frame components and the rotary type actuators were connected by the helical artificial muscle wire which was the shape memory alloy [20]. We used the BioMetal® Helix BMX50 to the helical artificial muscle wire (Available online at: [21]). The basic characteristics of the helical artificial muscle wire were as follows. The standard coil diameter of the artificial muscle wire was 0.2 mm where wire diameter was 0.05 mm. The practical maximum force produced 3 to 5 gf where kinetic displacement was 50%. The standard drive current was 50 to 100 mA where standard electric resistance was 3600 ohm/m.

Figure 3. Picture of the rotary type actuator.

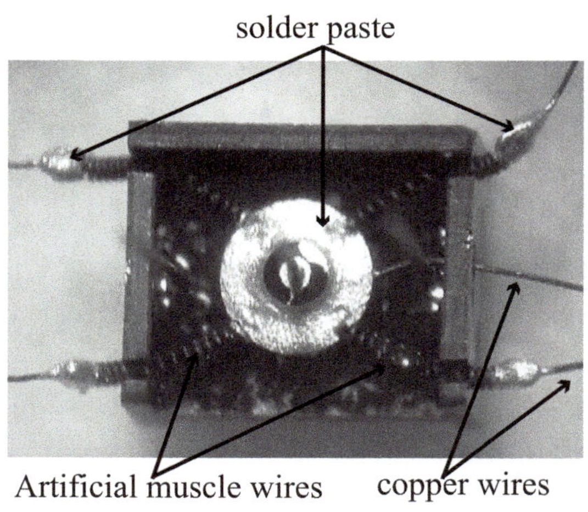

The picture of the fabricated MEMS microrobot was shown in Figure 4. The design size was 4.0, 2.7, 2.5 mm, width, length, height, respectively. The copper wires above the MEMS microrobot were two GND wires and eight signal wires. In the case of connecting the signal wires to the NNIC, the MEMS microrobot could locomote.

Figure 4. Picture of the fabricated MEMS microrobot.

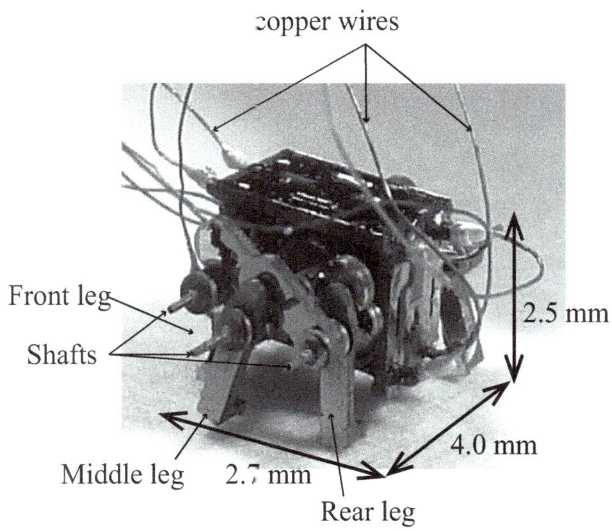

2.2. Locomotion Mechanisms

Ants can locomote smoothly on the uneven surface by using step pattern. On the other hand, many microrobots were weak in moving on the uneven surface. Therefore, microrobot which could locomote by step pattern was desired. We replicate the locomotion of an ant by using rotary type actuator and link mechanism. In this section, we will discuss about locomotion mechanisms of the MEMS microrobot.

Figure 5a shows the schematic diagrams of locomotion of the MEMS microrobot. The MEMS microrobot could move by the rotational actuator. The helical artificial muscle wire had a characteristic of changing length according to temperature. In the case of heating, the wire shrunk and in the case of cooling, the wire extended. In particular, when heating the helical artificial muscle wires from A to D, the MEMS microrobot would move forward. In contrast, heating the helical artificial muscle wires from D to A, the robot moved backward. The locomotion pattern is 180° phase shift on each side to represent the locomotion of an ant. The MEMS microrobot the locomotion method of ants (see Figure 5b. The touch with ground timing of the legs is the same.).

Figure 5. Locomotion of the MEMS microrobot.

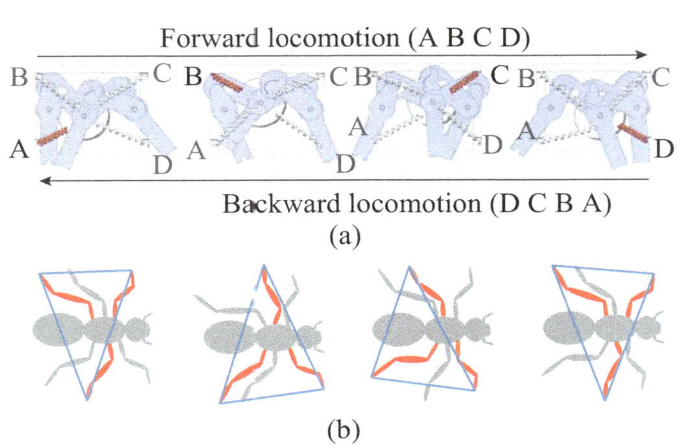

3. Neural Networks Integrated Circuit

It is well known that locomotion rhythms of living organisms are generated by central pattern generator (CPG). Previously, we proposed the CPG model using pulse-type hardware neuron model [22,23]. CPG model was board level circuit using surface-mounted components. The board level circuit was 10 cm square size. Therefore, it was impossible to integrate on the MEMS microrobot system. In addition, we proposed the packaged NNIC which could output the driving waveform of the robot. The packaged NNIC was externally connected to the robot system [23]. Proposing NNIC can mount on the MEMS microrobot system.

Figure 6 shows the circuit diagram of pulse-type hardware neuron model. The pulse-type hardware neuron model is a basic component of the NNIC. The pulse-type hardware neuron model has the same basic features of biological neurons such as threshold, refractory period, spatio-temporal summation characteristics, and enables the generation of continuous action potentials. The pulse-type hardware neuron model consists of a synaptic model and a cell body model. In this study, we use inhibitory synaptic model as a synaptic model. The synchronization phenomena of pulse-type hardware neuron models changes by the connection of the synaptic model. The cell body model connected by excitatory synaptic model cause the in-phase synchronization. The cell body model connected by inhibitory synaptic model cause the anti-phase synchronization. Therefore, we use the inhibitory mutual coupling to generate the driving pulses which can operate the actuators of MEMS microrobot (For more detail, see Saito [23]). The circuit parameters of synaptic model were as follows: $C_{ES} = C_{IS2} = 1$ pF, M_{ES1-3}, M_{IS1-5}: $W/L = 1$. The voltage source $V_{DD} = 5$ V. The circuit parameters of the cell body model were as follows: $C_G = 39$ μF, $C_M = 270$ nF, M_{C1}, M_{C2}: $W/L = 10$, M_{C3}: $W/L = 0.1$, M_{C4}: $W/L = 0.3$. The voltage source $V_A = 3.3$ V. The input voltage v_{ISin} of synaptic model were output voltage v_{Cout} of the other cell body model. The input current i_{Cin} of cell body model was output current i_{ISout} of synaptic model.

Figure 6. Circuit diagram of pulse-type hardware neuron model.

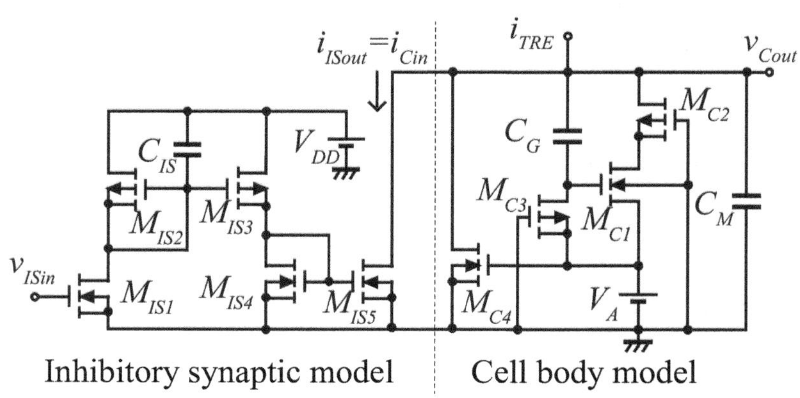

Using the anti-phase synchronization phenomena, we designed the NNIC. The connection diagram of designed NNIC was shown in Figure 7. The four cell body models are mutually coupled by 12 inhibitory synaptic models. Four output ports were extracted from NNIC. In addition, four trigger pulse input port were extracted to NNIC. According to the input timing of single external trigger pulse, NNIC could change the sequence of driving pulses.

Figure 7. Connection diagram of neural networks integrated circuit.

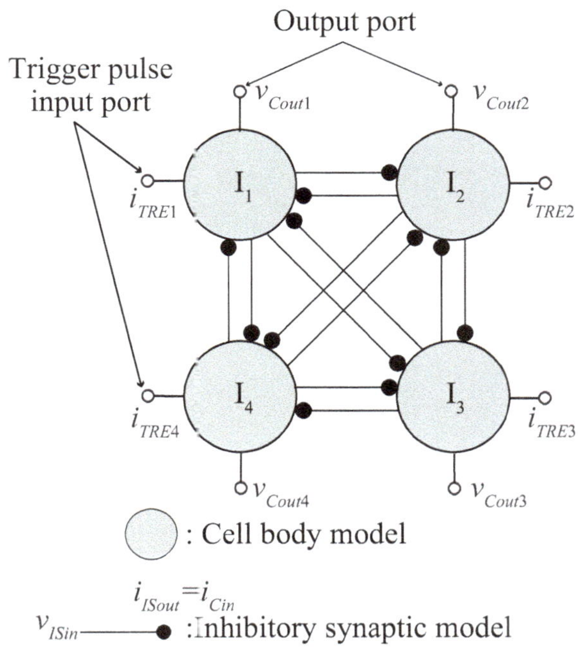

Figure 8 shows the layout pattern of NNIC. This IC chip was fabricated by Digian Technology, Inc. The design rule of the IC chip was four metal two poly CMOS 0.35 μm rule. The chip was a square 1.93 mm × 1.93 mm in size. The sizes of capacitors of cell body model are too large to implement to the CMOS IC chip. Therefore, the capacitors C_G and C_M were mounted externally of the bare chip.

Figure 8. Layout pattern of neural networks integrated circuit.

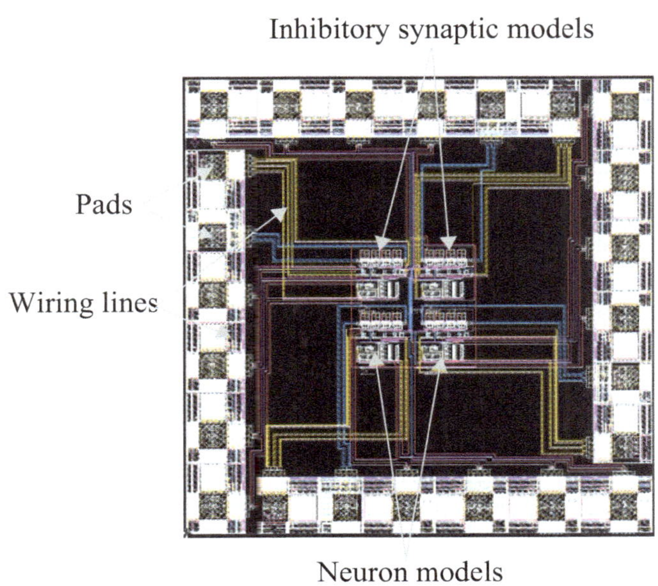

Figure 9 shows the example of output waveform of NNIC. NNIC is a coupled neural networks system which can generate the locomotion rhythms such as living organisms.

Figure 9. Example of output waveform of neural networks integrated circuit.

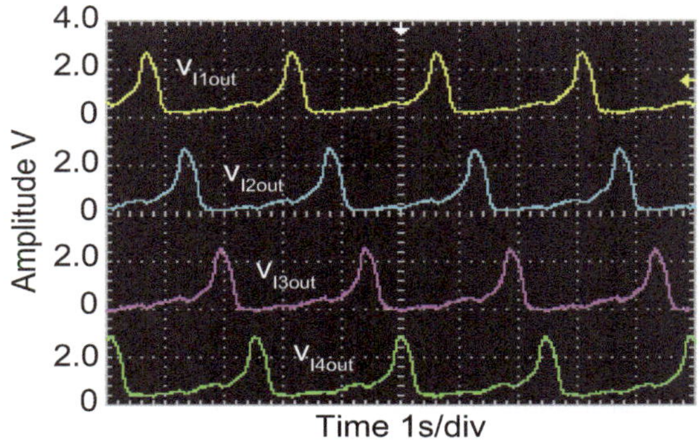

4. Results and Discussion

Four output ports were extracted from NNIC and they were connected to the artificial muscle wires. To heat the helical artificial muscle wires, we are required to input the pulse width 0.5 s, pulse period 2 s and pulse amplitude 2 V (such as shown in Figure 9). Therefore, the microrobot required 2 s to finish the 1 cycle locomotion. In the case of inputting shorter than pulse width 0.5 s, the artificial muscle wire could not shrunk because the joule heat by drive current was not enough. In contrast, in the case of inputting longer than pulse width 2 s, the artificial muscle wire could not be extended because the joule heat by drive current was too much and cooling must be difficult.

Figure 10 shows the example of locomotion of the MEMS microrobot. NNIC can output the driving pulse of locomotion which is necessary to actuate the MEMS microrobot. Thus, our NNIC is effective to generate the locomotion of the MEMS microrobot. The locomotion speed was 26.4 mm/min where the step width was 0.88 mm. (Examples of locomotion videos of MEMS microrobot are available at [24]).

Figure 10. Example of locomotion of the MEMS microrobot.

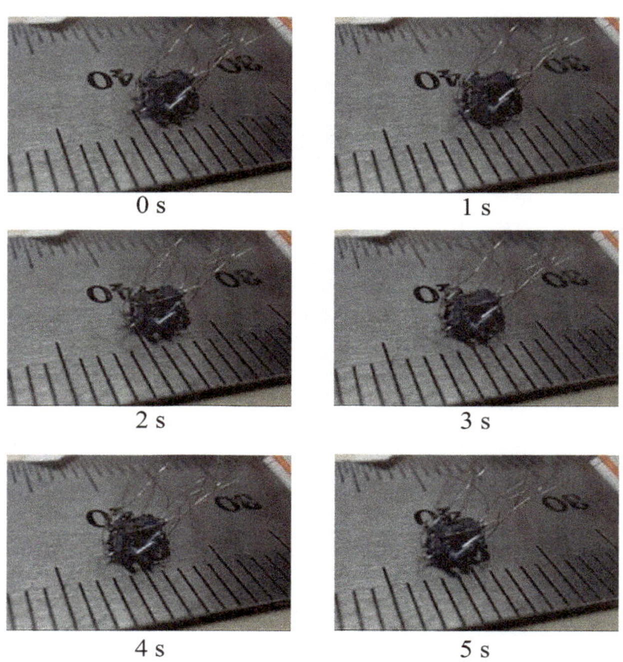

Figure 11 shows the peripheral circuit of the NNIC. The NNIC bare chip was fixed to the flame retardant type 4 (FR4) circuit board. Each pad of the NNIC bare chip was wire bonded to pad pattern of the FR4 circuit board by aluminum wire using ultrasonic wire bonding.

Figure 11. Neural networks integrated circuit bare chip with peripheral circuit.

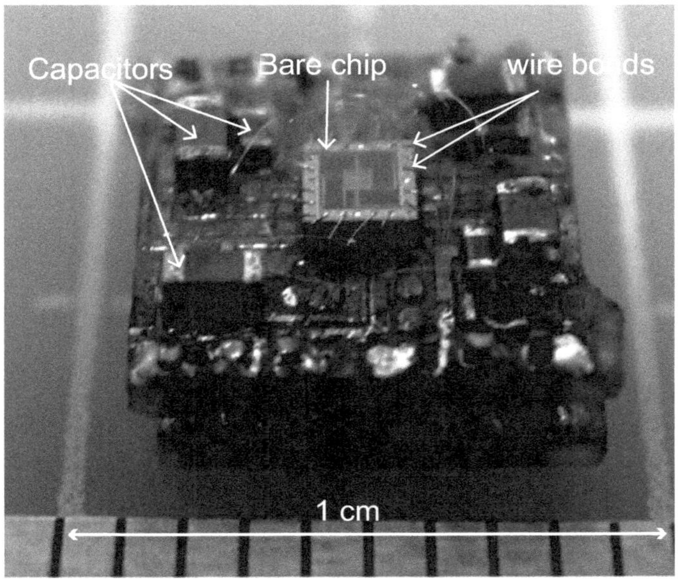

Figure 12 shows the complete system of the proposal microrobot system. The hexapod-type MEMS microrobot mounted the NNIC with peripheral circuit. The mechanical system and electrical system is connecting by enamel wire. We fix the enamel wire using solder paste.

Figure 12. Complete system of the proposal MEMS microrobot system.

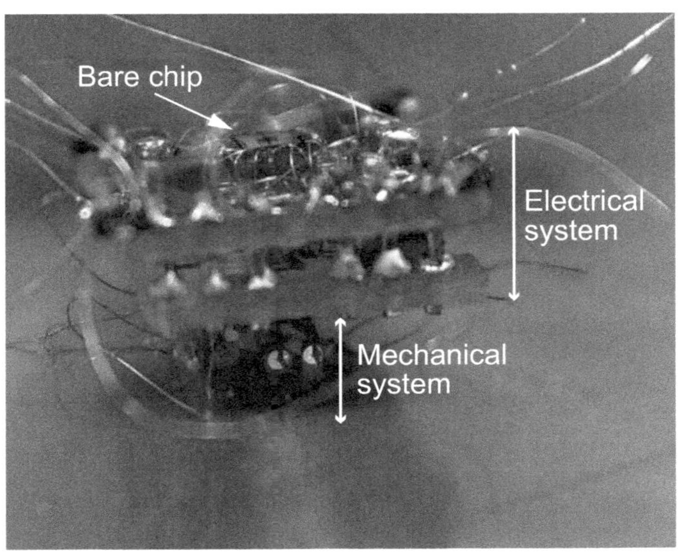

5. Conclusions

In this paper, we proposed the NNIC which is the driving waveform generator of the 4.0, 2.7, 2.5 mm, width, length, height size hexapod-type microrobot system. As a result, we developed the following conclusions.

(a) Neural networks integrated circuit could output the driving pulse which is necessary to actuate the MEMS microrobot. In particular our constructed NNIC can control the movements of MEMS microrobot. The locomotion speed of MEMS microrobot was 26.4 mm/min when the step width was 0.88 mm where the room temperature of measurement environment was 298 K. The power consumption of the microrobot system was 250 mWh. Our MEMS microrobot could locomote by step pattern. Therefore, our robot could locomote more smoothly on the uneven surface compared with the other microrobot.

(b) The NNIC with peripheral circuit was fixed to the FR4 circuit board. The size of the circuit board was 9.4, 7.5 mm width, length size. The weight of the circuit board was 0.31 g.

In the future, we will construct the neural networks integrated circuit with driver circuits inside the bare chip. Consequently, we will mount the IC chip (except the capacitors) directly on the frame of the MEMS microrobot.

Acknowledgments

The fabrication of the biomimetics MEMS microrobot was supported by the Research Center for Micro Functional Devices, Nihon University. This study was supported by Nihon University College of Science and Technology Project Research, JSPS KAKENHI (23760243, 25420226). We appreciate the support. The VLSI chip in this study has been fabricated by Digian Technology, Inc. This work is supported by VLSI Design and Education Center (VDEC), the University of Tokyo in collaboration with Synopsys, Inc., Cadence Design Systems, Inc. and Mentor Graphics, Inc.

Author Contributions

Study conception and design: Ken Saito and Fumio Uchikoba Acquisition of data: Kazuaki Maezumi, Yuka Naito, Tomohiro Hidaka, Kei Iwata, Yuki Okane and Hirozumi Oku Analysis and interpretation of data: Ken Saito and Minami Takato Drafting of manuscript: Ken Saito Critical revision: Ken Saito.

Conflicts of Interest

The authors declare no conflict of interest.

References

1. Shibata, T.; Aoki, Y.; Otsuka, M.; Idogaki, T.; Hattori, T. Microwave Energy Transmission System for Microrobot. *IEICE Trans. Electron.* **1997**, *E80-C*, 303–308.

2. Takeda, M. Applications of MEMS to Industrial Inspection. In Proceedings of the 14th IEEE International Conference on Micro Electro Mechanical Systems, Interlaken, Switzerland, 25 January 2001; pp. 182–191.
3. Habib, M.K.; Watanabe, K.; Izumi, K. Biomimetcs robots: From bio-inspiration to implementation. In Proceedings of the 33rd Annual Conference of the IEEE Industrial Electronics Society, Taipei, Taiwan, 5–8 November 2007; pp. 143–148.
4. Habib, M.K. Biomimetcs: Innovations and Robotics. *Int. J. Mechatron. Manuf. Syst.* **2011**, *4*, 113–134.
5. Baisch, A.T.; Sreetharan, P.S.; Wood, R.J. Biologically-inspired locomotion of a 2 g hexapod robot. In Proceedings of the 2010 IEEE/RSJ International Conference on Intelligent Robots and Systems (IROS), Taipei, Taiwan, 18–22 October 2010; pp. 5360–5365.
6. Tsuruta, K.; Mikuriya, Y.; Ishikawa Y. Micro sensor developments in Japan. *Sensor Rev.* **1999**, *19*, 7–42.
7. Donald, B.R.; Levey, C.G.; McGray, C.D.; Paprotny, I.; Rus, D. An untethered, electrostatic, globally controllable MEMS micro-robot. *J. Microelectromech. Syst.* **2006**, *15*, 1–15.
8. Edqvist, E.; Snis, N.; Mohr, R.C.; Scholz, O.; Corradi, P.; Gao, J.; Johansson, S. Evaluation of building technology for mass producible millimeter-sized robots using flexible printed circuit boards. *J. Micromech. Microeng.* **2009**, *19*, 1–11.
9. Suematsu, H.; Kobayashi, K.; Ishii, R.; Matsuda, A.; Sekine Y.; Uchikoba, F. MEMS Type Micro Robot with Artificial Intelligence System. In Proceedings of the International Conference on Electronics Packaging, Beijing, China, 10–13 August 2009; pp. 975–978.
10. Tang, W.C.; Nguyen, T.H.; Howe, R.T. Laterally driven poly silicon resonant microstructure. In Proceedings of the IEEE Micro Electro Mechanical Systems. An Investigation of Micro Structures, Sensors, Actuators, Machines and Robots, Fort Lauderdale, FL, USA, 20–22 February 1989; pp. 53–59.
11. Sniegowski, J.J.; Garcia, E.J. Surface-micromachined gear trains driven by an on-chip electrostatic microengine. *IEEE Electron Device Lett.* **1996**, *17*,366–368.
12. Asada, N.; Matsuki, H.; Minami, K.; Esashi, M. Silicone micromachined two-dimensional galvano optical scanner. *IEEE Trans. Magn.* **1994**, *30*, 4647–4649.
13. Suzuki, Y.; Tani, K.; Sakuhara, T. Development of a new type piezo electric Micromotor. *Transduceres* **1999**, *99*, 1748–1751.
14. Surbled, P.; Clerc, C.; Pioufle, B.L.; Ataka, M.; Fujita, H. Effect of the composition and thermal annealing on the transformation temperature sputtered TiNi shape memory alloy thin films. *Thin Solid Films* **2001**, *401*, 52–59.
15. Matsuoka, K. Mechanism of Frequency and Pattern Control in the Neural Rhythm Generators. *Biol. Cybern.* **1987**, *56*, 345–353.
16. Ikemoto, T.; Nagashino, H.; Kinouchi, Y.; Yoshinaga, T. Transitions in a Four Coupled Neural Oscillator Model. International Symposium on Nonlinear Theory and its Applications, Hilton Hawaiian Village, HI, USA, 29 November–3 December 1997; pp. 561–564.
17. Nakada, K.; Asai, T.; Amemiya, Y. An Analog CMOS Central Pattern Generator for Interlimb Coordination in Quadruped Locomotion. *IEEE Trans. Neural Netw.* **2003**, *14*, 1356–1365.

18. Saito, K.; Matsuda, A.; Saeki, K.; Uchikoba, F.; Sekine, Y. Synchronization of Coupled Pulse-Type Hardware Neuron Models for CPG Model. In *The Relevance of the Time Domain to Neural Network Models, Springer Series on Cognitive and Neural Systems*; Springer: Berlin, Germay, 2011; pp. 117–133.
19. Bhardwaj, J.K.; Ashraf, H. Advanced silicon etching using high-density plasmas. In Proceedings of the SPIE Micromachining and Micro fabrication Process Technology, Austin, TX, USA, 19 September 1995; Volume 2639, pp. 224–233.
20. Homma, D. Metal Artificial Muscle Bio Metal Fiber. *RSJ* **2003**, *21*, 22–24.
21. Toki Corporation's Homepage. Available online: http://www.toki.co.jp (accessed on 30 March 2014).
22. Okazaki, K.; Ogiwara, T.; Yang, D.; Sakata, K.; Saito, K.; Sekine, Y.; Uchikoba, F. Development of pulse control type MEMS micro robot with hardware neural network. *Artif. Life Robot.* **2011**, *16*, 229–233.
23. Saito, K.; Takato, M.; Sekine, Y.; Uchikoba, F. Biomimetics Micro Robot with Active Hardware Neural Networks Locomotion Control and Insect-Like Switching Behaviour. *Int. J. Adv. Robot. Syst.* **2012**, doi:10.5772/54129.
24. Micro Robot and Neural Networks Laboratory. Available online: http://www.eme.cst.nihon-u.ac.jp/~uchikoba/ (accessed on 30 March 2014).

8

Design Issues for Hexapod Walking Robots

Franco Tedeschi and Giuseppe Carbone *

Department DICEM, University of Cassino and Southern Lazio, Via di Biasio 43, Cassino 03043 (Fr), Italy; E-Mail: franco.tedeschi@unicas.it

* Author to whom correspondence should be addressed; E-Mail: carbone@unicas.it

---

Abstract: Hexapod walking robots have attracted considerable attention for several decades. Many studies have been carried out in research centers, universities and industries. However, only in the recent past have efficient walking machines been conceived, designed and built with performances that can be suitable for practical applications. This paper gives an overview of the state of the art on hexapod walking robots by referring both to the early design solutions and the most recent achievements. Careful attention is given to the main design issues and constraints that influence the technical feasibility and operation performance. A design procedure is outlined in order to systematically design a hexapod walking robot. In particular, the proposed design procedure takes into account the main features, such as mechanical structure and leg configuration, actuating and driving systems, payload, motion conditions, and walking gait. A case study is described in order to show the effectiveness and feasibility of the proposed design procedure.

Keywords: hexapod robots; walking machines; design procedure

---

1. Introduction

Legged hexapod robots are programmable robots with six legs attached to the robot body. The legs are controlled with a degree of autonomy so that the robot can move within its environments, to perform intended tasks. Hexapod robots can be suitable for terrestrial and space applications, and they can include features such as omnidirectional motion, variable geometry, good stability, access to diverse terrain, and fault tolerant locomotion [1].

One of the motivating factors often given for pursuing the development of hexapod robots is that they can climb over obstacles larger than the equivalent sized wheeled or trucked vehicle. In fact, the use of wheels or crawlers limits the size of the obstacle that can be climbed to half the diameter of the wheels. On the contrary, legged robots can overcome obstacles that are comparable with the size of the machine leg [2]. Hexapod walking robots also benefit from a lower impact on the terrain and have greater mobility in natural surroundings. This is especially important in dangerous environments like mine fields, or where it is essential to keep the terrain largely undisturbed for scientific reasons [3]. Hexapod legged robots have been used in exploration of remote locations and hostile environments such as seabed [4,5], in space or on planets [6,7] in nuclear power stations [8], and in search and rescue operations [9]. Beyond this type of application, hexapod walking vehicles can also be used in a wide variety of tasks such as forests harvesting, in aid to humans in the transport of cargo, as service robots and entertainment.

Despite the above referenced aspects, many challenges remain before hexapod walking robots can have a more widespread use. Some of their current disadvantages include higher complexity and cost, low energy efficiency [10], and relatively low speed. Walking robots are in fact complex and expensive machines, consisting of many actuators, sensors, transmissions and supporting hardware.

This paper gives a wide overview of the state of the art on six-leg walking robots aiming to identify the main design issues and constraints that influence the technical feasibility and performance of these systems. Then, a design procedure is proposed in order to systematically design a six-leg walking robot. In particular, the proposed design procedure takes into account mechanical structure and leg configuration, actuating and drive mechanisms, payload, motion conditions, walking gait and control system. A case study is described, referring to previous experiences at LARM in Cassino in order to show the effectiveness and feasibility of the proposed design procedure.

2. State of Art Overview

2.1. Early Designs

The first hexapods can be identified as robots based on a rigidly predetermined motion so that an adaptation to the ground was not possible. Early researches in the 1950s were focused on assigning the motion control completely by a human operator manually [11].

One of the first successful hexapod robot was constructed at University of Rome in 1972 (Figure 1a) as a computer-controlled walking machine with electric drives [12]. In the middle 70s, at the Russian Academy of Sciences in Moscow, a six-legged walking machine was developed with a mathematical model of motion control. It was equipped with a laser scanning range finder and was connected with a two-computer control system [13]. In 1976, Masha hexapod walking robot was designed at Moscow State University (Figure 1b). The robot had a tubular axial chassis, articulated legs with three DoFs [14]. The hexapod was able to negotiate obstacles using contact on the feet and a proximity sensor. Ohio State University in 1977 developed a six-legged insect-like robot system called "OSU Hexapod" [15]. This hexapod was kept tethered and was made to walk short distances over obstacles (Figure 1c).

In 1983, Carnegie-Mellon University developed a "Six-Legged Hydraulic Walker" [16], a first man-carrying hexapod capable of navigating rough terrain using different types of gaits. The hexapod used a combination of hydraulic feedback, computer control and human control and was about 2.5 m long and the same width. It weighted about 800 kg and was powered by a 13 kW gasoline engine.

Figure 1. Early hexapod design: (**a**) University of Rome's hexapod; (**b**) MASHA hexapod; (**c**) OSU hexapod; (**d**) ODEX I hexapod; (**e**) ASV hexapod.

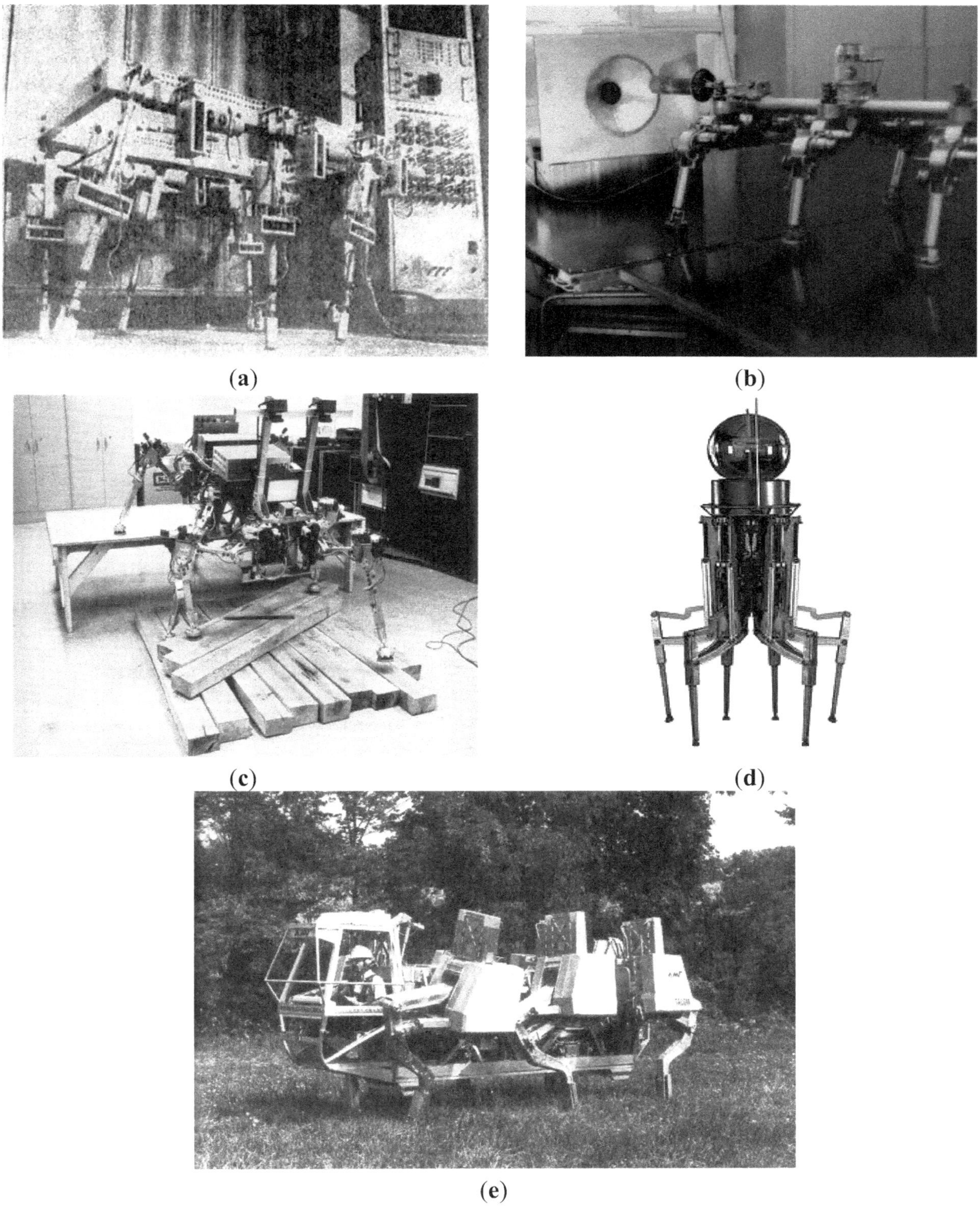

In 1984, Odetic Inc., California, USA, developed Odex I [17], a six-legged radially symmetric hexapod robot which used an onboard computer to play back pre-programmed motions (Figure 1d). Its onboard computer could be operated remotely and the robot moved under its own power. Using remote human control or the prerecorded motions, the hexapod could climb obstacles such as stairs or a pickup truck. Odex I weighed 136 kg; each leg was able to lift 180 kg. In 1985, the "NMIIIA" Hexapod Manned Rover was developed in Russia [18]. This hexapod was designed for investigating the walking propulsive device and control system. NMIIA had a mass of 750 kg; its load-carrying capacity was 80 kg; travel speed was 0.7 km/h. In 1989, the Ohio State University started the Adaptive Suspension Vehicle project [19]. The six-legged robot, shown in Figure 1e, used hydraulic actuation being powered by an internal combustion engine. A human was able to operate it through a joystick while the individual control of each leg was assured by a central computer. As main characteristics, its 250 kg payload capacity, and the possibility to surpass 1.8 m width ditches and climb vertical steps of maximum 1.65 m should be mentioned. A hexapod walking robot named Aquarobots was constructed in 1989 and used for underwater measurements of ground profiles for the construction of harbors [20]. A small hexapod robot named Genghis with 0.35 m length and 1 kg weight was developed in the same year [21]. The behavior of Genghis was not explicitly controlled, but was built by adding layers of control on top of existing simpler layer. This approach was different to the more traditional method of task decomposition.

Attila and Hannibal hexapod robots were built in the Mobot Lab in the early 1990s [22]; they were very sophisticated autonomous robots for their size, possessing over 19 degrees of freedom, more than 60 sensory inputs, eight microprocessors and real-time behavior. TUM Walking Machine was developed in 1991. The robot was designed and steered similar to a stick insect [23]; the control system was realized as a neural structure. AMBLER (Autonomous MoBiLe Exploration Robot) was a hexapod robot developed by the Jet Propulsion Laboratory during the mid-90s for operating under the particular constraints of planetary terrain [24]. The robot was about 5 m tall, up to 7 m wide, and weighed 2500 kg. While most robots bend their legs to step and walk, Ambler's legs remain vertical, while they swing horizontally, adopting a telescope like displacement to touch the ground.

2.2. Recent Developments

The two last decades have been characterized by a rapid development of control systems technology. Hexapod robots were equipped with various sensing systems. Artificial Intelligence systems were widely applied to the analysis of environment and motion of robots on a complex surface.

A series of bio inspired robots was developed at Case Western Reserve University (USA) at the end the 90s, such as, for example, Robot III that had a total of 24 DoFs. Robot III architecture was based on the structure of cockroach, trying to imitate their behavior [25]. In particular, each rear leg had three DoFs, each middle leg four DoFs and each front leg five DoFs. Similarly, Biobot was a biomimetic robot physically modeled as the American cockroach (Periplaneta Americana) and powered by pressurized air [26]. This hexapod had a great speed and agility. Each leg of the robot had three segments, corresponding to the three main segments of insect legs: coxa, femur, and tibia.

Hamlet was a hexapod robot constructed at the University of Canterbury, New Zealand [27]. Its legs were all identical and each had three revolute joints. Hamlet's application task was to study force and position control on uneven terrain.

In 2001, a project named RHex commenced [28]; RHex design comes from a multidisciplinary and multi-university DARPA funded effort that applies mathematical techniques from dynamical systems theory to problems of animal locomotion. Hexapod design consists of a rigid body with six compliant legs, each with one DoF (Figure 2a). Thus, RHex has only six motors that rotate the legs such as a wheel.

Figure 2. Recent developments in hexapod design: (**a**) RHex basic architecture; (**b**) Six-legged walking robot LAURON V from the FZI Research Center for Information Technology in Karlsruhe, Germany; (**c**) One configuration of the LEMUR prototype: the front limbs function as legs and arms, with integrated tools; (**d**) ATHLETE during experimental tests.

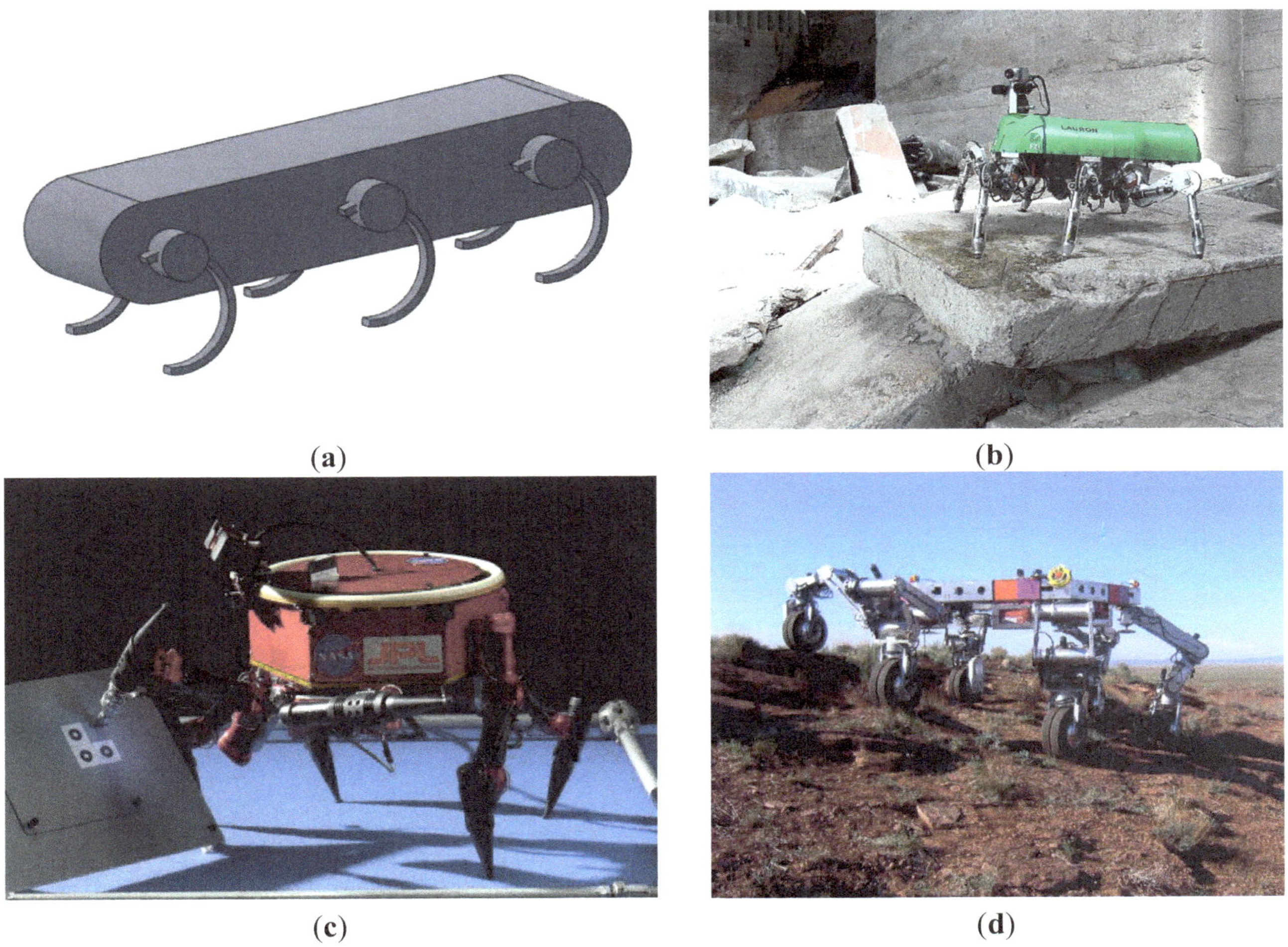

Several prototypes of Rhex have been developed. At present the project is still active.

Lauron V hexapod robot (Figure 2b) was the result of about 10 years of progressive improvement on the previous configurations Lauron I, II, III and IV. LAURON [29] is biologically-inspired by the stick insect. Like this insect, the robot has six legs fixed to a central body. Each of the six legs is actuated by four joints. Each foot has a three-axis force sensor, and each motor has a current sensor that detects forces opposing to its movement. At present the project is still active.

Gregor I reproduces the cockroach's agility where the locomotion control is based on the theory of the Central Pattern Generator [30]. Gregor I had a biological inspiration where each leg pair has a unique design. The front leg pair and the middle leg pair have three DoFs on each leg, and the rear leg pair has two DoFs. Another hexapod robot called Sprawlita [31] was developed following the basic principles of locomotion of cockroaches: self-stabilizing posture, different functions for the legs, passive visco-elastic structure, open-loop control and integrated construction. In 2005, the hexapod robot named BILL-Ant-p was developed [32]. The robot was based on ants' behavior and it is composed of three DoFs on each leg with six force-sensing feet, a three-DoF neck and head, and actuated mandibles with force-sensing for a total of 28 DoFs.

A series of hexapod named LEMUR (Limbed Excursion Mechanical Utility Robots robot) was developed by Jet Propulsion Laboratory [33] with the goals of using robots for repair and maintenance in near-zero gravity on the surface of spacecraft (Figure 2c). MARS (Multi Appendage Robotic System) was a hexapod mobile robotic research platform developed after the LEMUR project for similar applications by employing radial symmetry. MARS platforms were capable of walking in any direction without turning [34].

In 2004, a six-legged lunar robot called ATHLETE was developed by the Jet Propulsion Laboratory [35]. This robot had the ability to roll rapidly on rotating wheels over flat smooth terrain and walk carefully on fixed wheels over irregular and steep terrain. ATHLETE had a payload capacity of 450 kg, a diameter of around 4 m and a reach of around 6 m (Figure 2d).

AQUA was an amphibious hexapod robot developed with six independently-controlled leg actuators [36]. One of the most important features of this robot was the ability to switch from walking to swimming gaits as it is moving from a sand beach or surf-zone to deep water. The underwater walking robot CR200 was built as based on the concept of Crabster [37]. The application field was inspection of shipwrecks or scour and survey of seafloor in high current and turbid environments. The hexapod robot called RiSE is able to climb on a variety of vertical surfaces as well as demonstrating horizontal mobility [38]. A mechanism applied in RiSE uses compliant micro spines on its feet to reliably attach to textured vertical surfaces, in order to carry the payload while the robot climbs. COMET hexapods are a serious of robots designed to operate on extremely unstructured terrain [39]. The latest prototype, COMET IV, is a hydraulically driven hexapod involving walking with the force/impedance control, fully autonomous navigation with laser mapping and teleoperation. Mantis is a hexapod robot hydraulic powered developed by Micromagic Systems. It stands nearly 3 m tall and weighs about 2 tons; at present, it is one of the biggest hexapod robots in the world [40].

Many hexapod rovers for space exploration have been designed and built in the last years. A very rich reference of this vehicle and their design issues can be found in [41].

Given the wide variety of existing hexapod robots, several of the above described prototypes have been detailed in a specific table (Table 1). It provides a quick comparison tool for referring to the main features.

Table 1. Comparison of hexapod robots.

	Main Characteristics					Main Performance				
Robot Name	**Mass (Kg)**	**Length (cm)**	**Width (cm)**	**Height (cm)**	**Total DoFs**	**Max Speed (m/s)**	**Gait/Mobility**	**Power (W)**	**Year**	**Application Tasks**
Ambler	2700	500	500	700	12	0.007	Wave Free	1900	1989	Planetary exploration
ASV	3200	520	240	300	15	1.0	Wave Free	26000	1989	Navigate on uneven terrains
Hannibal	2.7	35	NA	20	19	0.04	Wave Free	NA	1989	Planetary exploration
Tum	23	80	40	100	18	0.3	Wave	500	1991	Hexapod following biological principles
Biobot	11	58	14	23	18	NA	NA	NA	2000	Locomotion over rough terrain
Hamlet	13	40	28	40	18	0.1	Wave Free	52	2001	Testing force and position control
Rhex	7	53	20	15	6	0.55	Wave Free	100	2001	Hexapod with reduced actuators
Sprawlita	0.27	16	NA	NA	12	0.35	Wave	NA	2002	Robots inspired to cockroaches
Lauron III	18	50	30	80	18	0.4	Wave Free	NA	1999–2003	Testing a hierarchical walk controller
Genghis	1	40	15	NA	12	0.04 0.12	Wave Free	NA	2004	Developing of a reactive controller
Aqua II	16.5	64	44	13	6	0.7 1	Land Water	200	2010	Underwater hexapod robot
Bill-Ant-p	2.3	47	33	16	22	0.004	Wave	25	2005	Biologically inspired legged robot
Gregor I	1.2	30	9	4	16	0.03	NA	25	2006	Robot inspired on cockroaches.
Athelete	850	2.75	2.75	2	36	2.78 0.016	Wheleed Wave	NA	2006	Navigate on rough soil of on the Moon
RiSe	2.8	41	NA	NA	12	NA	Wave	NA	2006	Hexapod climbing robots
Comet IV	2120	280	330	250	24	0.278	Wave	20600	2009–2011	Hexapod for multitasks on outdoor environment
CR200	600	250	200	130	18	0.5	Wave	20000	2013	Walk either on land or underwater in the turbulent surf zone
Mantis	1900	420	220	280	18	NA	Wave	42000	2013	Entertainment

NA means data not available.

2.3. Hexapod Robots' Performance Indices

The following indices have been proposed in the literature in order to compare legged robots which have different masses, shape and sizes:

- Duty factor
- Froude number
- Specific resistance
- Stability margin
- The duty factor β [42] is defined as:

$$\beta = \frac{\text{support period}}{\text{cycle time}} \tag{1}$$

Duty factor can be used to make the distinction between walks and runs, since we have $\beta \geq 0.5$ for walking and $\beta < 0.5$ for running.

According to [43], a Froude number can be calculated by:

$$\mathrm{F}_{r2} = \frac{V^2}{gh} \tag{2}$$

where V is the walking or running speed, g is the acceleration due to gravity, and h is the height of hip joint from the ground. V is a characteristic speed of the motion. According to [44], a possible way to estimate V can be by using the products (h times f) instead of the characteristic speed V since h is the characteristic height of the leg and f is the stride frequency. Thus, one can write for hexapod robots:

$$\frac{(hf)^2}{gh} = \frac{hf^2}{g} = \mathrm{F}_{r2} \tag{3}$$

Alexander [43] used the Froude number in order to characterize animal locomotion; he showed that animals of different sizes use similar gaits when they travel with equal Froude numbers. In particular, most animals change their gait from walking to running at a speed equivalent to a Froude number of F = 1.

The specific resistance is a dimensionless number that is used to evaluate the energy efficiency of a mobile robot. Gabrielli and von Karman discussed the performance of various vehicles using the power consumption per unit distance [45].

That is:

$$\varepsilon = \frac{E}{Mgd} \tag{4}$$

where E is the total energy consumption for a travel of distance d, M is the total mass of the vehicle, and g is the acceleration due to gravity. We can say that the specific resistance indicates how smooth the locomotion is.

For a dynamic walking robot, we can define a stability margin as the minimum distance of the ZMP to the boundaries of the support polygon, since the ZMP is the natural extension of a projected CoM on the ground [46]. For a legged robot on rough terrain, Messuri and Klein defined the energy stability margin as the minimum potential energy required to tumble the robot [47].

3. Design Considerations

Designing hexapod legged robots is far from trivial. A very numerous and a wide range of possibilities exist to design a hexapod as also described in the previous section. Designers must take several decisions which influence the operation and technical features. Some of the most important design issues and constraints according to [48] can be outlined as:

- the mechanical structure of robot body;
- leg architecture;
- max sizes;
- actuators and drive mechanisms;
- control architecture;
- power supply;
- walking gaits and speed;
- obstacle avoidance capability;
- payload;
- autonomy;
- operation features;
- cost.

The above mentioned design issues and constraints can be classified as design input (or key features) and design output (or main design characteristics) as shown in the scheme of Figure 3.

Figure 3. A scheme for preliminary layout design of hexapod walking robots.

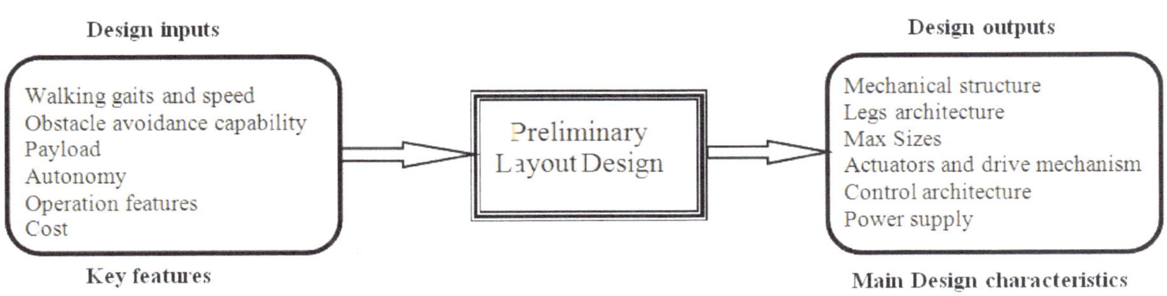

A survey on the state of the art shows that each hexapod walking robot design is almost unique. Hexapod can be developed in several configurations and every solution has its design criteria, specifications, shapes, advantages and disadvantages, but the literature is lacking a systematic design procedure for hexapod robots as referring to specific functional requirements.

A preliminary design is often a trade-off solution between design requirements and key features. Possible aids for the preliminary design process can be given by the Quality Function Deployment [49] in order to define the relationship between the design configuration and the robot capabilities or requirements. Based on this approach, we propose a novel simplified procedure that can be used to relate and prioritize preliminary layout design to key features through numerical ranking.

As a first step, the key features will be identified and then ranked by their importance to the project. Table 2 gives the ranking criteria proposed. Rankings values are recorded in the importance rating column of Table 3. Table 4 gives the criteria used to evaluate the importance of an engineering solution in fulfilling key features. The relationship matrix in Table 3 is completed by filling the row column intersection (key features—design requirements) according to Table 4 criteria.

Table 2. Ranking criteria of key features.

Importance of Key Features	Value
Not important to project success	1
Somewhat important to project success	2
Fairly important to project success	3
Very important to project success	4
Critical to project: design driver	5

Table 3. Cause and effect matrix design.

Main Design Characteristics (How) / Key Features (What)		Importance Rating R	Body Type	Legs Architecture	Actuators	Power Supply	
			1	2	3	4	j
Walking gaits and speed	1	R_1	K_{11}	K_{21}			
Obstacle avoidance	2	R_2	K_{12}	K_{22}			
Payload	3						
Autonomy	4						
Operation features	5						
Cost	6						
	i	R_i	K_{1i}	K_{2i}			K_{ji}
$Y_j = \sum_{i=1}^{n} K_{ji} \cdot R_i$			Y_1	Y_2	Y_3	Y_4	Y_j

Table 4. Relationship matrix values.

Engineering Solution Satisfies the Requirement	Value
By itself	9
In conjunction with one or two factors	3
In conjunction with many other factors	1
Does not satisfy the requirements	0

After filling in the relationship matrix, the importance rankings are multiplied by the relationship value and computed on each column of the matrix according to the rule of Table 3. The bottom of each column Y_j gives the importance rating of engineering requirements. Results are a Pareto of main design characteristics, which can be used as a starting point in the choice of hexapod architecture. This tool will help to focus the process development on the issues that are most critical to the proposed hexapod design.

A practical example of the proposed procedure is provided in Section 4. The proposed procedure is an aid to select the most favorable preliminary design solution among those under consideration. Non-conventional design solutions can be also considered into the cause and effect matrix, provided that proper relationship values can be estimated. Figure 4 depicted a flow-chart of proposed design in order to systematically consider the main design characteristics and key features.

Figure 4. Hexapod robot design flow-chart.

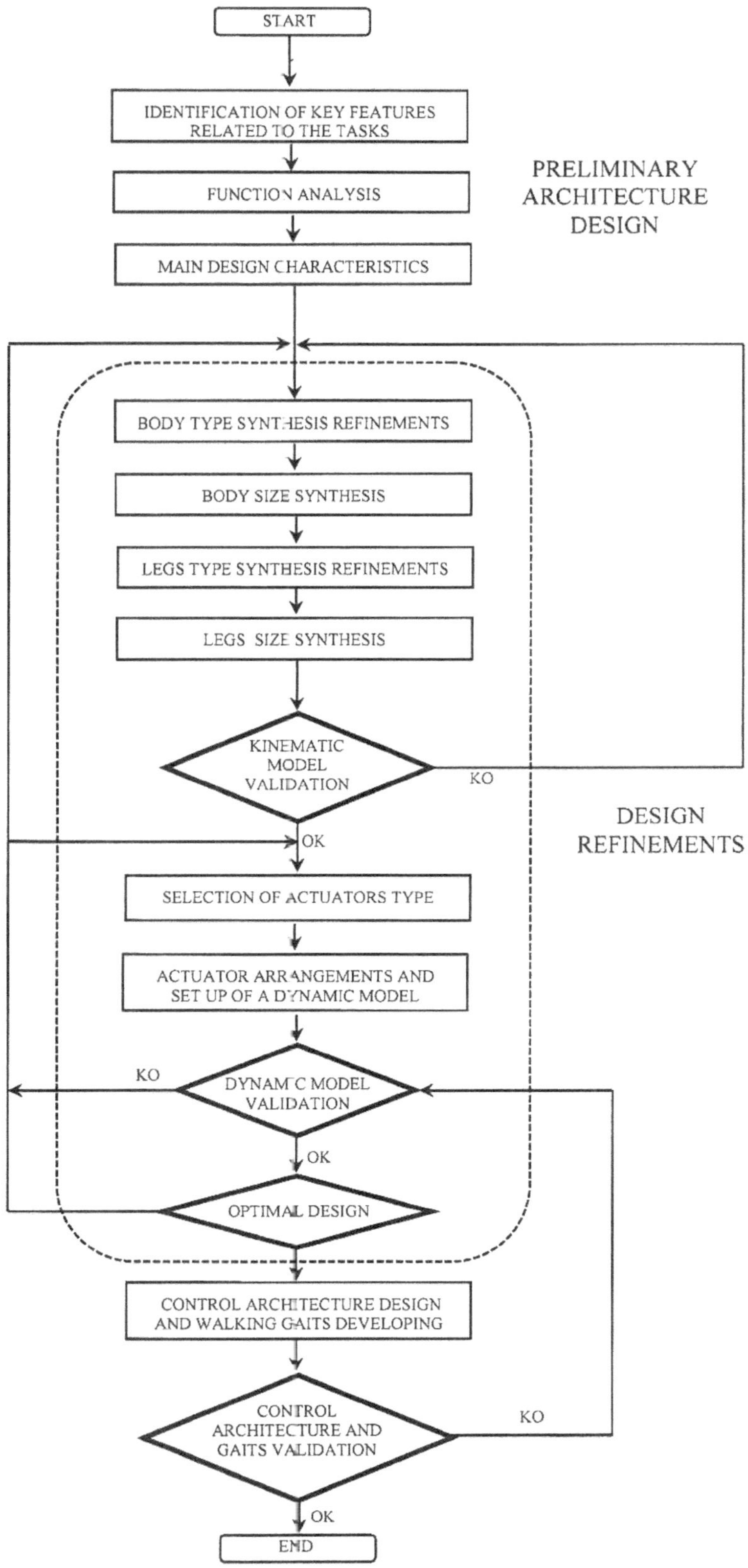

In the following, the main design characteristics and modeling issues are discussed in order to show how the proposed design procedure can be implemented. It is worth mentioning also that other design characteristics exist and additional features can be considered for specific applications. Thus, the most remarkable design characteristics have been considered as examples for filling in the data in the cause and effect matrix.

3.1. Robot Body Architecture

There are two basic architectures of hexapod robots: rectangular and hexagonal. The first one has six legs distributed symmetrically along two sides, each side having three legs. The second has legs distributed axi-symmetrically around the body, in a hexagonal or circular shape [50]. Many references can be found in the literature on rectangular six-legged robots. In [51], Lee *et al.* describe the longitudinal stability margin for rectangular hexapods. Also, the feasible walking gaits have been widely investigated and tested [52,53]. Bilateral symmetry may be better suited than radial symmetry to move along a straight line. Rectangular architectures require a special gait for turning action; generally, they need four steps in order to realize a turning action [50].

Hexagonal hexapod robots demonstrate better performances than rectangular robots for some aspects. As example hexagonal robots can have many kinds of gaits and can easily change direction—in fact true radial symmetry implies that all legs are equal and the body has no "front" or "rear"—there is thus no preferential direction for the motion. In [54], Preumont *et al.*, proved that hexagonal hexapods can easily steer in all directions and that they have a longer stability margin. In [55], Takahashi *et al.*, found that hexagonal robots rotate and move in all directions at the same time, better than rectangular ones, by comparing stability margin and stroke in wave gait. Chu and Pang in [56] proved theoretically that hexagonal hexapod robots have superior stability margin, stride and turning ability compared to rectangular robots.

3.2. Kinematic Architectures of Legs

Kinematics architecture depends on the factors related to the application in which the hexapod robot is required for, as for example the terrain form, the workspace, and the payload. Literature shows that there is a number of different leg types currently employed for hexapod walking robots. All have advantages and disadvantages. Figure 5a shows a schematic classification of hexapod legs types. At the first stage, one can choose between bio inspired and non-zoomorphic legs. Bio inspired leg configuration is motivated primarily by animal gait, such as reptiles, mammals or arachnid. The first one has legs and bodies for moving over rough and uneven terrains [57]. The principal characteristic of the Reptilian type is that the legs are placed on both ends of the protruding body and knees to the side of the base. Mammals' bodies are above the legs, which gives less support to the base and lower power consumption is needed to support the body, but it requires more stability than other types of animals [26]. In Aracnid configuration, legs extremities are situated on both sides, sticking the knees at the top of the spider's body. The orientation of the legs in respect to the body of the hexapod robot can be done with three configurations (Figure 5b): frontal, sagittal or circular. In the first one, the directions are perpendicular to the advancement of the legs' position, unlike the sagittal, which moves parallel to the robot legs, while in the circular arrangement the legs are positioned radially to the body of the system allowing the mechanism to move in any direction [58]. In the mammalian configuration, the legs are below the body and can place the knees in different positions depending on the application it requires (Figure 5c). Non zoomorphic legs can be hybrids such as in [59], telescopic such as in [24], or under-actuated such as in [28]. In [60], a solution named Roller-Walker is presented. The principle through which the robot propels itself during wheeled locomotion is the same as that of the skaters.

Figure 5. Type setting of hexapod legs' design.

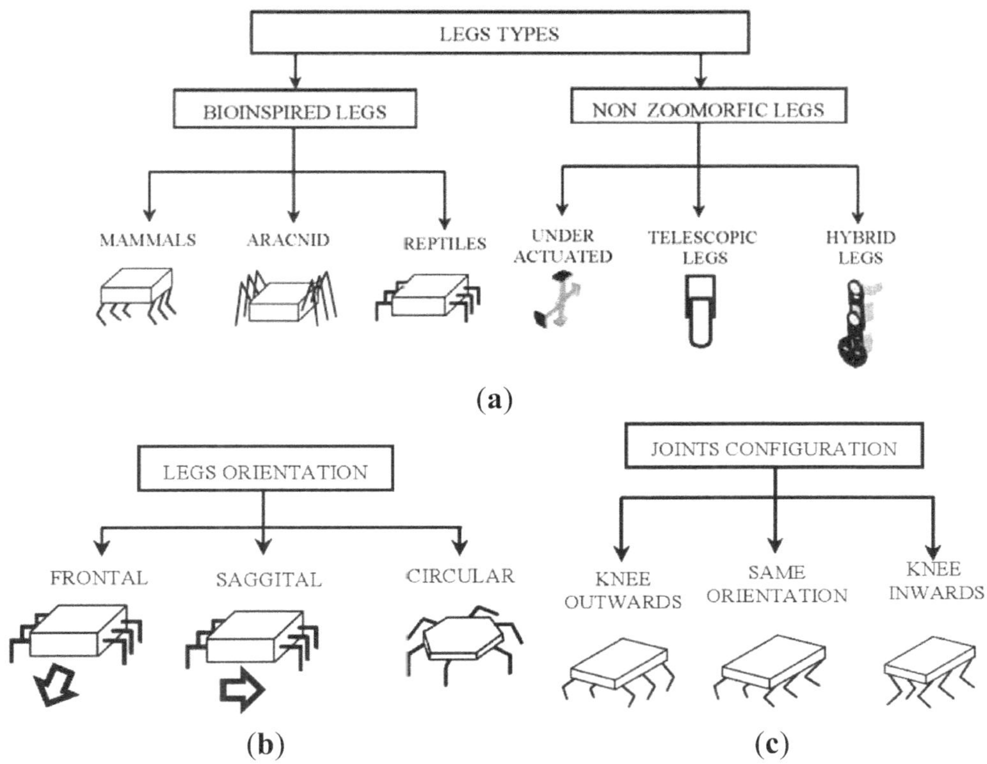

3.2.1. Actuator Types

Many kinds of actuators can be employed for operating hexapod walking robots. The majority of hexapods is actuated by electric rotating motors, as they are relatively cheap, easy to control and there are suitable technologies to store the energy Linear motors are able to generate considerable forces at very considerable speeds. However, these do not yet appear to have been utilized in many hexapods since they have a limited movable range to weight ratio. Pneumatics actuators have very low stiffness, inaccurate response, and low power to weight ratio. Pneumatics actuators, or air muscle, are able to offer a fast response time but they need an on-board air supply as bottles or compressors that are heavy pieces of equipment. Hydraulics actuators have high power/weight ratio; they are able to supply very high force, but suffer from the serious drawback of having to carry an engine to drive the pump. Hydraulics actuators are suitable for larger sized hexapod robots [61].

Unconventional actuators for hexapod robots can also be materials that can change shape through the direct application of electricity or a chemical agent. Ionic polymer-metal composites, for example, are materials that exhibit high strains under applied voltage differences allowing them to change from a sheet shape into curved shapes. Polyacrylonitrile is a form of artificial silk, classed as a gel polymer.

It can respond fast, but the activation method is a change in pH, which requires the fibers to be housed in a watertight bag. Another class of materials that can change shape under the application of electricity is the Shape Memory Alloys (SMA), such as a nickel-titanium alloy that exhibits extreme contraction when heated. Contraction rates are controlled by the heating and cooling; the major drawback resulting in slow response times and small operating forces. Thus, despite a great research interest and the building of some prototypes such as in [62], the uses of SMA as hexapod actuators have been very limited. Present trends indicate that they are more suited to micro robots.

3.2.2. Actuators Arrangements

Typically, specific actuator arrangements are developed in order to obtain maximum leg workspace with a minimal kinematic structure. Several types of geometrical arrangements such as in Figure 6a are recurrent in literature. The design consists of links connected through knee joints. The walking motion is accomplished by controlling the angle of the links to position the feet.

Figure 6. Actuators arrangement: (**a**) three DoF solution; (**b**) pinion-belt arrangement; (**c**) lead screw leg.

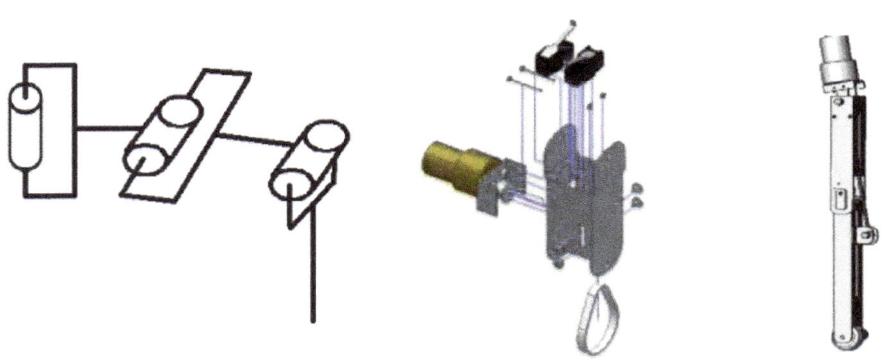

(**a**) (**b**) (**c**)

There are a number of different ways in which the joints can be actuated, which are referenced in [48]. Options include mounting the motor at the joint itself, or using a pulley and belt (Figure 6b) or lead screw (Figure 6c) to set the angle of the knee using an actuator mounted near the base of the leg [59]. The major drawback of last design is the necessity to actuate remote joints. On the other hand, latching the actuator at the knee joint adds various dynamic effects to the leg which have to be compensated by the controller. This adds complexity to the control algorithms needed to move the leg. It also requires more powerful motors at the hip joint to move the added mass of the leg. Remote actuation, in which the actuators are located at the base of the leg, eliminates some of these problems, at the cost of increasing the complexity of the mechanism. The coupling of the motion of the end effector relative to the actuators is another undesirable characteristic of this leg design.

Another potential leg design is modeled according to a typical mammalian leg with a four-bar linkage structure. The major drawback of this design is that the motions are highly coupled and the effective workspace is somewhat limited. Moreover, the entire weight of the robot is supported by the hip joint and they necessitate a powerful and expensive motor.

3.3. Modeling Issues

Computational programs dedicated to the kinematic and dynamic analysis of mechanical systems are a powerful alternative to the classic design methods. Simulation-based design can help in the development and the characterization of hexapod robots. Simulation software allows testing design performance and predicting component's behavior, prior to building a physical prototype. Simulation can also help the development of gait algorithms and the gait optimization.

Basically, the modeling of hexapod robots is composed of two steps. The first one concerns the kinematics of the robot motion: several types of software may be suitable for this purpose such as SolidWorks. The second stage of the model concerns dynamics and control of the actuators. This model is completed by the relevant forces encountered by walking robots, that is to say, the ground contact forces, the friction in the joints, the motor torques. Interaction of hexapod robots with the environment is a further step of the design study. Many references can be found in the literature as in [63]. Indeed, it is commonly acknowledged that high-geared actuators typically encountered in legged robots have very little sensitivity to inertia, and centrifugal and Coriolis forces. However, high-geared motors introduce flexibility, backlash and friction, which should be modeled to get reliable simulations. The development of gait algorithms can be initially based on pure kinematic robot models such as in [64,65] since hexapods are generally slow machines, and consequently inertial forces are not so important in the gait planning. A further aspect relates to gait optimization, which is focused, for example, on energy saving [66].

3.4. Optimal Design

The optimum design can be formulated as a multiobjective optimization problem in the form proposed in reference [67], as:

$$\min_{X} [\mathbf{F}(\boldsymbol{X})] = \min_{X} \{ \max_{i=1,\ldots,N} [w_i f_i(\boldsymbol{X})]\} \tag{5}$$

subjected to

$$\mathbf{G}(\boldsymbol{X}) < 0 \tag{6}$$

$$\mathbf{H}(\boldsymbol{X}) = 0 \tag{7}$$

where min is the operator for calculating the minimum of a vector function $\mathbf{F}(\boldsymbol{X})$; similarly max determines the maximum value among the N functions $[w_i f_i(\boldsymbol{X})]$ at each iteration; $\mathbf{G}(\boldsymbol{X})$ is the vector of constraint functions that describes limiting conditions, and $\mathbf{H}(\boldsymbol{X})$ is the vector of constraint functions that describes design prescriptions; $\boldsymbol{X}$ is the vector of design variables. The proposed optimization formulation uses the objective function $\mathrm{F}(\boldsymbol{X})$ at each iteration by choosing the worst-case value among all the scalar objective functions for minimizing it in the next iteration, as outlined in reference [68]. In particular, the worst-case value is selected in Equation (5) at each iteration as the objective function with maximum value among the N available objective functions. This approach for solving multi-objective problems with several objective functions and complex tradeoffs among them is known as the "minimax method". The "minimax method" is widely indicated in the literature for many problems, like, for example, for estimating model parameters by minimizing the maximum difference between model output and design specification, as outlined in references [69,70].

Since the formulated design problem is intrinsically highly non-linear, a solution is obtained when the numerical evolution from a tentative solution converges because of the iterative process to a solution that can be considered optimal within the explored range of the design variable domain. Therefore, a solution can be considered an optimal design as a local optimum, in general terms. This last remark makes clear the influence of a suitable formulation with computational efficiency for the involved criteria and constraints in order to have a design procedure, which is significant from an engineering viewpoint as well as being numerically efficient.

3.5. Hexapod Walking Robot Control

The control systems of recent hexapod robots are distributed hierarchical systems composed of a host computer, an onboard controller and various actuators and sensors. An example of hierarchical control architecture is shown in Figure 7. A schematic of hexapod basic locomotion control is shown in Figure 8. The user will set a desired position and walking gait, which is the input for the trajectory generator. The trajectory generator sends the leg coordinates to the inverse kinematic model for each leg. The inverse kinematic model re-calculates those given coordinates into its angular coordinates for all the joints. The outcomes of the inverse kinematics model are the joint angles which will then enter the control loop. The control at the joint level is almost always implemented using a PID control [71].

Figure 7. An example of hierarchical control architecture for Hexapod robot.

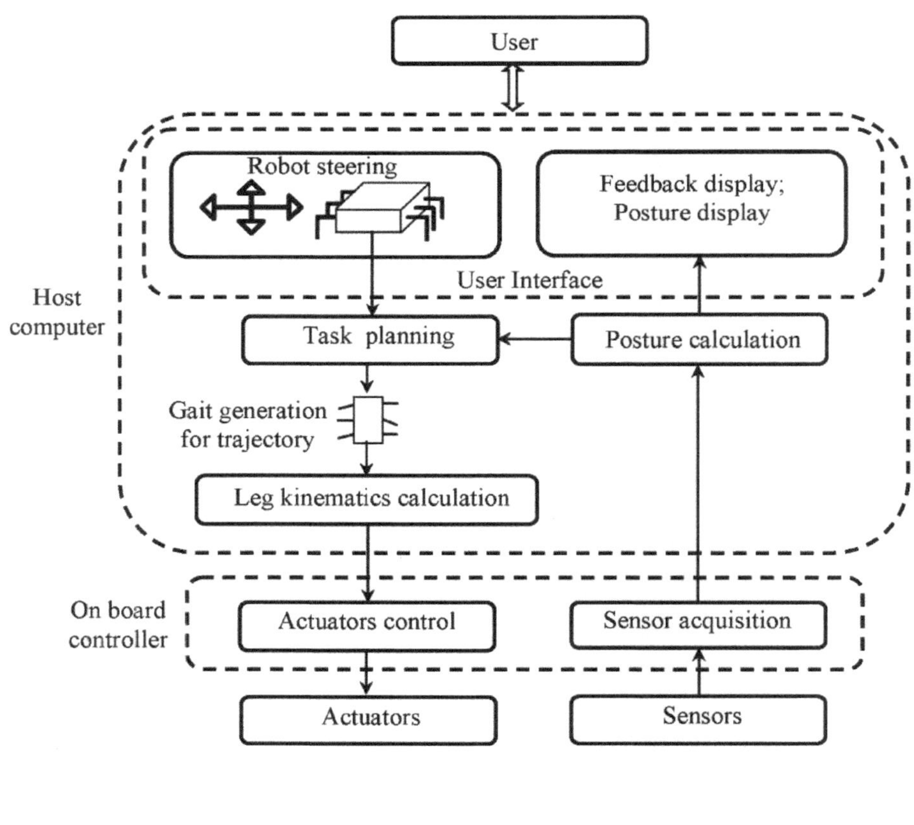

Figure 8. A scheme of hexapod robot basic locomotion control.

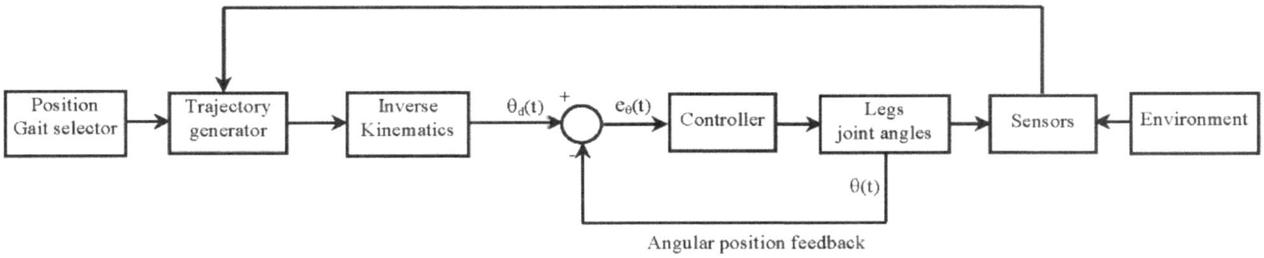

3.6. Gait Planning

A gait is a sequence of leg motions coordinated with a sequence of body motions for moving the overall body of the robot in the desired direction and for orientation from one place to another. A gait is described as periodic when similar states of the same leg during successive strokes occur at the same interval for all legs. Periodic gaits are suitable for smooth terrain and they have been studied by several investigators, such as [72]. Figure 9 shows the scheme of some periodical hexapodal gaits; white color indicates that the foot is in ground contact and the black color otherwise, according to [73]. Figure 9a reports the hexapod legs' description.

Figure 9. Gait diagram of hexapodal gaits: (**a**) legs description; (**b**) metachronal gait; (**c**) ripple gait; (**d**) tripod gait.

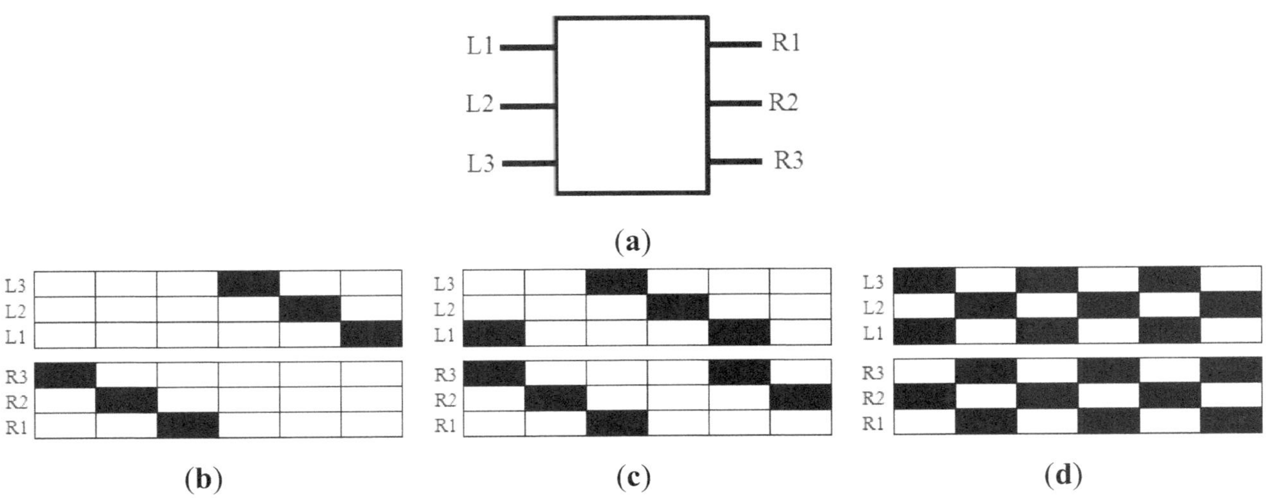

The metachronal gait illustrated in Figure 9b is adopted by the hexapod when it moves slowly. All legs on one side are moved forward in succession, starting with the rear-most leg. This is then repeated on the other side. Since only one leg is ever lifted at a time, with the other five down, the hexapod is always in a highly-stable posture. The ripple gait is used by the hexapod to move with a medium speed. Gait diagram representation is shown in Figure 9c. Tripod gait (Figure 9d) is a regular, periodic gait where $\beta = 0.5$, and where the anterior and posterior legs on one side lift in time with the contralateral middle leg, forming alternating tripods. This is a gait suitable for high speed walking over relatively flat ground.

One problem with periodic gaits is that they require synchronization between the expected and actual times that legs make contact with the ground. Over uneven ground, the time of footfall becomes unpredictable, breaking the rigorously defined leg phasing and compromising stability. The free gait, on the other hand, is much more effective on rough terrain with obstacles. The free gait is when each leg can move on a free chosen interval and its own stroke algorithm. Some examples can be found in the literature such as [74–77]. The difficulty with these approaches is the computational complexity. Genetic Algorithms (GA) are mostly used for free gait implementation [78]. Another way for using GA is a fuzzy logic controller, a potential tool for handling imprecision and uncertainty [79]. Also, using the techniques of soft computing will improve the leg speed; this is done by on-line trajectory generation [80]. Most of the studies focused on the control at the leg level and leg coordination using neural networks and fuzzy logic are based also on hybrid force/position control.

4. A Case of Study for Preliminary Lay-Out Design

This section describes the preliminary design process for a hexapod robot aimed for inspection and analysis of historical sites of architectonic interest, with low-cost and easy-operation features.

Key feature requirements for the specific application can be summarized as follows:

- low cost (<1000 Euros);
- user-friendly operation, also for non-expert users (e.g., architects);
- possibility to negotiate a large variety of obstacles, such as
  - a step with maximum height of 60 mm;
  - a crest with maximum width of 100 mm and maximum height of 60 mm;
  - a ditch with a maximum width of 60 mm;
- the robot must be able to move inside archaeological and/or architectural sites by carrying surveying devices and by avoiding damage of the surface and other parts of the site;
- the robot can be operating wirelessly, in environments that cannot be reached or that are unsafe for human operators;
- operating speed on regular terrain should be >0.1 m/s on regular terrain and ≥0.01 m/s on uneven terrains as based on previous experience in architectonical survey.

According to the above referenced criteria, Table 5 shows a matrix related to preliminary design. As a first step, the key features are ranked by their importance to the project, according to Table 2. Each key feature has been detailed in several aspects. For example, walking speed has been described in terms of performances on regular terrain and uneven terrain. The cost has been divided according to two factors: total budget and possibility of use for commercial components. Regarding the walking gait, three types of gaits have been considered. In a similar way, all key features have been recorded in the rows of the matrix. Rankings are recorded in the second column of Table 5. Similarly, each design requirement has been divided into several categories. For example, the body type category reports two solutions: the first one is a rectangular shape and the second one is a hexagonal shape. The leg type issue has been split into three categories: configuration, orientation and knee. Configuration of legs has been categorized in mammals, reptile, spider and hybrid. Orientation of legs can be sagittal, frontal, and circular. In the same way, the knee item has been divided into three categories: outwards, same orientation and inwards. After the rankings have been chosen, the relationship matrix has been completed according to criteria given in Table 3 and the above mentioned design issues.

In the following, we explain, for example, how to fill row 5 of the matrix regarding the step avoidance category. Step avoidance $H_{max} \geq 60$ mm is a very important key feature, thus its importance ranking is 4. Regarding the shape, rectangular or hexagonal, the score is 3 since each of the solutions satisfies the obstacle avoidance requirement only in conjunction with one or two features. Mammalians' configuration score is 9, since this solution demonstrated very good ability for obstacle traversing due to the high center of gravity. Reptile leg score is 1 since this solutions satisfies the requirements only in conjunction with many other factors. In the same way, spider leg configuration score is 1. Referring to [81,82], the hybrid legs score is 9, similarly to mammalian configuration. As regards the legs' orientation, sagittal score is 1 since this solution is not effective in satisfying the obstacle avoidance requirement. Frontal orientation score is 3 since this solution satisfies the requirements only in conjunction with one or two factors. In the same way, circular orientation score is 3. Knee orientation score is 3 for the outwards solution, since it can satisfy the requirements only in conjunction with one or two factors. In the same way, orientation solution score is 9 due to the high center of gravity. Inward solution score is 1 since this solutions satisfies the requirements only in conjunction with many other factor. Regarding the actuators, the score for each item is 1 since each of them satisfies the requirement only in conjunction with many other factors. The same procedure holds for filling all the matrix rows.

After filling in the relationship matrix, the importance rankings have been multiplied by the relationship value and the total is given at the bottom of each column of the matrix. Thus, the value at the bottom of the column gives the importance rating of the considered main design characteristics. These evaluations must be confirmed with simulation and trade-off studies; however, only the highest ranked items are subject to design refinements. This will significantly reduce the required time investment for the design.

Final results in Table 5 show that the most significant engineering requirements are rectangular body shape, hybrid legs, frontal configuration, same knee orientation, and electrical actuators. In fact, the above mentioned characteristics have obtained the highest ranking in their category of the matrix (value 218, 294, 174, 200, 292, respectively). According to these results, the preliminary hexapod architecture has been selected. Based on the results of this preliminary design procedure, a 3D CAD model has been developed as shown in Figure 10. Design refinements and a prototype will be developed in the near future as final validation of the proposed approach.

Figure 10. A CAD model of a novel hexapod.

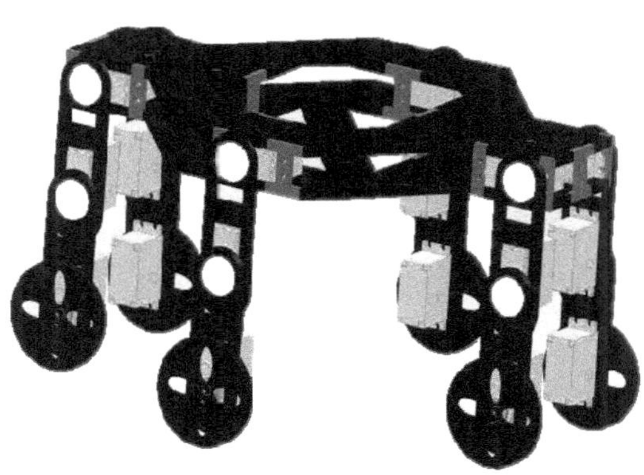

Table 5. Example of application of the proposed procedure on hexapod design.

Main Design Characteristics (How) / Key Features (What)		Rating of Importance R	Body Type		Legs Type Synthesis												Actuators		
					Configuration				Orientation			Knee							
			Rectangular Shape	Hexagonal Shape	Mammalian Legs	Reptile Legs	Spider Legs	Hybrid Legs	Sagittal	Frontal	Circular	Outwards	Same Orientation	Inwards			Electrical	Pneumatic	Hydraulics
			1	2	3	4	5	6	7	8	9	10	11	12			13	14	15
Walking speed																			
Regular terrain ≥ 0.1 m/s	1	5	3	3	1	1	1	9	3	9	3	3	3	3			9	3	3
Uneven terrain ≥ 0.01 m/s	2	3	3	3	1	1	1	3	3	9	3	3	3	3			9	3	3
Low Cost																			
Cost ≤ 1000 Euro	3	5	3	3	3	3	3	3	1	1	1	1	1	1			9	1	1
Use of commercial components	4	4	3	1	9	9	9	9	3	3	3	3	3	3			9	3	1
Obstacle avoidance																			
Step H_{max} ≥ 60 mm	5	4	3	3	9	1	1	9	1	3	3	3	9	1			1	1	1
Crest H_{max} ≥ 60 mm W ≥ 100 mm	6	4	3	3	9	1	1	9	1	3	3	3	9	1			1	1	1
Ditch W_{max} ≥ 100 mm	7	4	3	3	9	1	1	9	1	3	3	3	9	1			1	1	1
Operation																			
Wireless	8	5	3	3	3	1	1	1	1	1	1	1	1	1			9	1	1
Omnidirectional Steering	9	4	3	3	1	3	3	3	3	3	9	3	3	3			3	3	3
Autonomy ≥ 1 h	10	5	1	1	1	1	1	1	1	1	1	1	1	1			9	0	0
Walking Gait																			
Tripod gait	11	3	9	9	3	3	3	9	3	3	3	3	3	3			1	1	1
Metachronal gait	12	3	9	3	3	3	3	3	9	3	3	3	3	3			1	1	1
Free gaits	13	1	3	3	1	1	1	9	3	1	1	1	3	3			1	1	1
Ground clearance > 15 cm	14	3	9	3	9	3	1	3	1	1	1	1	1	1			1	1	1
Load carrying capacity ≥ 0.5 kg	15	5	3	3	1	1	1	1	1	1	1	1	1	1			3	3	3
$Y_j = \sum_{i=1}^{n} K_{ji} \cdot R_i$		218	174	242	126	120	294	122	174	150	126	200	104	292			95	87	87

5. Conclusions

This paper gives an overview of the state of the art on six-leg walking robots. Careful attention is paid to the main design issues and constraints that influence the technical feasibility and performance of these systems. A design procedure is outlined in order to systematically design a six-leg walking robot. In particular, the proposed design procedure takes into account the mechanical structure and leg configuration, the actuating and drive mechanisms, payload, motion conditions, and walking gait, providing a useful tool for the systematic choice of main design characteristic. A case study has been reported in order to show the effectiveness and engineering feasibility of the proposed procedure.

Acknowledgments

Authors wish to acknowledge the fruitful discussions and support given by Prof. Marco Ceccarelli, director of LARM Laboratory.

Author Contributions

Franco Tedeschi has systematized the proposed design procedure within his PhD studies under the supervision of Giuseppe Carbone. Basic concepts and design features of "Cassino Hexapod" prototype series have been proposed by Giuseppe Carbone.

Conflicts of Interest

The authors declare no conflict of interest.

References

1. Chàvez-Clemente, D. Gait Optimization for Multi-legged Walking Robots, with Application to a Lunar Hexapod. Ph.D. Thesis, Stanford University, California, CA, USA, 2011.
2. Carbone, G.; Ceccarelli, M. Legged robotic systems. In *Cutting Edge Robotics*; Kordic, V., Lazinica, A., Merdan, M., Eds.; InTech: Vienna, Austria, 2005; pp. 553–576.
3. Cigola, M.; Pelliccio, A.; Salotto, O.; Carbone, G.; Ottaviano, E.; Ceccarelli, M. Application of robots for inspection and restoration of historical sites. In Proceedings of the International Symposium on Automation and Robotics in Construction of the Conference, Ferrara, Italy, 11–14 September 2005; p. 37.
4. Jun, B.H.; Shim, H.; Kim, B.; Park, J.Y.; Baek, H.; Yoo, S.; Lee, P.M. Development of seabed walking robot CR200. In Proceedings of the OCEANS'13 MTS/IEEE of the Conference, San Diego, CA, USA, 23–26 September 2013; pp. 1–5.
5. Georgiades, C. Simulation and Control of an Underwater Hexapod Robot. M.D. Thesis, McGill University, Montreal, QC, Canada, 2005.
6. Bares, J.; Hebert, M.; Kanade, T.; Krotkov, E.; Mitchell, T.; Simmons, R.; Whittaker, W. Ambler: An autonomous rover for planetary exploration. *IEEE Comput.* **1989**, *26*, 6–18.
7. Preumont, A.; Alexandre, P.; Doroftei, I.; Goffin, F. A conceptual walking vehicle for planetary exploration. *Mechatronics* **1997**, *7*, 287–296.
8. Bartholet, T.; Crawson, R. *Robot Applications for Nuclear Power Plant Maintenance*; EPRI Report-NP-3941, Research Report Center: Palo Alto, CA, USA, 1985.
9. Oku, M.; Yang, H.; Paio, G.; Harada, Y.; Adachi, K.; Barai, R.; Nonami, K. Development of hydraulically actuated hexapod robot COMET-IV-The 1st report: System design and configuration. In Proceedings of the 2007 JSME Conference on Robotics and Mechatronics, Akita, Japan, 26–28 May 2007.
10. Gregorio, P.; Ahmadi, M.; Buehler, M. Design, control, and energetics of an electrically actuated legged robot. *Syst. Man Cybern. B IEEE Trans.* **1997**, *27*, 626–634.

11. Schneider, A.; Schmucker, U. Force sensing for multi-legged walking robots: Theory and experiments part 1: Overview and force sensing. In *Mobile Robotics, Moving Intelligence*; Buchli, J., Ed.; Pro Literatur Verlag ARS: Germany; Austria, 2006; pp. 125–174.
12. Peternella, M.; Salinari, S. Simulation by digital computer of walking machine control system. In Proceedings of the 5th IFAC Symposium on Automatic Control in Space of the Conference, Genova, Italy, June 1973.
13. Okhotsimski, D.; Platonov, A. Control algorithm of the walking climbing over obstacles. In Proceedings of the 3rd International Joint Conference on Artificial Intelligence, Stanford, CA, USA, 20 August 1973.
14. Gurfinkel, V.; Gurfinkel, E.; Devjanin, E.; Efremov, E.; Zhicharev, D.; Lensky, A.; Schneider, A.; Shtilman, L. Investigation of robotics. In *Six-legged Walking Model of Vehicle with Supervisory Control*; Nauka Press: Moscow, Russia, 1982; pp. 98–147.
15. McGhee, R. Control of legged locomotion systems. In Proceedings of the 18th Automatic Control Conference, San Francisco, CA, USA, 3–8 December 1977; pp. 205–215.
16. Raibert, M. *Legged Robots that Balance*; MIT Press: Cambridge, London, 1986; pp. 180–201.
17. Byrd, J.; de Vries, K. A six-legged telerobot for nuclear applications development. *Int. J. Robot. Res.* **1990**, *9*, 43–52.
18. Efimov, V.; Kudriasev, M.; Titov, A. Investigation of robotics systems. In *A Physical Similar of Motion Walking Apparatus*; Nauka Press: Moscow, Russia, 1982; pp. 86–91.
19. Song, S.M.; Waldron, K. *Machines that Walk: The Adaptive Suspension Vehicle*; MIT Press: Cambridge, London, 1989; pp. 283–299.
20. Akozono, J.; Iwasaki, M.; Asakura, O. Development on a walking robot for underwater inspection. In Proceedings of the International Conference on Arabidopsis Research (ICAR' 89), Columbus, OH, USA, 13–15 June 1989.
21. Brooks, R.A. A robot that walks; emergent behaviors from a carefully evolved network. *Neural Comput.* **1989**, *1*, 253–262.
22. Bares, J.; Hebert, M.; Kande, T.; Krotkov, E.; Mitchell, T.; Simmons, R.; Whittaker, W. Ambler: An autonomous rover for planetary exploration. *IEEE Comput.* **1989**, *22*, 18–26.
23. Angle, C.A.; Brooks, R.A. Small planetary rovers. In Proceedings of the IEEE International Workshop on Intelligent Robots and Systems, Ibaraki, Japan, 3–6 July 1990; pp. 1–5.
24. Pfeiffer, F.; Eltze, J.; Weidermann, H. Six-legged technical walking considering biological principles. *Robot. Autom.* **1995**, *14*, 223–232.
25. Nelson, G.M.; Quinn, R.D.; Bachmann, R.J.; Flannigan, W.C.; Ritzmann, R.E.; Watson, J.T. Design and simulation of a cockroach-like hexapod robot. In Proceedings of the 1997 IEEE International Conference on Robotics and Automation, Albuquerque, NM, USA, 25 April 1997; pp. 1106–1111.
26. Delcomyn, F.; Nelson, M.E. Architectures for a biomimetic hexapod robot. *Robot. Auton. Syst.* **2000**, *30*, 5–15.
27. Fielding, M.R.; Dunlop, R.; Damaren, C.J. Hamlet: Force/Position controlled hexapod walker—Design and systems. In Proceedings of the IEEE Conference on Control Applications, Mexico City, Mexico, 5–9 September 2001; pp. 984–989.

28. Saranli, U.; Buehler, M.; Koditschek, D.E. RHex—A simple and highly mobile hexapod robot. *Int. J. Robot. Res.* **2001**, *20*, 616–631.
29. Roennau, A.; Heppner, G.; Pfozter, L.; Dillman, R. Lauron V: Optimized leg configuration for the design of a bio-inspired walking robot. In Proceedings of the 16th International Conference on Climbing and Walking Robots and the Support Technologies for Mobile Machines, Sydney, Australia, 14–17 July 2013; pp. 563–571.
30. Arena, P.; Fortuna, L.; Frasca, M.; Patanè, L.; Pavone, M. Implementation and experimental validation of an autonomous hexapod robot. In Proceedings of the IEEE International Symposium on Circuits and Systems, Kos, Greece, 21–24 May 2006.
31. Cham, J.G.; Bailey, S.A.; Clark, J.E.; Full, R.J.; CutKosky, M.R. Fast and robust: Hexapedal robots via shape deposition manufacturing. *Int. J. Robot. Res.* **2002**, *21*, 869–882.
32. Lewinger, W.A.; Branicky, M.S.; Quinn, R.D. Insect-inspired, actively compliant hexapod capable of object manipulation. In Proceedings of the 8th International Conference on Climbing and Walking Robots (CLAWAR' 2005), London, UK, 13–15 September 2005; pp. 65–72.
33. Kennedy, B.; Okon, A.; Aghazarian, H.; Garrett, M.; Huntsberger, T.; Magnone, L.; Robinson, M.; Townsend, J. The lemur II-Class robots for inspection and maintenance of orbital structures: A system description. In Proceedings of the 8th International Conference on Climbing and Walking Robots (CLAWAR' 2005), London, UK, 13–15 September 2005; pp. 1069–1076.
34. Showalter, M. *Hexapod Work Space and Walking Algorithm*; VDM Verlag: Saarbrücken, Germany, 2009; pp. 20–34
35. Hauser, T.; Bretl, K.; Latombe, J.C.; Harada, W. Motion planning for legged robots on varied terrain. *Int. J. Robot. Res.* **2008**, *27*, 1325.
36. Georgiadis, C.; German, A.; Hogue, A.; Liu, H.; Prahacs, C.; Ripsman, A.; Sim, R.; Torres, L.A.; Zhang, P.; Buehler, M.; Dudek, G.; Jenkin, M.; Milios, E. AQUA: An aquatic walking robot. In Proceedings of the IEEE/RSJ International Conference on Intelligent Robots and Systems (IROS), Sendai, Japan, 28 September–2 October, 2004.
37. Kang, H.; Shim, H.; Jun, B.H.; Lee, P.M. Design of underwater link and joint for the development of multi-legged underwater robot CR200. In Proceedings of the MTS/IEEE OCEANS 2012, Yeosu, Korea, 21–24 May 2012; pp. 1–6.
38. Asbeck, A.T.; Kim, S.; McClung, A.; Parness, A.; Cutkosky, M.R. Climbing walls with microspines. In Proceedings of the IEEE International Conference Robotics and Automation, Orlando, FL, USA, 15–19 May 2006.
39. Nonami, K.; Huang, Q.J. Humanitarian mine detection six-legged walking robot COMET-II with two manipulators. In Proceedings of the 4th International Conference on Climbing and Walking Robots (CLAWAR' 2001), Karlsruhe, Germany, 24–26 September, 2001; pp. 989–996.
40. Mantis homepage. Available online: http://www.mantisrobot.com (accessed on 24 March 2014).
41. Genta, G. *Introduction to the Mechanics of Space Robots*; Springer: Berlin, Germany, 2012; pp. 381–423.
42. Song, S.M.; Choi, B.S. The optimally stable ranges of 2n-Legged wave gaits. *IEEE Trans. Syst. Man Cybern. B* **1990**, *20*, 888–902.
43. Alexander, R. McN. *Exploring Biomechanics—Animals in Motion*; W.H. Freeman Press: New York, NY, USA, 1992.

44. Alexander, R. McN. The gait of bipedal and quadrupedal animals. *Int. J. Robot. Res.* **1984**, *3*, 49–59.
45. Gabrielli, G.; von Karman, T. What price speed—Specific power required for propulsion of vehicles. *Mech. Eng.* **1950**, *72*, 775–781.
46. McGhee, R.B.; Frank, A.A. On the stability properties of quadruped creeping gaits. *Math. Biosci.* **1968**, *3*, 331–351.
47. Messuri, D.A.; Klein, C.A. Automatic body regulation for maintaining stability of a legged vehicle during rough-terrain locomotion. *IEEE J. Robot. Autom.* **1985**, *1*, 132–141.
48. Zielinska, T. Autonomous walking machines, discussion of the prototyping problems. Bulletin of the Polish academy of sciences. *Tech. Sci.* **2010**, *58*, 443–451.
49. Rashid, M.Z.A.; Aras, M.S.M.; Radzak, A.A.; Kassim, A.M.; Jamali, A. Development of hexapod robot with manoeuvrable wheel. *Int. J. Adv. Sci. Tech.* **2012**, *49*, 119–136.
50. Ding, X.; Wang, Z.; Rovetta, A.; Zhu, J.M. Locomotion analysis of hexapod robot. In *Climbing and Walking Robots*; In Tech: Rijeka, Croatia, 2010; pp. 291–310.
51. Lee, T.T.; Liao, C.M.; Chen, T.K. On the stability properties of hexapod tripod gait. *IEEE J. Robot. Autom.* **1988**, *4*, 427–434.
52. Waldron, K.J.; McGhee, R.B. The adaptive suspension vehicle. *IEEE Control Syst. Mag.* **1986**, *6*, 7–12.
53. Wang, X.J. A Study of Locomotion and Force Planning for Multilegged Walking Robots. Ph.D. Thesis, Huazhong University of Science & Technology, Wuhan, China, 2005; pp. 95–100.
54. Preumont, A.; Alexadre, P.; Ghuys, D. Gait analysis and implementation of a six leg walking machine. In Proceedings of the Fifth International Conference on Advanced Robotics. Robots in Unstructured Environments (ICAR' 91), Pisa, Italy, 19–22 June 1991; Volume 2, pp. 941–945.
55. Takahashi, Y.; Arai, T. Development of multi-limb robot with omnidirectional manipulability and mobility. In Proceedings of the 2000 IEEE-RSJ International Conference on Intelligent Robots and Systems, Takamatsu, Japan, 31 October–5 November 2000; Volume 3, pp. 2012–2017.
56. Chu, S.K.-K.; Pang, G.K.-H. Comparison between different models of hexapod robot in fault-tolerant gait. *IEEE Trans. Syst. Man Cybern.* **2002**, *32*, 752–756.
57. Silva, M.F.; Tenreiro Machado, J.A. A historical perspective of legged robots. *J. Vib. Control* **2007**, *13*, 1447–1486.
58. Billah, M.M.; Ahmed, M.; Farhana, S. Walking hexapod robot in disaster recovery developing algorithm for terrain negotiation and navigation. In *New Advanced Technologies*; Lazinica, A.; In Tech: Rijeka, Croatia, 2010.
59. Nava Rodríguez, N.E.; Carbone, G.; Ceccarelli, M.; Moreno Lorente, L.E. Design evolution of cassino hexapod robot. In Proceedings of the 10th Biennial ASME Conference on Engineering Systems Design and Analysis ESDA2010, Istanbul, Turkey, 12–14 July 2010.
60. Endo, G.; Hirose, S. Study on Roller-Walker: System integration and basic experiments. *Robot. Autom.* **1999**, *3*, 2032–2037.
61. Nonami, K.; Barai, R.K.; Irawan, A.; Daud, M.R. *Hydraulically Actuated Hexapod Robots*; Springer: London, UK, 2014; pp. 78–104.
62. Conrad, J.M.; Mills, J.W. The history and future of stiquito: A hexapod insectoid robot. In *Artificial Life Models in Hardware*; Springer: London, UK, 2009; pp. 1–20.

63. Rahnejat, H. Multi-body dynamics: Historical evolution and application. *J. Mech. Eng. Sci.* **2000**, *214*, 149–173.
64. Salmi, S.; Halme, A. Implementing and testing a reasoning-based free gait algorithm in the six-legged walking machine MECANT. *Control Eng. Pract.* **1996**, *4*, 487–492.
65. Silva, M.F.; Tenreiro Machado, J.A. A Kinematic and dynamic performance analysis of artificial legged systems. *Robotica* **2008**, *26*, 19–39.
66. De Santos, P.G.; Garcia, E.; Ponticelli, R.; Armada, M. Minimizing energy consumption in hexapod robots. *Adv. Robot.* **2009**, *23*, 681–704.
67. Grace, A. *Optimization Toolbox User's Guide*; The MatlabWorks Inc.: Denver, CO, USA, 1992.
68. Vanderplaats, G. *Numerical Optimization Techniques for Engineers Design*; McGraw-Hill: NewYork, NY, USA, 1984.
69. Eldar, Y.C. Minimax estimation of deterministic parameters in linear models with a random model matrix. *IEEE Trans. Signal Process.* **2006**, *54*, 601–612.
70. Medanic, J.; Andjelic, M. Minimax solution of the multiple-target problem. *IEEE Trans. Autom. Control* **1972**, *17*, 597–604.
71. Franklin, G.F. *Feedback Control of Dynamic Systems*; Addison-Wesley Longman Publishing Co.: Boston, MA, USA, 2002.
72. Collins, J.J.; Stewart, I. Hexapodal gaits and coupled nonlinear oscillator models. *Biol. Cybern.* **1993**, *68*, 287–298.
73. Song, S.M.; Waldron, K.J. An analytical approach for gait study and its applications on wave gaits. *Int. J. Robot. Res.* **1987**, *6*, 60–71.
74. Ferrell, C. *Robust Agent Control of an Autonomous Robot with Many Sensors and Actuators*; MIT AI Lab: Cambridge, MA, USA, Technical Report 1443, 1993; pp. 1–100.
75. McGhee, R.B.; Iswandhi, G.I. Adaptive locomotion of a multilegged robot over rouch terrain. *IEEE Trans. Syst. Man Cybern.* **1979**, *9*, 176–182.
76. Shih, T.; Tsai, C.S.; Her, I. Comparison of alternative gaits for multiped robots with severed legs. *Int. J. Adv. Robot. Syst.* **2012**, *9*, 1–8
77. Silva, M.F.; Tenreiro Machado, J.A.; Lopes, A.M. Quasi-periodic Gaits in Multi-legged Robots. In Proceedings of the 5th International Conference on Climbing and Walking Robots and the Support Technologies for Mobile Machines (CLAWAR' 2002); Paris, France, 25–27 September 2002; pp. 733–740.
78. Goldberg, D.E. *Genetic Algorithms in Search, Optimization, and Machine Learning*; Addison-Wesley Longman Publishing Co.: Upper Saddle River, NJ, USA, 1989; pp. 110–130.
79. Kosko, B. *Neural Networks and Fuzzy Systems: A Dynamical Systems Approach to Machine Intelligence*; Prentice-Hall, Inc.: Upper Saddle River, NJ, USA, 1992; pp. 179–193.
80. Gonzalez de Santos, P.; Garcia, E.; Estremera, J. *Quadrupedal Locomotion: An Introduction to the Control of Four-Legged Robots*; Springer: London, UK, 2006.
81. Carbone, G.; Ceccarelli, M. A mechanical design of a low-cost easy-operation anthropomorphic wheeled leg for walking machines. *Int. J. Robot. Manag.* **2004**, *9*, 3–8.

82. Carbone, G.; Shrot, A.; Ceccarelli, M. Operation strategy for a low-cost easy-operation cassino hexapod. *Appl. Bionics Biomech.* **2007**, *4*, 149–156.

9

Adaptive Neuro-Fuzzy Technique for Autonomous Ground Vehicle Navigation

Auday Al-Mayyahi *, William Wang and Phil Birch

Department of Engineering and Design, University of Sussex, Famer, Brighton, BN1 9QJ, UK;
E-Mails: w.j.wang@sussex.ac.uk (W.W); p.m.birch@sussex.ac.uk (P.B.)

* Author to whom correspondence should be addressed; E-Mail: auday.almayyahi@sussex.ac.uk

External Editor: Huosheng Hu

Abstract: This article proposes an adaptive neuro-fuzzy inference system (ANFIS) for solving navigation problems of an autonomous ground vehicle (AGV). The system consists of four ANFIS controllers; two of which are used for regulating both the left and right angular velocities of the AGV in order to reach the target position; and other two ANFIS controllers are used for optimal heading adjustment in order to avoid obstacles. The two velocity controllers receive three sensor inputs: front distance (FD); right distance (RD) and left distance (LD) for the low-level motion control. Two heading controllers deploy the angle difference (AD) between the heading of AGV and the angle to the target to choose the optimal direction. The simulation experiments have been carried out under two different scenarios to investigate the feasibility of the proposed ANFIS technique. The simulation results have been presented using MATLAB software package; showing that ANFIS is capable of performing the navigation and path planning task safely and efficiently in a workspace populated with static obstacles.

Keywords: ANFIS; autonomous ground vehicle; navigation; obstacle avoidance; static environment

---

1. Introduction

Autonomous ground vehicles are an important key of industrial automation. Such vehicles could be utilized in different applications such as monitoring, transportation and many other potential applications. The path planning and navigation problems are one of area of current research. A considerable attention has been paid in recent years to deal with such problems. The autonomous vehicle must be able to gather and extract information from its surrounding using sensors. Thus, these sensors are fed to a controller to plan and execute its mission within its environment without human intervention [1].

The path planning methods are classified into either the global planning method or the local planning method [2]. In the global planning method, the autonomous vehicle requires the environment to be completely priori-known and the terrain should be static. In contrast, local path planning means the environment might be only partially known or completely unknown. In the local planning method, the autonomous ground vehicle utilises the received sensory information during its local navigation [3]. The autonomous ground vehicle must use sensors to perceive its surroundings and plan the motion.

In the current literature, many techniques have been used for solving the navigation and path planning problems. Some of the popular methods are the Visibility Graph algorithm [4], Voronoi Diagram [5], Bug Algorithm and the Potential Field method. For instance, in the potential field method, there is an attractive potential pulling the robot towards the goal point and a repulsive potential pushing the robot away from the obstacles over free space. However, the potential field method is prone to becoming stuck within local search minima.

In recent years, with the rapid development of modern computing techniques, artificial intelligence techniques have been applied widely in solving path planning and navigation problems. These techniques include genetic algorithm [6,7]. In [8], an obstacle avoidance control algorithm for mobile robots based on fuzzy controller was presented. The environment information surrounding the robot detected by the ultrasonic sensors is fuzzified, and then input into a fuzzy control system. The output of the fuzzy control system is used to drive the robot. The simulation results show that the algorithm can help the robot to avoid obstacles safely and is certainly feasible.

In [9], fuzzy logic controllers using different membership functions are developed and used to navigate mobile robots. First, a fuzzy controller was used with four types of input members, two types of output members and three parameters each. Next, two types of fuzzy controllers were developed having same input members and output members with five parameters each. Each robot had an array of ultrasonic sensors for measuring the distances of obstacles around it and an infrared sensor for detecting the bearing of the target. In [10], a fuzzy logic system was designed for a path planning in an unknown environment for a mobile robot. The ultrasonic sensors were employed for detecting the distance and positions of obstacles. An angular velocity control for left and right wheels was implemented by a fuzzy logic system. This included another new rule table that was induced from the consideration of the distance to obstacles and the angle between the robot and the goal.

In [11], an obstacle avoidance approach for an e-puck robot by using fuzzy logic controller was introduced. The inputs from eight IR sensors and the output of the motor speed were used to construct the fuzzy logic rules. A Webots Pro Simulation software was used to model and program the e-puck robot and fuzzy algorithm and to test an environment. Chi and Less [12], showed an artificial neural networks was applied to learn the environment from sensory data to move along a collision free trajectory. In [13], the navigation of non-holonomic robot was discussed using neuro-fuzzy. The ANFIS parameters were discovered offline by using suitable datasets and the obtained parameters were fed into the robot.

Joshi and Zaveri [14], showed a neuro-fuzzy based system for the behaviour based control of a mobile robot for reactive navigation. Systems transform sensors' input to yield wheel velocities using behaviour based fuzzy reasoning. Singh *et al.* [15], analysed an adaptive neuro-fuzzy technique for the navigation of a mobile robot in unknown environment. The results improved the validity of the proposed technique to determine the optimal time and create a collision free path of the mobile robot. In [16], a hybrid neuro-genetic fuzzy technique was proposed for choosing the collision free path form a set of paths. Fuzzy logic was used for obstacle avoidance when a neural network is unable to choose a path due to blockage from obstacles in a number of sets of directions.

From the literature survey, it can be clearly seen that several issues have not been addressed yet. Firstly, the consideration of the physical dimensions of the autonomous ground vehicle (length and width); authors normally assume the vehicle is a point in a workplace. Secondly, the online sensing information from the three directions (front, right, left) during the navigation. Thirdly, the problem of the complexity of the navigation environment. Therefore, in this paper, all aforementioned issues have been involved to create a new problem formulation for solving the navigation problem in a complex static environment contains multiple obstacles with different shapes and sizes.

The contribution of the proposed approach can be understood as the implementing a new ANFIS model which has the ability to avoid becoming stuck into local minimum and generates a smooth trajectory between the starting and the target points in unknown environment based local planning method. In other words, the ANFIS model successfully and efficiently guide the vehicle through the workspace. Moreover, the performance of the ANFIS controller is verified by changing the poses of obstacles.

The work in this paper is divided in two stages: firstly, the target reaching controller, and secondly, the obstacle avoidance controller. The target reaching controller was implemented using two ANFIS controllers. Each controller received the same input, which was the angle difference between the autonomous ground vehicle's direction and the target's angle. The output of first ANFIS controller was used to control the right angular velocity, and the second ANFIS controller was used to control the left angular velocity. The obstacle avoidance controller was also implemented also using two ANFIS controllers. Each controller received three inputs that represent the front, right and left distances. These distances represent sensory information between the autonomous vehicle and obstacles. The paper is organized as follows: Section 2 describes a kinematic model of an autonomous ground vehicle. The architecture of neuro-fuzzy inference system is presented in Section 3. The design of adopted four ANFIS controllers and autonomous ground vehicle platform are described in Section 4. To validate the proposed approach, simulation results are conducted and discussed in Section 5. Finally, the conclusions are presented in Section 6.

2. The Kinematic Model of an Autonomous Ground Vehicle

The schematic diagram of a two-dimensional plane for an autonomous ground vehicle is shown in Figure 1. The initial Cartesian coordinates for this vehicle are x_o in x-axis and y_o in y-axis. Similarly, target point coordinates are denoted as x_t and y_t respectively. The vehicle's current position is coordinate P_c. The vehicle has four fixed standard wheels and is differentially driven by skid steer motion. The wheels have the same radius, r. The driving wheels are separated by a distance L. In general, the posture of the vehicle in the two dimensional plane at any instant is defined by the systems Cartesian coordinates and heading angle, θ, with respect to a global frame reference, $Pc = (x_c, y_c, \theta)$. In this paper, the wheels' radius equals $r = 0.1$ m, and the distance between the driving wheels equals $L = 0.3$ m. It is assumed that the autonomous ground vehicle is subject to the kinematic constraints such as the contact between the wheels and the ground is pure rolling, and non-slipping [17]. Such constraints cause some challenge in the motion planning.

Figure 1. The schematic diagram of the autonomous ground vehicle.

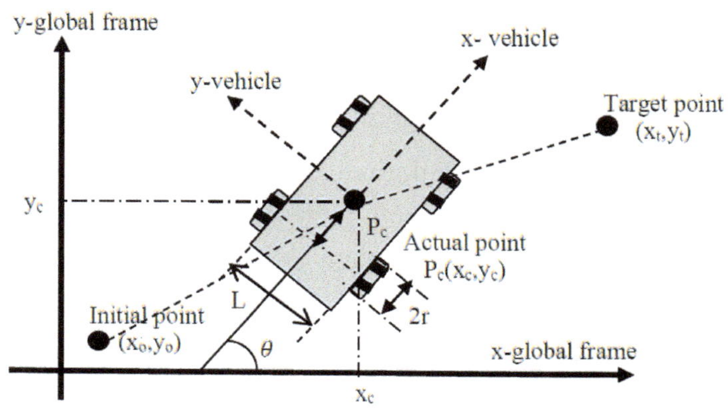

A set of relationships for the autonomous ground vehicle can be defined as:

a. No slip constraint

$$\dot{x_c}(t)\cos\theta(t) - \dot{y_c}(t)\sin\theta(t) = a\,\dot{\theta}(t) \tag{1}$$

b. Pure rolling constraint

$$\dot{x_c}(t)\cos\theta(t) + \dot{y_c}(t)\sin\theta(t) + L\,\dot{\theta}(t) = r\,\dot{\emptyset}_r(t) \tag{2}$$

$$\dot{x_c}(t)\cos\theta(t) + \dot{y_c}(t)\sin\theta(t) - L\,\dot{\theta}(t) = r\,\dot{\emptyset}_l(t) \tag{3}$$

$$\begin{bmatrix} v_x(t) \\ v_y(t) \\ \dot{\theta}(t) \end{bmatrix} = \begin{bmatrix} \frac{r}{2} & \frac{r}{2} \\ 0 & 0 \\ -\frac{r}{L} & \frac{r}{L} \end{bmatrix} \begin{bmatrix} w_l(t) \\ w_r(t) \end{bmatrix} \tag{4}$$

$$v(t) = r.\frac{w_r(t) + w_l(t)}{2} \tag{5}$$

$$\dot{\theta}(t) = r.[\frac{w_r(t) - w_l(t)}{L}] \tag{6}$$

$$\dot{x}(t) = v(t)\cos\theta(t) \tag{7}$$

$$\dot{y}(t) = v(t)\sin\theta(t) \tag{8}$$

$$w(t) = \dot{\theta}(t)\ (9) \tag{9}$$

$$\mathrm{w_r}(\mathrm{t}) = \dot{\phi}_r(t) \tag{10}$$

$$\mathrm{w_l}(\mathrm{t}) = \dot{\phi}_l(t) \tag{11}$$

$$\text{or} \begin{bmatrix} \dot{x}(t) \\ \dot{y}(t) \\ \dot{\theta}(t) \end{bmatrix} = \begin{bmatrix} cos\theta & 0 \\ Sin\theta & 0 \\ 0 & 1 \end{bmatrix} \begin{bmatrix} V \\ w \end{bmatrix} \tag{12}$$

where,

$\mathrm{v_x}(\mathrm{t})$ = longitudinal velocity of the moving vehicle [m/s]
$\mathrm{v_y}(\mathrm{t})$ = lateral velocity of the moving vehicle [m/s]
θ = moving vehicle orientation [degree]
$\mathrm{w_r}(\mathrm{t})$ = angular velocity of right wheel [rad/s]
$\mathrm{w_l}(\mathrm{t})$ = angular velocity of left wheel [rad/s]
$\mathrm{w}(\mathrm{t})$ = angular velocity of vehicle [rad/s]
r = wheel radius [m]
a = the distance between the centre of mass and driving wheels axis in x-direction [m]
L = the distance between each driving wheel and the vehicle axis of symmetry in y-direction [m].

3. The Architecture of Adaptive Neuro-Fuzzy Inference System (ANFIS)

The architecture of the ANFIS technique, consists of the fuzzy inference system and neural network with given input and output data pairs. This technique is a self-tuning and adaptive hybrid controller that uses learning algorithms. In other words, this technique gives the fuzzy logic capability to adapt the membership function parameters that best allow the associated fuzzy inference system to track the given input and output data parameters of ANFIS model. In order to process a fuzzy rule by neural networks, it is necessary to modify the standard neural network structure accordingly. Figure 2 depicts the architecture model of ANFIS. This model is called a first-order Takagi-Sugeno-fuzzy model [18]. For simplicity, it is assumed that the ANFIS model has two inputs k_1 and k_2, and one output f.

Figure 2. The architecture of an adaptive neuro-fuzzy inference system (ANFIS).

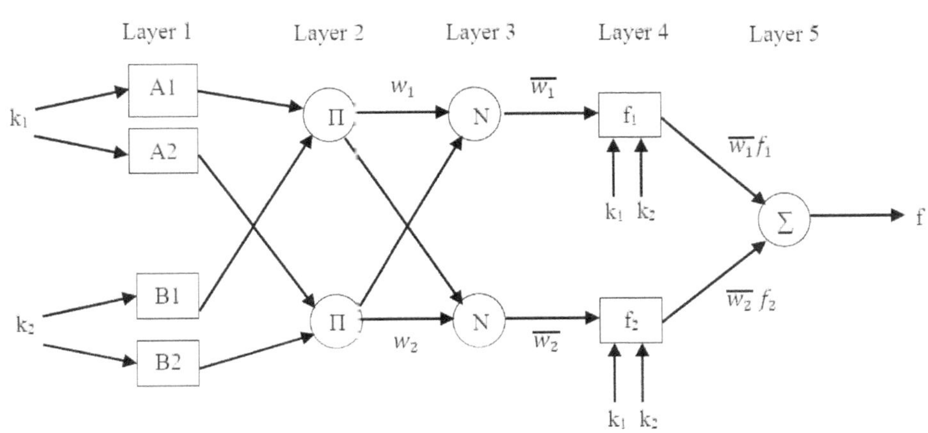

This system is broken down into five layers:

Layer 1: Every node i in this layer is an adaptive node with a node function

$$O_{1,i} = \mu_{A\_i}(x), for\ i = 1,2 \tag{13}$$

$$O_{1,i} = \mu_{B_{i-2}}(y), for\ i = 3,4 \tag{14}$$

where k_1 (or k_2) is the input to node i and A_i (or B_{i-2}) is a linguistic label (such as "small" or "large") associated with this node. In other words, $O_{1,i}$ is the membership grade of a fuzzy set A (= A_1 , A_2 , B_1 or B_2) and it specifies the degree to which the given input k_1 (or k_2) satisfies the quantifier A. Here the membership function for A is assumed be triangular shaped membership function:

$$\mu_A(x) = max\left[min\left\{\frac{x - a_i}{b_i - a_i}, \frac{c_i - x}{c_i - b_i}\right\}, 0\right] \tag{15}$$

where $\{a_i, b_i, c_i\}$ is the parameter set. As the values of these parameters change, the bell-shaped function varies accordingly, thus exhibiting various forms of membership function for fuzzy set A. Parameters in this layer are referred to as premise parameters.

Layer 2: Every node in this layer is a fixed node labelled Π, whose output is the product of all the incoming signals:

$$O_{2,i} = w_i = \mu_{A\_i}(x)\mu_{B_i}(y), i = 1,2 \tag{16}$$

Each node output represents the firing strength of a rule. In general, any other T-norm operators that perform a fuzzy AND can be used as the node function in this layer.

Layer 3: Every node in this layer is a fixed node labelled N. The ith node calculates the ratio of the ith rule's firing strength to the sum of all of the rules firing strengths:

$$O_{3,i} = \overline{w_i} = \frac{w_i}{w_1 + w_2}\ i = 1,2 \tag{17}$$

The output of this layer is called the normalised firing strengths.

Layer 4: Every node *i* in this layer is an adaptive node with a node function:

$$O_{4,i} = \overline{w_i}f_i = \overline{w_i}(p_i x + q_i y + r_i), i = 1,2 \tag{18}$$

where $\overline{w_i}$ is the normalised firing strength from layer 3 and $\{p_i, q_i, r_i\}$ is the parameter set of this node. Parameters in this layer are referred to as consequent parameters.

Layer 5: The single node in this layer is a fixed node labelled ∑, which computes the overall output, f, as the summation of all incoming signals:

$$f = O_{5,1} = \sum_i \overline{w_i}\, f_i = \frac{\sum_i w_i f_i}{\sum_i w_i}\ i = 1,2 \tag{19}$$

The first and fourth layers are adaptive layers in the ANFIS architecture. The modifiable parameters are called premise parameters in the first layer and consequent parameters in the fourth layer. The task of learning is to tune all modifiable parameters to make the ANFIS match the training data. In this paper, the trainable parameters of ANFIS, *i.e.*, premise parameters and consequent parameters $\{a_i, b_i, c_i\}$ and $\{p_i, q_i, r_i\}$ are adjusted to make the ANFIS output match the training data. The adopted hybrid learning technique combines the least square method and gradient descent method in ANFIS Toolbox. The hybrid learning algorithm is composed of a forward pass and a backward pass. The least squares method (forward pass) is used to optimize the consequent parameters with the premise parameters fixed. Once the optimal consequent parameters are found, the backward pass starts immediately. The gradient descent method (backward pass) is used to adjust optimally the premise parameters corresponding to the fuzzy sets in the input domain. The output of the ANFIS is calculated by employing the consequent parameters found in the forward pass. The output error is used to adapt the premise parameters by means of a standard back propagation algorithm. It has been proven that this hybrid algorithm is highly efficient in training the ANFIS.

4. Design of the Autonomous Ground Vehicle's ANFIS Controller

In this section, the proposed ANFIS controllers are discussed in detail. There are four ANFIS controllers which have been designed to accomplish the navigation task; firstly, two ANFIS controllers for achieving the target reaching task, secondly, the other two ANFIS controllers for performing the obstacle avoidance mission.

All four controllers will be combined through a switch block for choosing which controller will be activated. For instance, if there are no obstacles in the vehicle's path, the target reaching controller will be activated. Otherwise, if the vehicle senses an obstacle, the obstacle avoidance controller will be activated. The switching between these two controllers will be decided according to an obstacle sensing signal, OS, from an environment model. This signal is generated in accordance to measured distances (front, right, and left) from sensory information. If the vehicle does not sense an obstacle in its path, this OS parameter will indicate '0' if there is no an obstacle and '1' if the vehicle senses an obstacle near to its platform. Thus, the output of the switching block will be the angular velocities of the left and right wheel of the autonomous ground vehicle. These velocities are fed into the vehicle model in order to obtain the instantaneous vehicle's posture through the movement of the vehicle. Figure 3 depicts the structure of the proposed ANFIS controllers based navigation systems with the autonomous ground vehicle and the implemented workspace.

Figure 3. The structure of the proposed ANFIS and the autonomous ground vehicle.

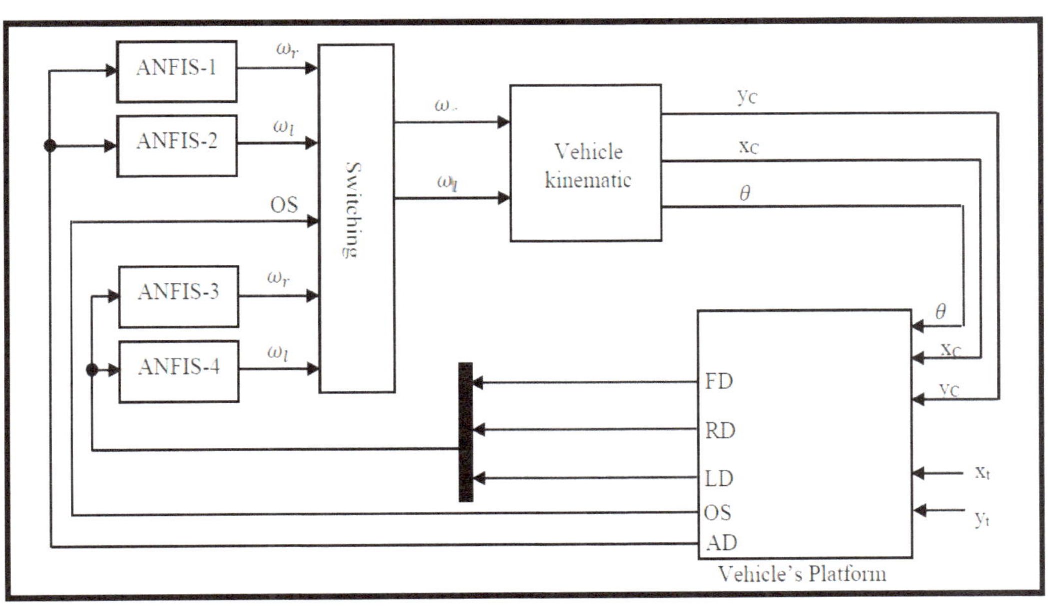

4.1. Autonomous Ground Vehicle Platform

The autonomous ground vehicle utilized in this paper is equipped with four wheels. The right and left back wheels drive independently. Two encoders are fixed on the back axis for the left and right wheel in order to measure the instantaneous position of the vehicle. The platform is created by using the Matlab-Simulink software package. This software has a vast diversity of tools that used for technical computing and interface graphical implementations. The implementation of this platform is illustrated in Figure 4. As show in the figure the vehicle has three degrees of freedom that are represented by the vehicle's posture (x_c, y_c, θ) in the global coordinate system. The s-function is a Matlab Simulink block was used to create the platform. As illustrated in the Figure 4, the Simulink block receives six inputs and produces five outputs. The platform parameters are explained below:

Remark 1. The front, right and left distances (FD, RD, and LD) represent the shortest distance between the vehicle and obstacles in the front, right and left directions respectively. The sensory information is modelled by assuming these three sensors are placed on a vehicle's platform, and each senor carries the information for three directions of the platform. These sensor outputs change depending on the distance between the instantaneous positions of the vehicle.

Remark 2. The angle difference (AD) represents the difference between the vehicle's heading and the target point.

Remark 3. Obstacle sensing (OS) signal is generated in accordance to the measured distances (front, right, left) from the sensory information. If the vehicle does not sense an obstacle in its path, this OS parameter will indicate '0' if there is no an obstacle, and '1' if the vehicle senses an obstacle near to its platform.

Remark 4. The clock timer is used for measuring the simulation running time that the vehicle elapsed to reach the destination in the platform.

Figure 4. The Schematic diagram for the autonomous ground vehicle platform.

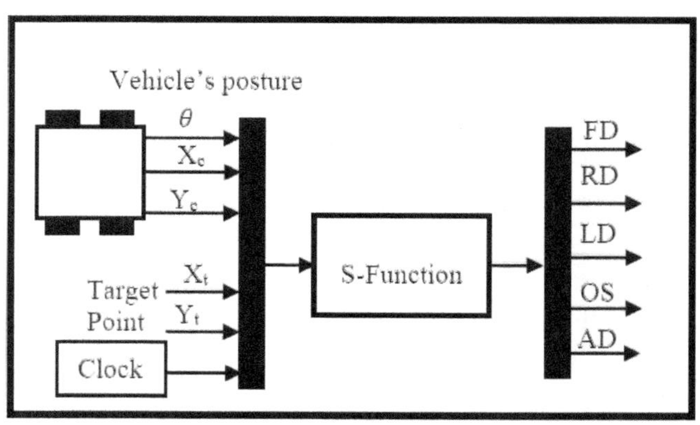

4.2. Target Reaching ANFIS Controller

To achieve the target reaching mission, two ANFIS controllers are implemented to drive the right and left angular velocities. The input of this controller is the angle difference, AD. This angle represents the difference between the vehicle's heading and the target's point. The calculation of this angle is depicted in Figure 5. The output of the first controller is the right angular velocity and for the second controller is the left angular velocity. In this study, a total of 40 sets of data range were selected for implemented the target-reaching controller for the purpose of training the ANFIS. The training data for input and outputs ranges are given in Table 1. This training adjusts the membership parameters to implement the required model. The ANFIS learning information for predicting the angular velocity for the left and the right wheels are as follows: number of nodes, 52; number of linear parameters, 12; number of nonlinear parameters, 36; total number of parameters, 48; number of fuzzy rules, 12. Training results shows that the average error for predicting angular velocity is 0.15631 rad/s with an epoch number of 200. The epoch value is chosen after a number of iteration for obtaining the minimum value of error between the input and local set points. The relation between the average error and epoch number is given in Figure 6. Figure 7a,b shows the output training data for obstacle avoidance ANFIS initially before the training and after the training process is completed.

Figure 5. Coordinates for moving to a target point.

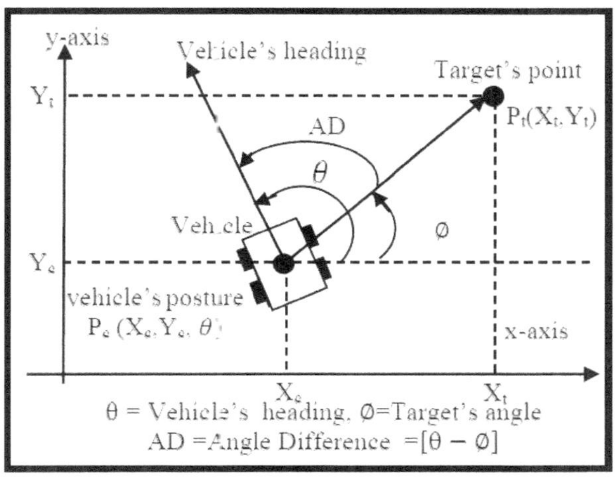

Table 1. Training data of target reaching ANFIS controller.

Item Number	Angle Difference	Right Angular Velocity	Left Angular Velocity
1	0	80	80
2	−44.007	40.882	−37.352
3	−29.214	54.031	2.093
4	−14.226	67.353	42.061
5	−4.319	76.160	68.480
6	0.505	78.652	79.550
7	1.975	74.731	78.243
8	1.809	75.173	78.391
9	0.530	78.584	79.528
10	−0.009	79.991	79.974
11	4.4693	68.081	76.027
12	13.668	43.550	67.850
13	25.192	12.819	57.606
14	26.956	8.116	56.038

Table 1. *Cont.*

Item Number	Angle Difference	Right Angular Velocity	Left Angular Velocity
15	27.114	7.694	55.898
16	18.5223	30.607	63.535
17	9.047	55.873	71.957
18	2.774	72.602	77.534
19	−0.295	79.737	79.211
20	6.759	61.975	73.991
21	15.705	38.117	66.039
22	29.555	1.186	53.728
23	43.460	−35.895	41.368
25	55.161	−40	40
26	62.435	−40	40
27	14.562	41.167	67.055
28	4.477	68.058	76.019
29	−5.788	74.854	64.563
30	−13.107	68.348	45.046
31	−20.435	61.835	25.505
32	−27.755	55.328	5.985
33	−85.460	40	−40
34	−37.315	46.830	−19.507
35	−22.376	60.110	20.330
36	−22.843	59.694	19.082
37	−15.563	66.165	38.496
38	−7.999	72.888	58.66
39	−2.812	77.500	72.500
40	0.2973	79.207	79.735

Figure 6. Relationship between training error and epoch's number for the target-reaching ANFIS controller.

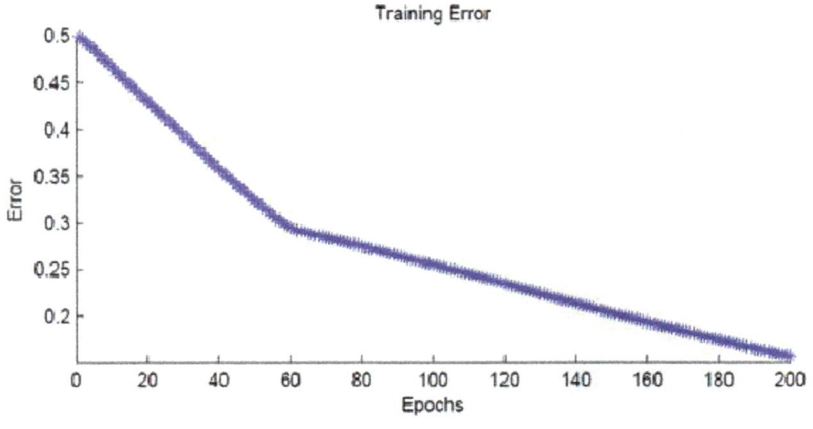

Figure 7. The output data for target reaching ANFIS controller. (**a**) The initial data before the training (**b**). The matched data after the ANFIS is trained.

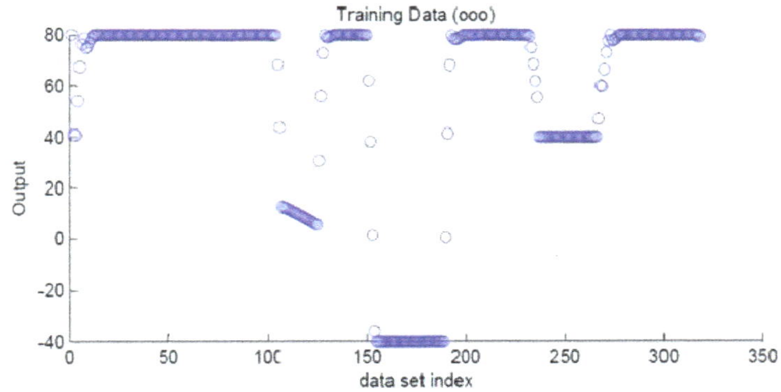

(**a**)

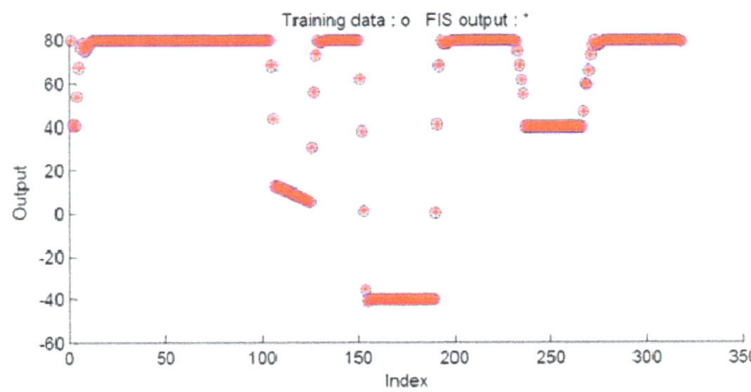

(**b**)

The number of membership functions was chosen for the angle difference input is ten. This number gives a reasonable amount of error as shown in Figure 6. The type of membership function is triangular. The relationship of surface view between angle difference and the output for the first and second ANFIS controllers is illustrated in Figure 8a, b respectively.

Figure 8. The surface view for target reaching ANFIS controller. (**a**) The first ANFIS controller. (**b**) The second ANFIS controller.

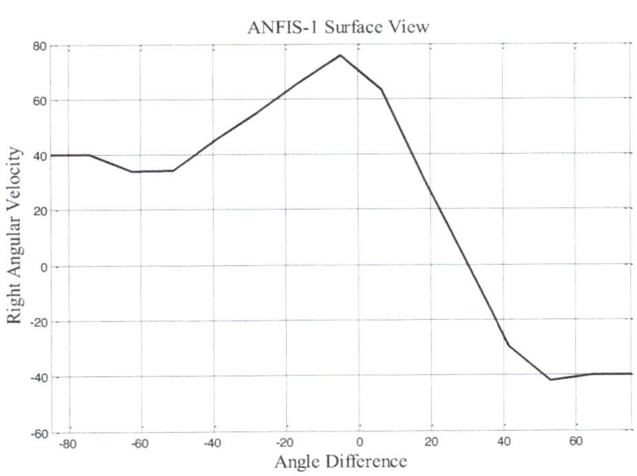

(**a**)

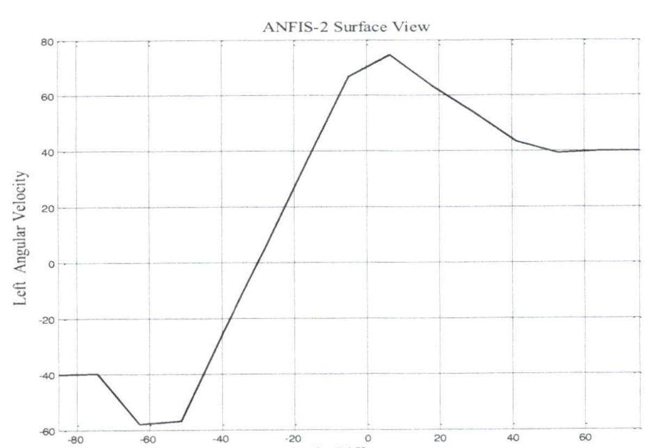

(**b**)

4.3. Obstacle Avoidance ANFIS Controller

The essential mission of the autonomous ground vehicle is to have a collision free path during the navigation. The other two ANFIS controllers have been implemented to comprise the third and fourth ANFIS controllers. These controllers are utilised to steer the vehicle's orientation when the vehicle be in the vicinity an obstacle. Each ANFIS has three inputs which are the front, right and left distance and one output which is either the right or left angular velocity. The received sensory information is depicted in Figure 9 which represents the left, right and front distances as inputs for each ANFIS. The output of the third and fourth ANFIS controller is right angular velocity (w_l) and left angular velocity (w_r) respectively. Each controller regulates the vehicle's velocities (left or right) depending upon the distances between the obstacles and the current posture of the autonomous ground vehicle.

Figure 9. Schematic diagram for the sensory information.

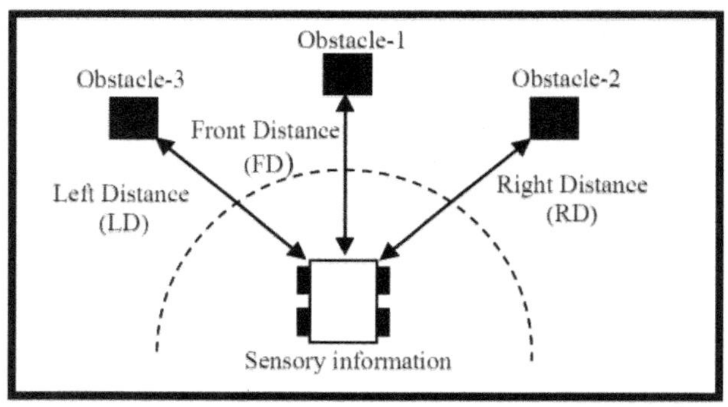

For implementing the obstacle avoidance ANFIS controller, the 21 sets of data range were selected for the training purpose. These training data ranges for inputs and outputs are given in Table 2. This training as mentioned previously adjusts the membership parameters to implement the required model. The ANFIS learning information for predicting the angular velocity for the left and the right wheels are as follows: number of nodes, 286; number of linear parameters, 125; number of nonlinear parameters, 45; total number of parameters, 170; number of fuzzy rules, 125. Training result shows that the average error for predicting angular velocity is 0.329231 rad/s with an epoch number of 200. Again, the relationship between the average error and epoch number is given in Figure 10. Figure 11a, b shows the output training data for obstacle avoidance ANFIS initially before the training and after the training process is completed.

Table 2. Training data of obstacle avoidance ANFIS controller.

Item Number	Front Distance	Right Distance	Left Distance	Right Angular Velocity	Left Angular Velocity
1	0.100	0.100	0.100	−40	−40
2	0.800	0.800	0.800	80	80
3	0.800	0.399	0.800	79.99	79.999
4	0.800	0.339	0.800	67.999	43.999
5	0.800	0.279	0.800	55.999	7.999
6	0.800	0.249	0.800	49.99	−10.006
7	0.547	0.800	0.800	80	80

Table 2. *Cont.*

Item Number	Front Distance	Right Distance	Left Distance	Right Angular Velocity	Left Angular Velocity
8	0.4873	0.309	0.800	61.997	25.992
9	0.397	0.279	0.800	54.922	8.351
10	0.4269	0.219	0.800	43.997	−28.008
11	0.515	0.219	0.800	43.967	−28.098
12	0.800	0.249	0.800	49.853	−10.440
13	0.800	0.308	0.800	61.644	24.932
14	0.800	0.800	0.392	75.580	78.526
15	0.800	0.800	0.333	40.013	66.671
16	0.786	0.800	0.303	22.230	60.743
17	0.800	0.800	0.304	22.418	60.806
18	0.800	0.800	0.304	22.803	60.934
19	0.800	0.800	0.335	41.084	67.028
20	0.800	0.800	0.395	77.330	79.110
21	0.800	0.800	0.455	80	80

Figure 10. Relationship between training error and epoch's number for obstacle avoidance ANFIS.

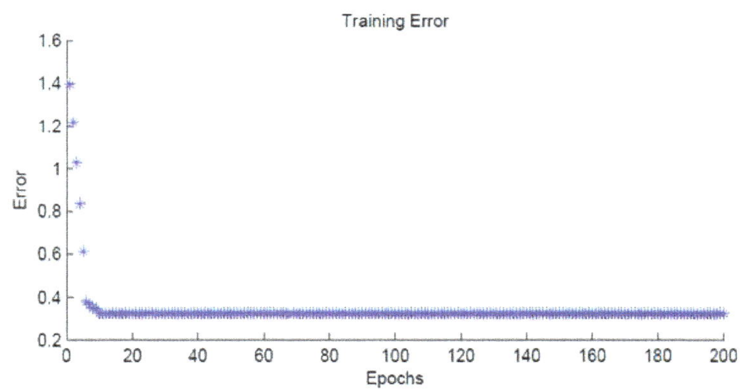

Figure 11. The output data for obstacle avoidance ANFIS. (**a**) The initial data before the training. (**b**) The matched data after the ANFIS is trained.

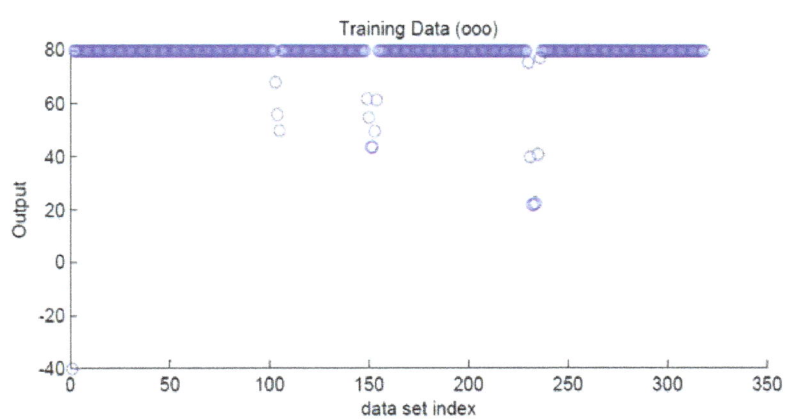

(a)

Figure 11. *Cont.*

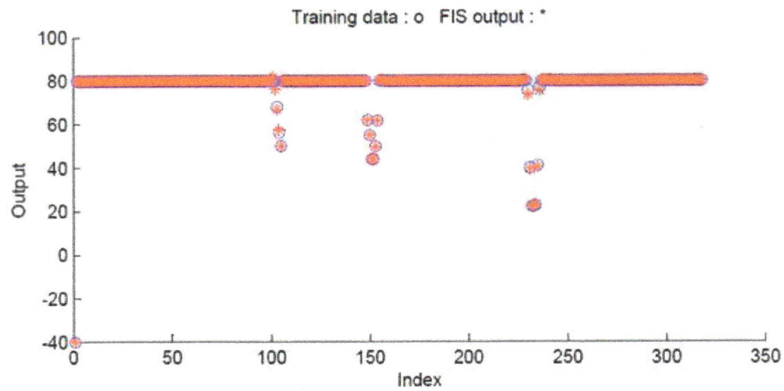

(b)

For obstacle avoidance ANFIS controller, the number of membership functions that was chosen for each input is five. Thus, the amount of error was reasonable with this number of membership functions. The type of the membership function is triangular. The surface view between the inputs and the output for the third ANFIS controllers is illustrated in Figure 12.

Figure 12. The surface view for obstacle avoidance ANFIS controller. (**a**) The surface view between input-1, input-2 and the output. (**b**) The surface view between input-1, input-3 and the output. (**c**) The surface view between input-2, input-3 and the output.

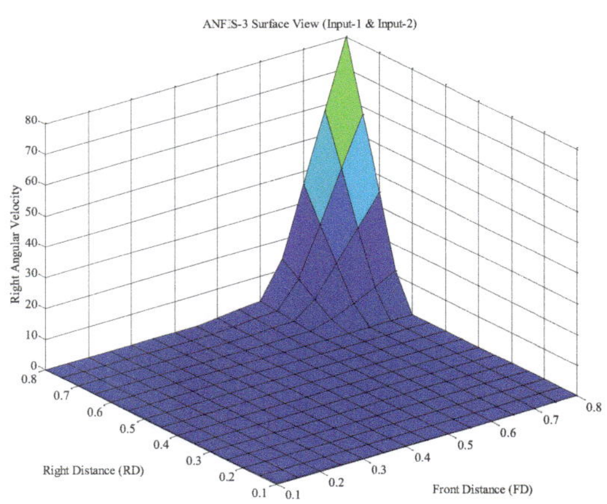

(a)

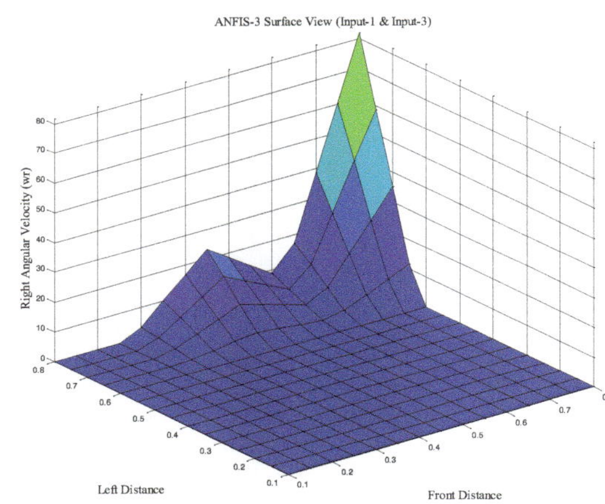

(b)

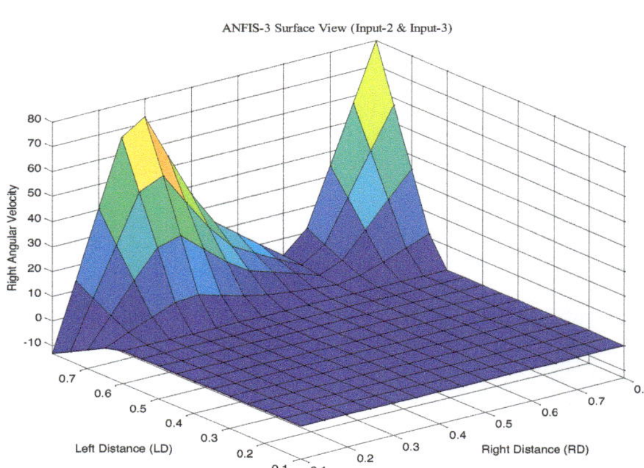

(c)

For the fourth obstacle avoidance ANFIS controller, the surface view of the parameters is described in Figure 13.

Figure 13. The surface view for obstacle avoidance ANFIS controller. (**a**) The surface view between input-1, input-2 and the output. (**b**) The surface view between input-1, input-3 and the output. (**c**) The surface view between input-2, input-3 and the output.

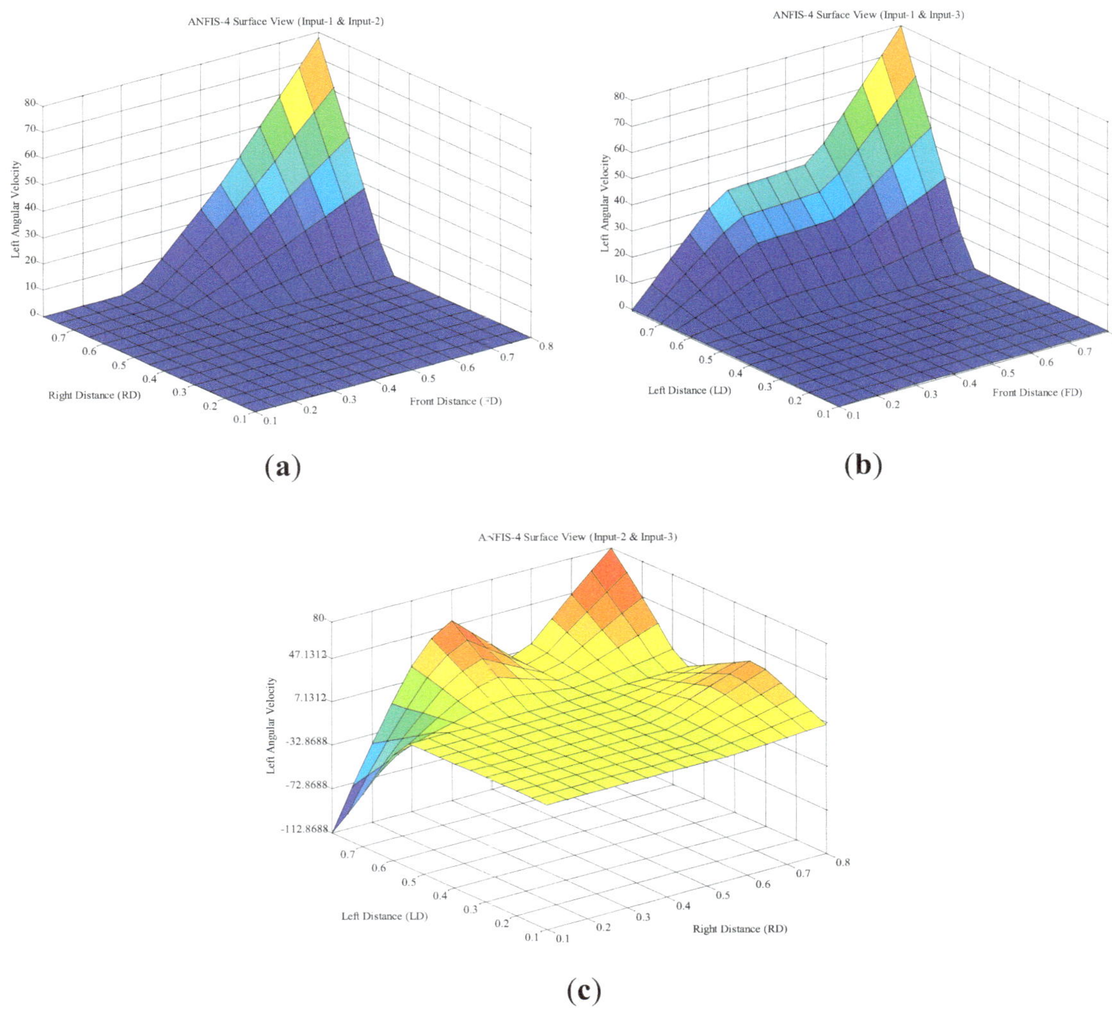

5. Simulation Results

To validate the proposed ANFIS controller, two scenarios have been adopted and carried out in this paper using the MATLAB-SIMULINK environment. In the first scenario, an artificial workspace has been created with seven identical static obstacles that were placed in different positions in the workspace. An initial position and the target point of an autonomous ground vehicle can be set arbitrarily in the workspace. Figure 14 illustrates the workspace for the autonomous vehicle platform filled with seven static obstacles. The workspace dimensions are fixed by four corner points having coordinates (–2, –2), (18, –2), (18, 18), (–2, 18) to combine the workspace grid. The dimensions of the obstacles described by their peripheral vertices are given in Table 3. The Cartesian coordinates of the initial and target points are P_s (0, 0) and P_t (15, 15) respectively.

By running the implemented SIMULINK model, the autonomous ground vehicle started moving from a start position towards to the destination position. As can be observed from the Figure 14, the vehicle avoided obstacles successfully and safely. The decision is made to correctly change the vehicle's direction when it approached the obstacles, and finally, the target is reached. The simulation results of angular velocity for both right and left wheels are shown in Figure 15. Figure 16 represents the angular position of the autonomous ground vehicle. The linear velocity profile is shown in Figure 17. These figures illustrate the changing in the position and velocity of the vehicle when it confronts an obstacle and during the vehicle's turning.

Table 3. Obstacles description on the workspace grid.

Obstacle No.	Peripheral Vertices Coordinates
1	(2, 2), (2, 3.5), (3.5, 3.5), (3.5, 2)
2	(8, 4), (8, 5.5), (9.5, 5.5), (9.5, 4)
3	(6, 7), (6, 8.5), (7.5, 8.5), (7.5, 7)
4	(10, 8), (10, 9.5), (11.5, 9.5), (11.5, 8)
5	(12, 12), (12, 13.5), (13.5, 13.5), (13.5, 12)
6	(3.5, 12.5), (3.5, 14), (5, 14), (5, 12.5)
7	(8, 14), (8, 15.5), (9.5, 15.5), (9.5, 14)

Figure 14. Navigation platform contains five identical static obstacles.

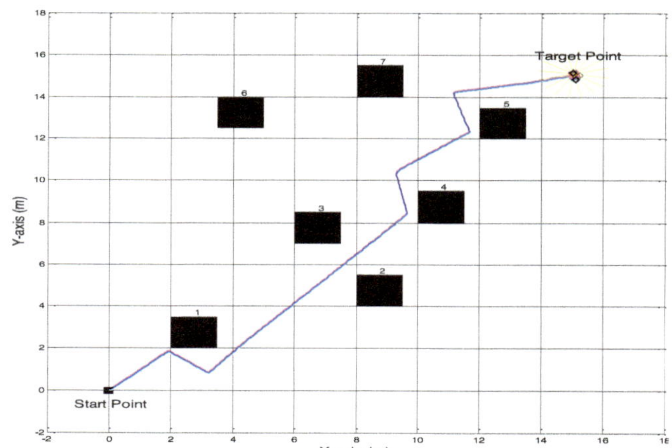

Figure 15. The angular velocity for both of the right and left wheel angular velocities.

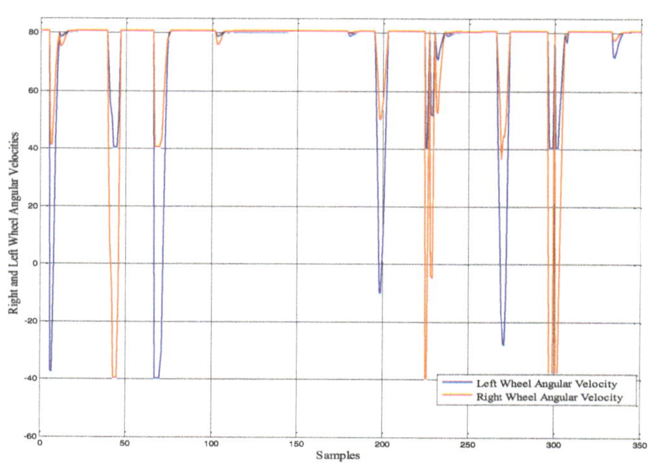

Figure 16. The angular position for autonomous ground vehicle.

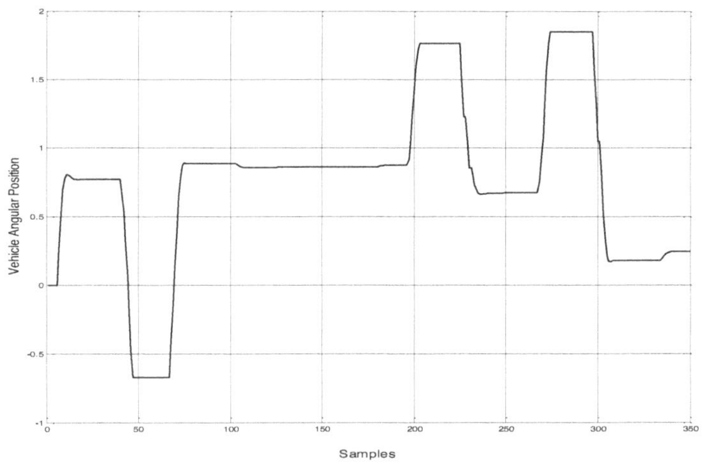

Figure 17. The linear velocity profile for autonomous ground vehicle.

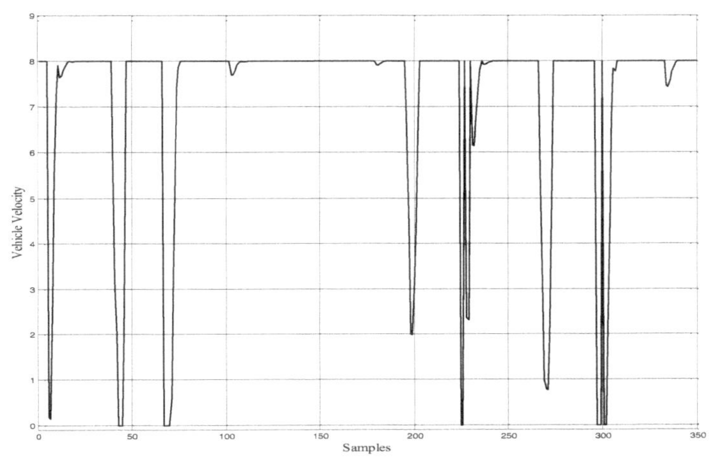

In a more realistic situation, environments might contain obstacles that have different shapes and sizes. In case of increasing the number of obstacles, the frequency of the switching between the obstacle avoidance ANFIS controller and the target reaching ANFIS controller will be increased most the time. This makes the behaviour of the whole system more challenging and the environment more complicated. The performance of the platform was tested in such a case to validate the response of the system in different situations.

In this second scenario, we considered the vehicle navigating in the case that the obstacles have different sizes and shapes. The dimensions of the obstacles are described by their peripheral vertices are given in Table 4. The number of obstacles was increased to eleven to make a more complex environment. The simulation results showed that the trained obstacle avoidance ANFIS controller made the vehicle traverses all the obstacles in the vehicle's route by making decisions to change the vehicle's headings when it approached an obstacle. Decisions are made upon the controllers' inputs to manipulate the left and right angular velocities. A new feasible trajectory was generated by the autonomous ground vehicle after roving around the obstacles as shown in Figure 18.

The simulation results of angular velocity for both of right and left wheels are shown in Figure 19. Figure 20 represents the angular position of the autonomous ground vehicle. The linear velocity profile is shown in Figure 21. These figures illustrate the changing in the position and velocity of the vehicle when it wanders around the obstacles.

Table 4. Obstacles description on the workspace grid.

Obstacle No.	Peripheral Vertices Coordinates	Shape Type
1	(3.5, 3.5), (5, 5), (2.5, 4), (4, 2.5)	Square
2	(8, 0), (9, 1.5), (10, 0)	Triangle
3	Centre (9, 4) and Radius = 0.75m	Circle
4	(12, 1), (12, 4), (14, 4), (14, 1)	Rectangle
5	(0, 6), (0, 8), (2, 8), (2, 6)	Square
6	(3.5, 3.5), (5, 5), (15.5, 4)	Triangle
7	(14.5, 8), (16, 8), (15.5, 6), (14, 6)	parallelogram
8	Centre (10, 10) and Radius = 1m	Circle
9	Centre (3, 13) and Radius = 0.75m	Circle
10	(12, 12), (12.5, 5), (13.5, 13), (14, 12)	Trapezoid
11	(8, 14), (8, 15.5), (9.5, 15.5), (9.5, 14)	Square

Figure 18. Navigation platform contains eleven static obstacles with different sizes and shapes.

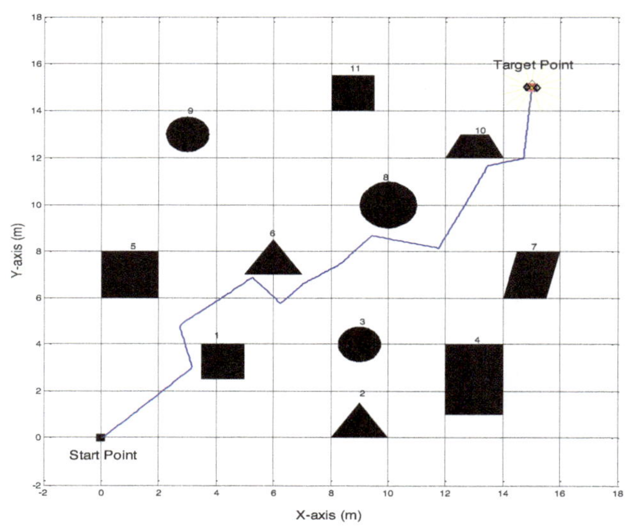

Figure 19. The angular velocity for both of the right and left wheel angular velocities.

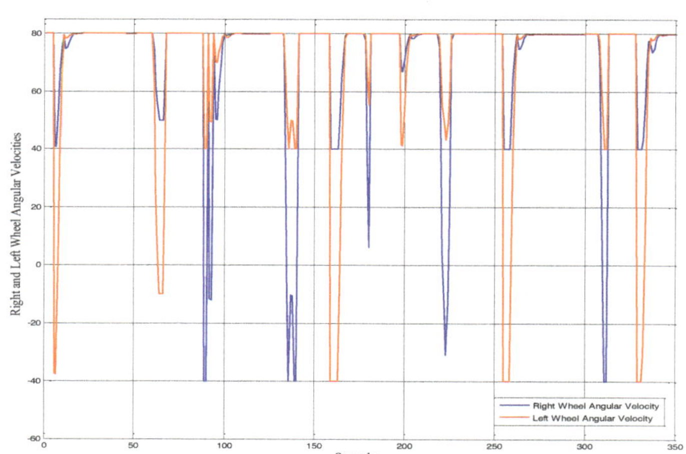

Figure 20. The angular position for autonomous ground vehicle.

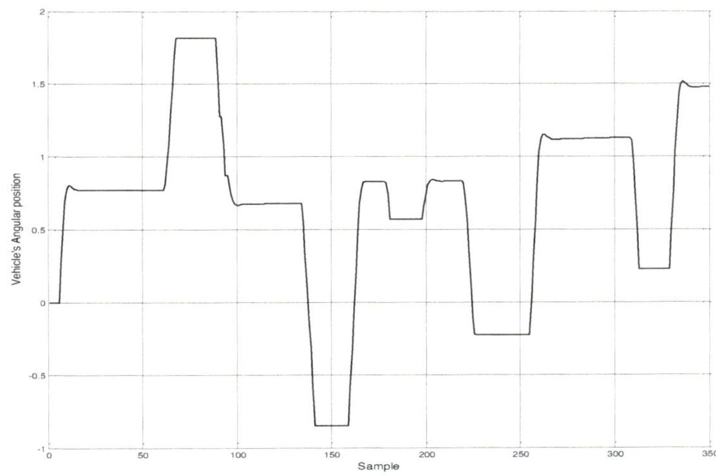

Figure 21. The linear velocity profile for autonomous ground vehicle.

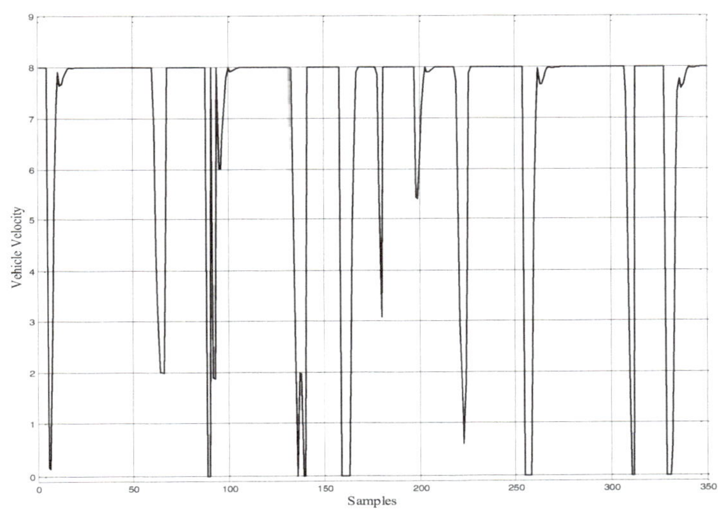

Figure 22. The Error rate between the target point and vehicle's actual point. (**a**) First scenario. (**b**) Second scenario.

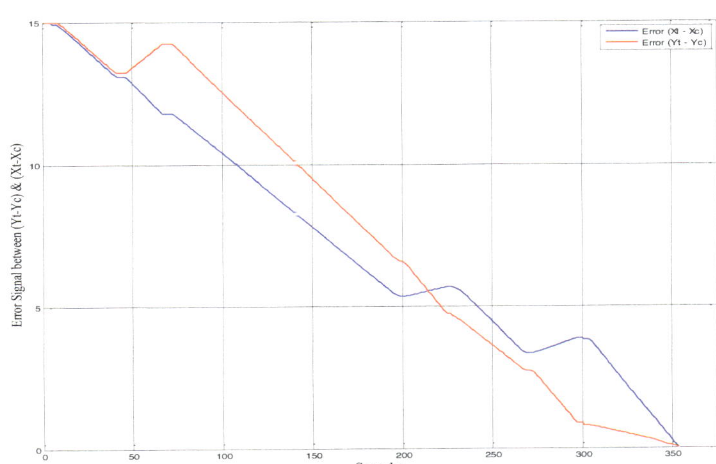

(**a**)

Figure 22. *Cont.*

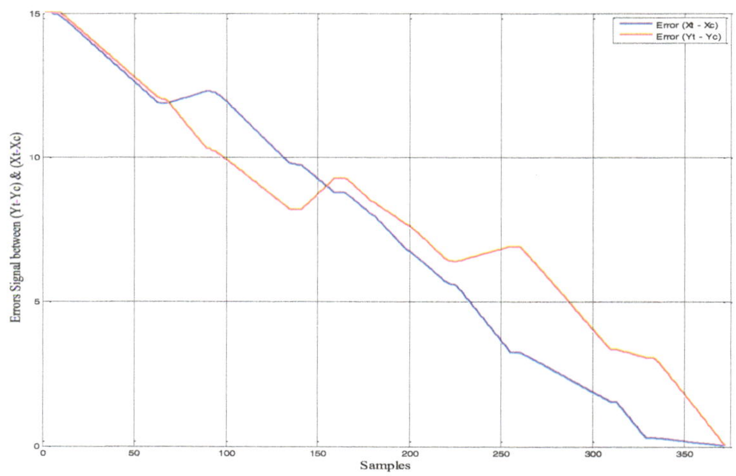

(b)

The error signals for both of x and y coordinates between the target point ($x_t = 15$, $y_t = 15$) and the vehicle's actual points during the navigation was obtained. These signals of the autonomous ground vehicle for the first and second scenarios are shown in Figure 22a,b respectively. The results show the behaviour of the vehicle as it reached its destination. Before the vehicle started moving the error rate was maximum and when the vehicle arrived at its target the error rate became zero.

6. Conclusion

In this paper, an adaptive neuro-fuzzy inference system has been implemented to control the angular velocities of the wheels of an autonomous ground vehicle. These velocities guide the vehicle safely to reach the destination in a static environment without colliding with obstacles presented on its way. The proposed system utilises four ANFIS controllers. Firstly, two controllers for the target reaching which to ensure that the vehicle reaches its destination point; Secondly, two controllers for guiding the vehicle to avoid the collision with obstacles. In these four controllers, the right and left angular velocities will derive the vehicle's heading during the navigation.

The validation of the proposed method has been demonstrated by considering two scenarios; firstly, seven identical static obstacles are placed randomly with the workspace; secondly, eleven obstacles have different sizes and shapes. In case of reducing obstacle numbers significantly, the obstacle avoidance ANFIS controller did not activate until the vehicle confronted an obstacle. Therefore, the target reaching ANFIS controller was active most of the time. This case will be similar in the situation if a path is free of obstacles so the vehicle will take straight line that connects between both start and target points.

In contrast, in multiple obstacles simulation, the obstacle avoidance ANFIS controller was activated frequently in order to avoid obstacles. After each avoiding point the vehicle switched to the target reaching ANFIS controller. In both considered scenarios, the autonomous ground vehicle was capable of avoiding obstacles safely and reaching the target with a feasible and smooth online generated path between the initial and the target points. The simulation experiments have been conducted using MATLAB-Simulink software package.

In the future, the proposed ANFIS control strategy will be implemented in a real AGV to further verify its feasibility and performance.

Acknowledgements

The authors would like to express their gratefulness for the sponsor, ministry of higher education and scientific research in Iraq for funding the research programme in the United Kingdom. The authors also would like to thank the University of Basrah, Iraq, for their support.

Author Contributions

Auday Al-Mayyahi made contribution in implementing a new ANFIS model which has the capability of avoiding stuck into the local minimum and generates a smooth trajectory between the starting and the target points in unknown environment based on the local planning method. He also has demonstrated the ANFIS model can efficiently guide the vehicle through the workspace. Moreover, the performance of the ANFIS controller is verified by changing the poses of obstacles. William Wang made contributions in advising the methodology and the control strategy, also made corrections to the manuscript. Phil Birch revised the manuscript widely and made the reading proof.

Conflicts of Interest

The authors declare no conflict of interest.

References

1. Nonami, K.; Kartidjo, M.; Yoon, K.-J.; Budiyono, A. *Autonomous Control Systems and Vehicles: Intelligent Unmanned Systems*; Springer Tokyo Heidelberg: Tokyo, Japan, 2013; p. 306.
2. Sedighi, K.H.; Ashenayi, K.; Manikas, T.W.; Wainwright, R.L. Autonomous Local Path Planning for a Mobile Robot Using a Genetic Algorithm. In Proceedings of the IEEE Congress on Evolutionary Computation, Portland Marriott Downtown, Portland, OR, USA, 19–23 June 2004; pp. 1338–1345.
3. Wang, M.; Liu, J.N.K. Fuzzy Logic Based Robot Path Planning in Unknown Environment. In Proceedings of the IEEE International Conference on Machine Learning and Cybernetics, Guangzhou, China, 18–21 August 2005; Volume 1, pp. 813–818.
4. De Berg, O.S.M.; van Kreveld, M.; Overmars, M. *Computaional Geometry: Algorithms and Applications*; Germany: Springer-Verlag: Berlin, Germany, 2000.
5. Garrido, S.; Moreno, L.; Blanco, D.; Jurewicz, P. Path planning for mobile robot navigation using voronoi diagram and fast marching. *Int. J. Robot. Autom.* **2011**, *2*, 42–64.
6. L-Taharwa, I.A.; Sheta, A.; Al-Weshah, M. A mobile robot path planning using genetic algorithm in static environment. *J. Comput. Sci.* **2008**, *4*, 341–344.
7. Samadi, M.; Othman, M.F. Global Path Planning for Autonomous Mobile Robot Using Genetic Algorithm. In Proceedings of the IEEE International Conference on Signal-Image Technology & Internet-Based Systems, Kyoto, Japan, 2–5 December 2013; pp. 726–730.

8. Cui, S.; Su, X.; Zhao, L.; Bing, Z.; Yang, G. Study on Ultrasonic Obstacle Avoidance of Mobile Robot Based on Fuzzy Controller. In Proceedings of the IEEE International Conference on Computer Application and System Modeling, TaiYuan, China, 22–24 October 2010, pp. 233–237.
9. Pradhan, S.K.; Parhi, D.R.; Panda, A.K. Fuzzy logic techniques for navigation of several mobile robots. *Appl. Soft Comput.* **2009**, *9*, 290–304.
10. Li, X.; Choi, B. Design of Obstacle Avoidance System for Mobile Robot Using Fuzzy Logic Systems. *Int. J. Smart Home* **2013**, *7*, 321–328.
11. Hajar, S.; Mohammad, A.; Jeffril, M.A.; Sariff, N. Mobile Robot Obstacle Avoidance by Using Fuzzy Logic Technique. In Proceedings of the 3rd IEEE International Conference on System Engineering and Technology, Shah Alam, Kuala Lumpur, 19–20 August 2013; pp. 331–335.
12. Chi, K.-H.; Lee, M.-F.R. Obstacle Avoidance in Mobile Robot Using Neural Network. In Proceedings of the IEEE International Conference on Consumer Electronics, Communications and Networks, XianNing, China, 16–18 April 2011; pp. 5082–5085.
13. Deshpande, S.U.; Bhosale, S.S. Adaptive neuro-fuzzy inference system based robotic navigation. In Proceedings of the IEEE International Conference on Computational Intelligence and Computing Research, Enathi, India, 26–28 December 2013; pp. 1–4.
14. Joshi, M.M.; Zaveri, M.A. Neuro-Fuzzy Based Autonomous Mobile Robot Navigation System. In Proceedings of the 11th International Conference Control, Automation, Robotics and Vision, Singapore, 7–10 December 2010; pp. 384–389.
15. Singh, M.K.; Parhi, D.R.; Pothal, J.K. ANFIS Approach for Navigation of Mobile Robots. In Proceedings of the IEEE International Conference on Advances in Recent Technologies in Communication and Computing, Kottayam, Kerala, India, 27–28 October **2009**; pp. 727–731.
16. Khelchandra, T.; Huang, J.; Debnath, S. Path planning of mobile robot with neuro-genetic-fuzzy technique in static environment. *Int. J. Hybrid Intell. Syst.* **2014**, *11*, 71–80.
17. Fierro, R.; Lewis, F.L. Control of a nonholonomic mobile robot using neural networks. *IEEE Trans. Neural Networks* **1998**, *9*, 589–600.
18. Jang, J.R. ANFIS: Adaptive network based fuzzy inference system. *IEEE Trans. Syst. Man Cybern.* **1993**, *23*, 665–685.

10

Rotation Matrix to Operate a Robot Manipulator for 2D Analog Tracking Objects Using Electrooculography

Muhammad Ilhamdi Rusydi [1,2], Takeo Okamoto [1], Satoshi Ito [1] and Minoru Sasaki [1,*]

[1] Department of Mechanical Engineering, Gifu University, 1-1 Yanagido, Gifu City 501-1193, Japan; E-Mails: q3812204@edu.gifu-u.ac.jp (M I.R.); okatake0331@gmail.com (T.O.); satoshi@gifu-u.ac.jp (S.I.)

[2] Department of Electrical Engineering, Andalas University, Limau Manis, Padang City 25163, Indonesia

* Author to whom correspondence should be addressed; E-Mail: sasaki@gifu-u.ac.jp

Abstract: Performing some special tasks using electrooculography (EOG) in daily activities is being developed in various areas. In this paper, simple rotation matrixes were introduced to help the operator move a 2-DoF planar robot manipulator. The EOG sensor, NF 5201, has two output channels (Ch1 and Ch2), as well as one ground channel and one reference channel. The robot movement was the indicator that this system could follow gaze motion based on EOG. Operators gazed into five training target points each in the horizontal and vertical line as the preliminary experiments, which were based on directions, distances and the areas of gaze motions. This was done to get the relationships between EOG and gaze motion distance for four directions, which were up, down, right and left. The maximum angle for the horizontal was 46°, while it was 38° for the vertical. Rotation matrixes for the horizontal and vertical signals were combined, so as to diagonally track objects. To verify, the errors between actual and desired target positions were calculated using the Euclidian distance. This test section had 20 random target points. The result indicated that this system could track an object with average angle errors of 3.31° in the x-axis and 3.58° in the y-axis.

Keywords: gaze motions; directions; analog; distance; rotation matrixes

1. Introduction

Nowadays, in daily activity, people are used to working with computers or machines to improve their quality of work. However, for some paralysis patients, armless persons [1], older adults [2] and locked-in syndrome patients, they face some barriers in controlling devices, because of their limitations. To solve this problem, people have tried to discover alternative ways to transmit and receive signals between humans and machines [3].

One of the methods to handle this problem is through bio-signal technology for communication. There are many kinds of bio-signals, such as electromyography (EMG) [4], electroencephalographic (EEG) [5] and electrooculography (EOG) [6]. This study used the EOG signal to connect human and machine.

EOG is recorded as a weak electric potential existing around the eye. This signal comes from eye activities, caused by the difference of the potential between cornea and retina, since cornea has a more positive polarity than retina [7]. The advantage of EOG is the linearity between its potential and the gaze distance. Nevertheless, the range of linear area in the horizontal (±45°) was usually wider than vertical linearity (±30°) [8]. Three basic parts in the human-machine interface using EOG were EOG amplifying and acquisition, EOG pattern recognition and the control command. Beside the linearity, compared to EEG, EOG was easy to detect, and the amplitude is relatively high [9].

Many methods were generated by researchers to detect and recognize eye movement based on the EOG signal to control some applications. Carlos G Pinhero *et al.* in his review [3] provided some examples of EOG function, such as an alphanumeric/symbol/number writing system, cursor control and generating Morse code. Controlling robot movement using EOG had been reported by [10].

Human and computer interface systems accessed by EOG were proposed to control functions in a monitor. In [11], EOG was classified by a fuzzy algorithm into four basic directions (right, left, up and down). Operators could use this system to activate some functions, such as a game, TV, an eye test, and others. Although the condition of the operators also influenced the EOG signal, such as nervousness, blinking, exaggerated gestures and speech, this research verified that the horizontal movement had a wider range than vertical movement.

While, in [12,13], an artificial neural network using training samples had the ability to learn a nonlinear characteristic and to build a model of it, this advantage was used to develop a prosthetic eye moving with a natural action [14]. A neural network was also used to classify EOG from people with good or bad health conditions [15].

The threshold method also effectively distinguished EOG based on gaze motion distances. This method, reported by [16,17], successfully divided gaze motion into four thresholds, each in vertical and horizontal movement. Then, these were all associated with the velocities of the robot manipulator for each direction (right, left, up and down). The robot velocity was controlled by EOG with four different levels of speed using the four thresholds. The operator would blink to mark targets if the robot passed them.

EOG signals were easily interrupted by some interferences in real circumstance, and the waveforms belonging to the different trials and subjects were also changed. Many studies tried to solve this problem; one of them [18] used the combination of spectral entropy, the LPCC coefficient and dynamic time warping (DTW). This method was also used to recognize seven actions of gaze motion (looking up, down, left and right; blinking twice, three times and four times).

Using EOG for assistive technology was also proposed. In [19], a wheelchair was successfully controlled by EOG. In the linear region, the EOG potential varied between 50 and 3500 μV. Operators could use four commands to control the robot. The commands were to move forward, backward, turn right and turn left. In a similar function [1], EOG was also useful in assisting the discrimination between three different distances between operators and objects. The operators also found that there was no difference between the signal when the operators sat down or stood up. In another study [2], an eye-gaze input system became the favorite for adults compared to young people. They felt that the EOG signal was easier to generate than moving their hands to operate a computer, due to declined motor functions.

The EOG signal used in this study was not only developed to control a two-degrees of freedom (DoF) planar robot manipulator's direction, but also its position. Operators looked at targets on a monitor; then, this robot moved to points in the real world belonging to the EOG potential. Therefore, there were conversions from gaze motion to the pixel coordinate, followed by the conversion from the pixel coordinate to the joint angle.

This research has developed a communication interface between human and robot using the EOG signal. Rotation matrixes are proposed in this research to improve the performance of the gaze motion tracking system using EOG in diagonal movements. The EOG signal controlled the analog movements of a robot manipulator in a two-dimensional area. This signal generated by eye movements was converted to the pixel coordinates on a monitor display. The gaze distance of the pixel was used to calculate the target positions. The positions in the pixel are converted to the robot coordinates to calculate the end-effector position of the robot. Using inverse kinematics, the angles of two joints were determined.

2. Electrooculography

This study combines a bio-signal and a control system to move a flexible robot manipulator by using gaze motion. Therefore, in its simplest form, the input of the system is a bio-signal and the output is an end-effector movement. The bio-signal sensor was produced by NF Instrument and has an amplifier, a processor and four electrodes, as shown in Figure 1. These electrodes consist of two positive channels (Ch1 and Ch2), a ground electrode and a reference electrode or minus channel.

In previous research [20], the processor had embedded filters inside. The filters were a 60-Hz low-pass filter, a 1.6-Hz high-pass filter and 60-Hz hum filter. Figure 2 shows the signals before and after being filtered. The characteristics to differentiate the up gaze motion and blink signal were also introduced. The ratio between the maximum and absolute minimum value of EOG from Ch1 was used to ignore blink signals. This ratio for the blink signal is always greater than one. On the other hand, the ratio for the up gaze motion is smaller than one. Figure 3 shows the signals for the blinking and up gaze motion.

Figure 1. The NF Instrument as the electrooculography (EOG) sensor.

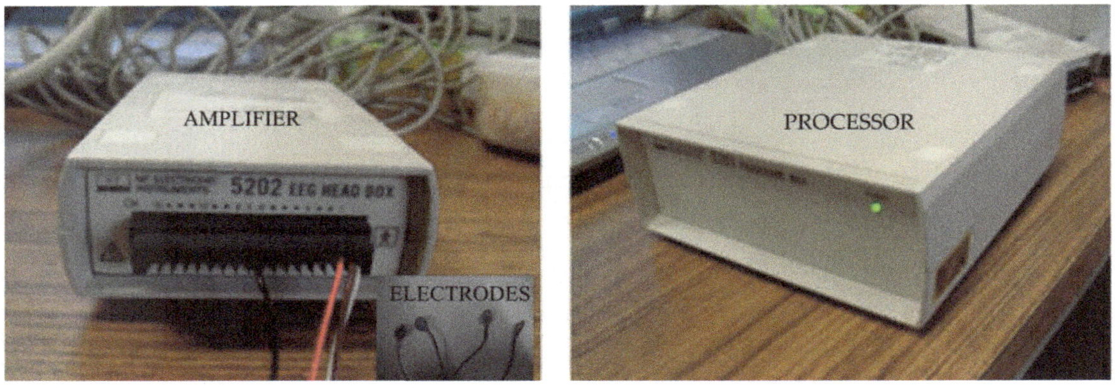

Figure 2. Example of the EOG signal: (**a**) before being filtered; (**b**) after being filtered.

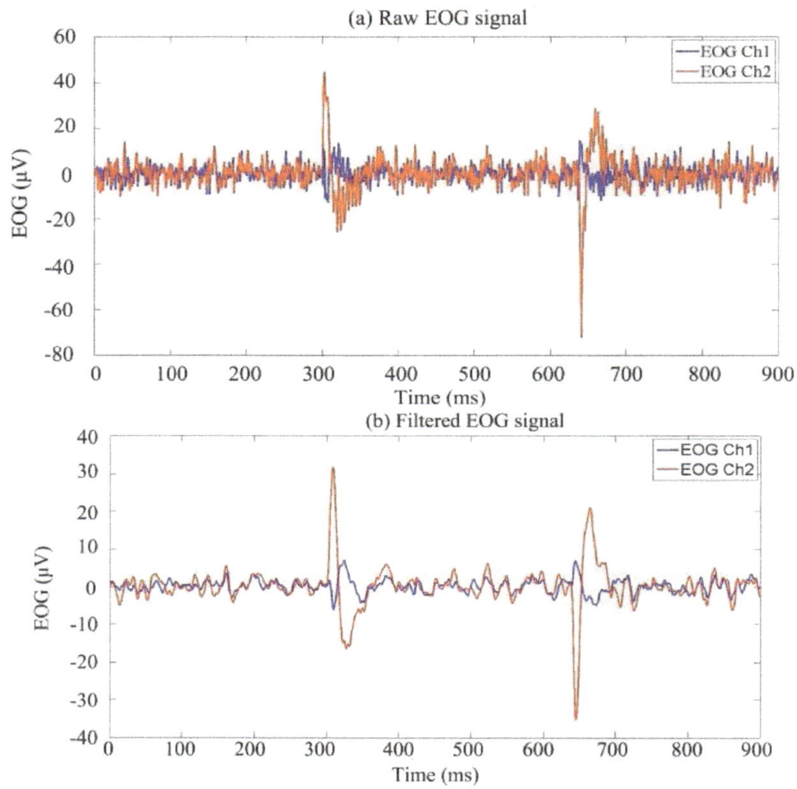

Figure 3. The EOG signal for the blinking and up gaze motion.

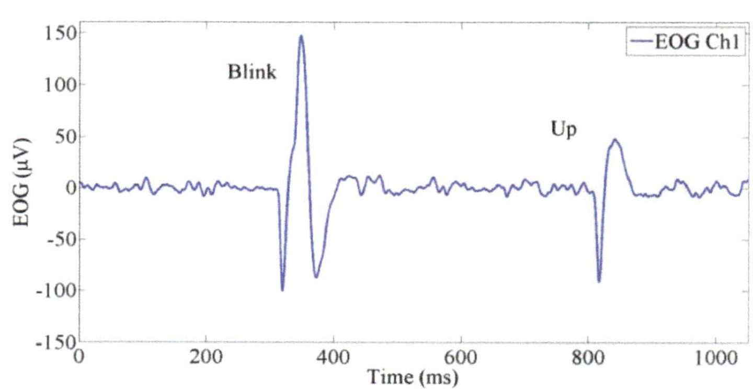

EOG is a weak signal that naturally occurs because of the potential difference between the cornea and retina. This distinction arises due to eye activities, as well as gaze motions [21] and blinks [22]. The EOG signal can be used in a dark environment, as long as there is eye activity distinguishing the electric potential between the two poles.

EOG is also a susceptible signal that is easily influenced by certain factors. Electrode positions are very important in this case, as well as the operator condition and the environment in which the research is being carried out [20,23]. Furthermore, unnecessary facial movements have impacts on the EOG value and pattern [24]. Figure 4 illustrates the polarity of human eyes and the positions of electrodes.

There are two patterns of EOG signals from the NF Instrument. These are a negative signal (−) and a positive signal (+), as shown in Figure 5. Two thresholds were used to detect when a gaze motion occurred: a negative signal passed the negative threshold (th −) and a positive signal passed the positive threshold (th +). It was simple to detect gaze motions in four basic directions (right, left, up and down) using the signal pattern combinations from Ch1 and Ch2. As shown in Table 1, there are four signal combinations from Ch1 and Ch2 to detect four directions. One is a left gaze motion if Ch1 has a negative signal and Ch2 has a positive signal. A contrary condition happens with the right gaze motion. In the same way, the up gaze motion and down gaze motion can be distinguished. These combinations will be explored later to detect the target points.

Figure 4. Polarity of eye and the electrode positions.

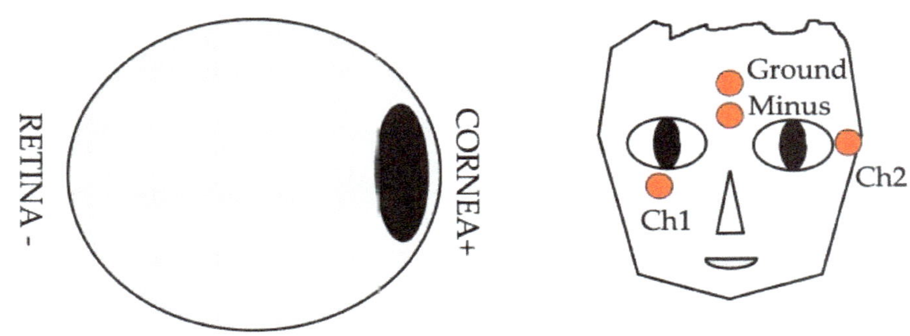

Figure 5. Two patterns of EOG signal using the NF Instrument.

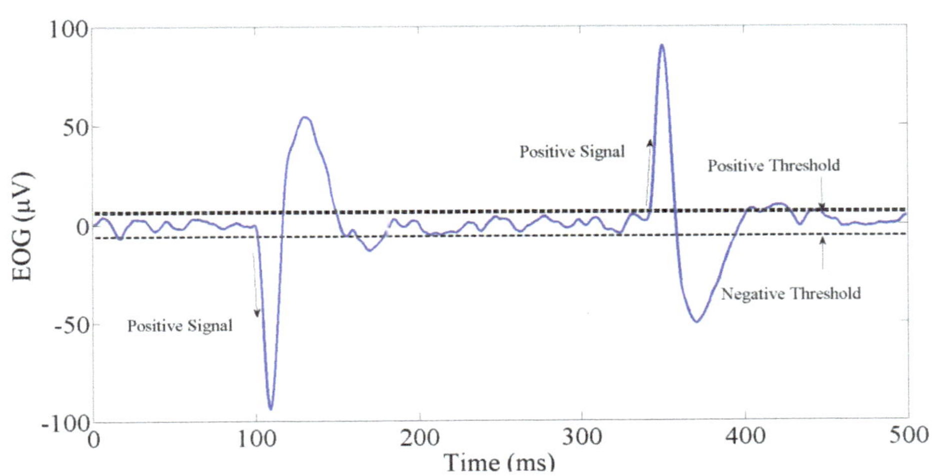

Table 1. Signal combinations for basic directions.

Channel	Gaze Direction			
	Left	Right	Up	Down
Ch1	(−)	(+)	(−)	(+)
Ch2	(+)	(−)	(−)	(+)

3. Design of a Two-DoF Robot Manipulator

Operators drove the robot end-effector to some positions that they desired using gaze motion. Therefore, the working area of the robot was designed to cover the target area. To assess the accuracy of the system, some targets were spread on a monitor display. The target positions were called actual positions, and the positions generated from the EOG were called desired positions. The size of the monitor was 34 cm (horizontal) × 27 cm (vertical) or 1020 pixels × 720 pixels. Figure 6 shows the location of the target area based on the robot workspace.

Figure 6. Target area and robot workspace.

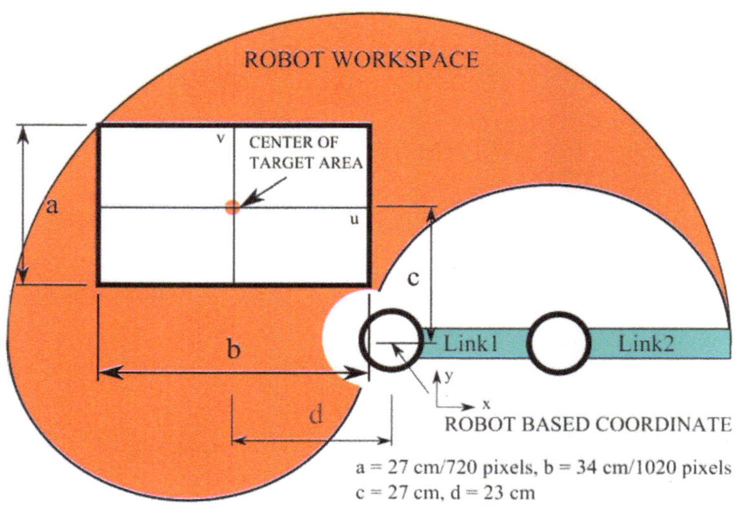

A two-DoF planar robot manipulator was already designed to cover this area. This robot had two joints and two links. The length of each links was 30 cm. The first joint had a 0° to 180° rotation angle and the second joint a rotation angle of 0° to 120°. Considering the torque of the end effector, the HSR-5980SG motor, which had 24 kg/cm of torque and a weight of 70 g, was chosen as the first joint and the HS-5085MG motor as the second joint had 3.6 kg/cm of torque and a weight of 21.9 g. The joints were controlled by an Arduino microcontroller and connected by serial communication to an EOG computer.

The output from the EOG was the pixel coordinates (u, v) of the gaze motion. Then, these pixels were converted into (*Px*, *Py*) in robot coordinates. Every pixel in u and v could be calculated as cm by using Equations (1) and (2). Offset coordinates also occurred, since the center of the target area is not exactly the same as the robot-based coordinates. The effects of the offset in the horizontal direction were 23 cm and 27 cm in the vertical direction.

$$X_{cm/pixel} = 34\ \text{cm}/1020\ \text{pixel} \tag{1}$$

$$y_{cm/pixel} = 27 \text{ cm/720 pixel} \quad (2)$$

The end-effector of the robot moved to the target position (*Px*, *Py*). *Px* was the end-effector position in the *x*-axis, and *Py* was the end-effector position in the y-axis. The correlation between the end-effector position (*Px*, *Py*) and the joint angles (α and β) was determined by inverse kinematics. Joint 1 rotated with the α angle, and Joint 2 rotated with the β angle. The link lengths were Link1 and Link2, which were equal, so that they were named as "Links". Consequently, the angles of Joint 1 and Joint 2 for the end-effector position (*Px*, *Py*) are given by Equations (3–10), for which, sin α and sin β were calculated by the trigonometry property, $\sin\theta = \sqrt{1-\cos^2\theta}$. Figure 7 shows the robot and its joint coordinates.

Figure 7. End-effector position, *P* = (*Px*, *Py*).

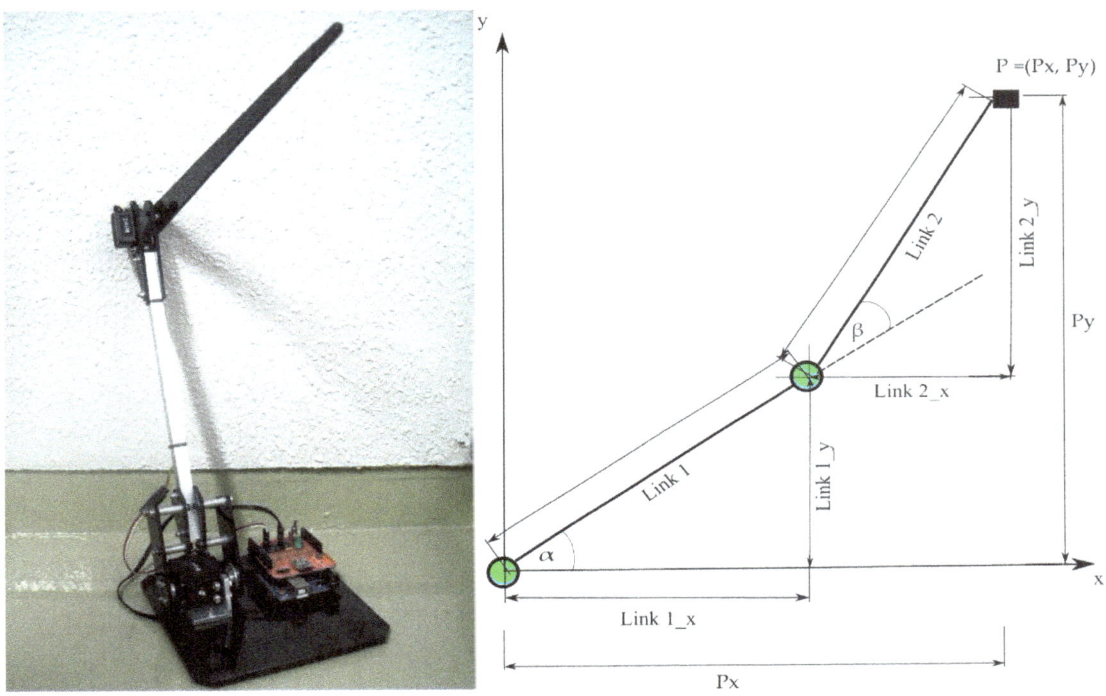

$$Px = \text{Link1}\_x + \text{Link2}\_x = \text{Link1}\times\cos\alpha + \text{Link2}\times\cos(\alpha+\beta) \quad (3)$$

$$Py = \text{Link1}\_y + \text{Link2}\_y = \text{Link1}\times\sin\alpha + \text{Link2}\times\sin(\alpha+\beta) \quad (4)$$

$$\cos\beta = \left(\frac{P^2x + P^2y - \text{Link1}^2 - \text{Link2}^2}{2\times\text{Link1}\times\text{Link2}}\right)$$

$$\text{Link1} = \text{Link2} = \text{Link}$$

$$\cos\beta = \left(\frac{P^2x + P^2y - 2\times\text{Link}^2}{2\times\text{Link}^2}\right) \quad (5)$$

$$\cos\alpha = \left(\frac{Px*\left(\text{Link1} + \text{Link2}\times\cos\beta\right) + Py\times\text{Link2}\times\left(\sqrt{1-\cos^2\beta}\right)}{P^2x + P^2y}\right)$$

$$\cos\alpha = \left(\frac{\text{Link} \times \left(Px \times (1+\cos\beta) + Py \times \left(\sqrt{1-\cos^2\beta} \right) \right)}{P^2x + P^2y} \right) \tag{6}$$

$$\sin\beta = \sqrt{1 - \left(\frac{P^2x + P^2y - 2 \times \text{Link}^2}{2 \times \text{Link}^2} \right)} \tag{7}$$

$$\sin\alpha = \sqrt{1 - \left(\frac{\text{Link} \times \left(Px \times (1+\cos\beta) + Py \times \left(\sqrt{1-\cos^2\beta} \right) \right)}{P^2x + P^2y} \right)^2} \tag{8}$$

$$\beta = \text{atan2}(\cos\beta, \sin\beta) \tag{9}$$

$$\alpha = \text{atan2}(\cos\alpha, \sin\alpha) \tag{10}$$

4. Methodology

After discussing the EOG sensor and the robot manipulator, this section explains the experiment setup, which includes the position of the operators and the positions of the targets for training. First, the similarity of EOG signals in four areas each for horizontal and vertical gaze motions was checked for each operator. Then, a preliminary experiment was done to get the relationship between the vertical gaze motion and the EOG signal from Ch1, as well the relationship between the horizontal gaze motion and the EOG signal from Ch2. This step gave the linear equations of EOG. The products of these equations were the pixel positions of the targets. Based on the signal combinations, these equations were improved by rotation matrixes, so that the system was well operated, not only in horizontal or vertical direction, but also in the diagonal gaze motion for a free gaze motion distance. Finally, three healthy operators tried to operate the robot using their EOG signal. Figure 8 briefly shows the main process for achieving the goal.

Figure 8. The main process of the research.

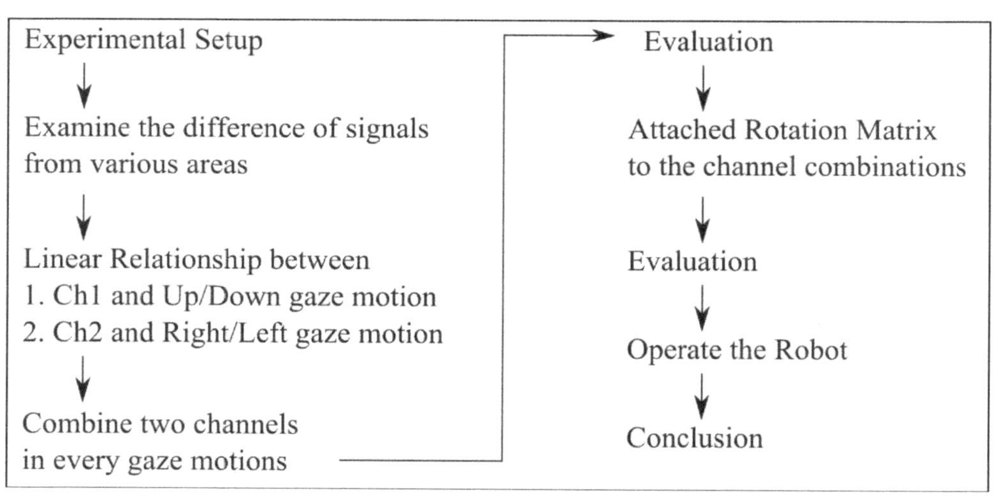

There were ten training targets in total, 40 cm in front of the operators, which were five training targets in the horizontal line and five training targets in the vertical line, as shown in Figure 9a,b. The training targets were used to determine the linear relationships between the EOG and the gaze motion. The training targets were numbered from 1 to 5. Target Number 3 was exactly located in the middle of display. This point was a reference point for the head position. This position was fixed to reduce the noise from the head movement. The normal distance between the operator and the target area is 40 cm, as shown in Figure 9b.

Figure 9. A monitor in front of operator as the target location. (**a**) Target points for training. (**b**) The normal distance between eyes and the target.

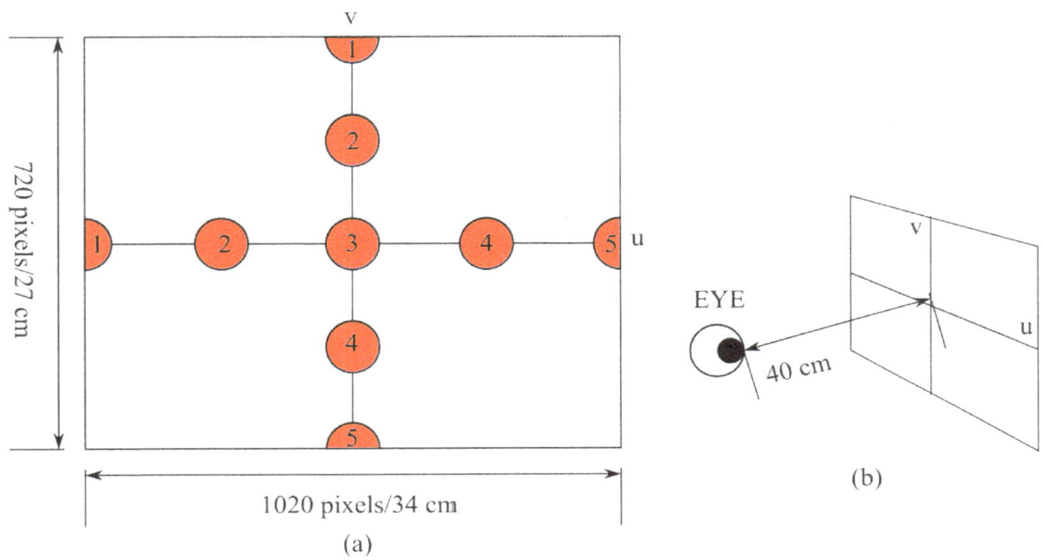

Linear Relationship between EOG and Gaze Motion

The training experiments depended on directions, distances and areas. The directions were up, down, right and left. There were four gaze distances, each in the horizontal and vertical directions. The distances were grouped into 1-point, 2-point, 3-point and 4-point gaze motions. In the vertical case, the shortest distance was 180 pixels, whereas the longest distance was 720 pixels. On the other hand, the shortest distance in the horizontal direction was 255 pixels, and the longest distance was 1020 pixels, as shown in Table 2. In this experiment, the area for horizontal movement was approximately ±46°, and the area for the vertical movement was approximately 38°.

Table 2. The distance for four types of movement distances.

Direction	**Gaze Distance**			
	1-Point	**2-Point**	**3-Point**	**4-Point**
Vertical	180 pixels/9°	360 pixels/19°	540 pixels/28°	720 pixels/38°
Horizontal	255 pixels/12°	510 pixels/23°	765 pixels/35°	1020 pixels/46°

The gaze distances had some different working areas: 1-point movement had four areas; 2-point movement had three areas; 3-point movement had two areas; and 4-point movement had only one area.

Area 1 for the 1-point gaze distance was *1–2*, which refers to the gaze motions from training Target 1 to training Target 2 or *vice versa*. Table 3 summarizes the working area types.

The feature of the EOG that was used in this research was the absolute integral of EOG ($\text{int\_}EOG_{Chi,i=1,2}$), formulated by Equations (11) and (12). This variable was needed to establish the relationships between EOG and gaze distance (pixel). It was counted first when there was a signal bigger than the positive threshold (th +) or smaller than the negative threshold (th −), until the full wave was formed. The integral value from Ch1 belonged to up or down gaze motions, and Ch2 was concerned with the right or left gaze motions.

Table 3. Types of working areas.

Gaze Distance	Area				Pixel Distance (u,v)	
	1	2	3	4	**Horizontal**	**Vertical**
1-point	*1–2*	*2–3*	*3–4*	*4–5*	(255, 0)	(0, 180)
2-point	*1–3*	*2–4*	*3–5*		(510, 0)	(0, 360)
3-point	*1–4*	*2–5*			(765, 0)	(0, 540)
4-point	*1–5*				(1020, 0)	(0, 720)

The Ch1 signal was only associated with vertical gaze motion, and the signal from Ch2 was for the horizontal gaze motion. Operators performed all possibilities for the movements belonging to the training targets. Figure 10a–d shows the average of the absolute integral of EOG signals for up, down, right and left gaze motions.

$$\text{int\_}EOG_{Ch1} = \left| \int_{\Omega_+} EOG_{Ch1}(\text{t})\,dt \right| + \left| \int_{\Omega_-} EOG_{Ch1}(\text{t})\,dt \right| \tag{11}$$

$$\begin{gathered} \text{int\_}EOG_{Ch2} = \left| \int_{\Omega_+} EOG_{Ch2}(\text{t})\,dt \right| + \left| \int_{\Omega_-} EOG_{Ch2}(\text{t})\,dt \right| \\ \Omega_+ = \{t : EOG_{Chi}(\text{t}) > th+\} \\ \Omega_- = \{t : EOG_{Chi}(\text{t}) < th-\} \\ i = 1,2 \end{gathered} \tag{12}$$

Before making a conclusion about the linear relationship between gaze motion and the EOG, checking the signal difference for all areas, especially in 1-point movement, was necessary. This step was just to make sure that the EOG signals in the same gaze motion distances had equal values for those areas, so that all of the areas could be used to examine the linear relationship with the gaze motion distance.

An ANOVA was used to compare the signals in four areas for the 1-point gaze motion distance. The level of confidence was 95% or $\alpha = 0.05$, and the null hypothesis was that there was no difference between the EOG signals from Area 1, Area 2, Area 3 and Area 4 or that the EOG signal from Area 1 was equal to Area 2, Area 3 and Area 4. There were three operators, and they looked 20 times to 1-point right and 20 times to 1-point left in each area in the horizontal case, as well as to up and down for the vertical test.

Table 4 shows that all of the values of *F_crit* were smaller than *F*. That means that there was no difference between the EOG signal among the four areas as long as the gaze motion distances were the same. Therefore, if an operator looked to any target points in all possible areas, the values of the EOG were not different as long as the gaze distances were same. This evidence showed that the EOG signal from these operators could be used for further experiments.

Figure 10. Target points for training and normal distance between eyes and target.

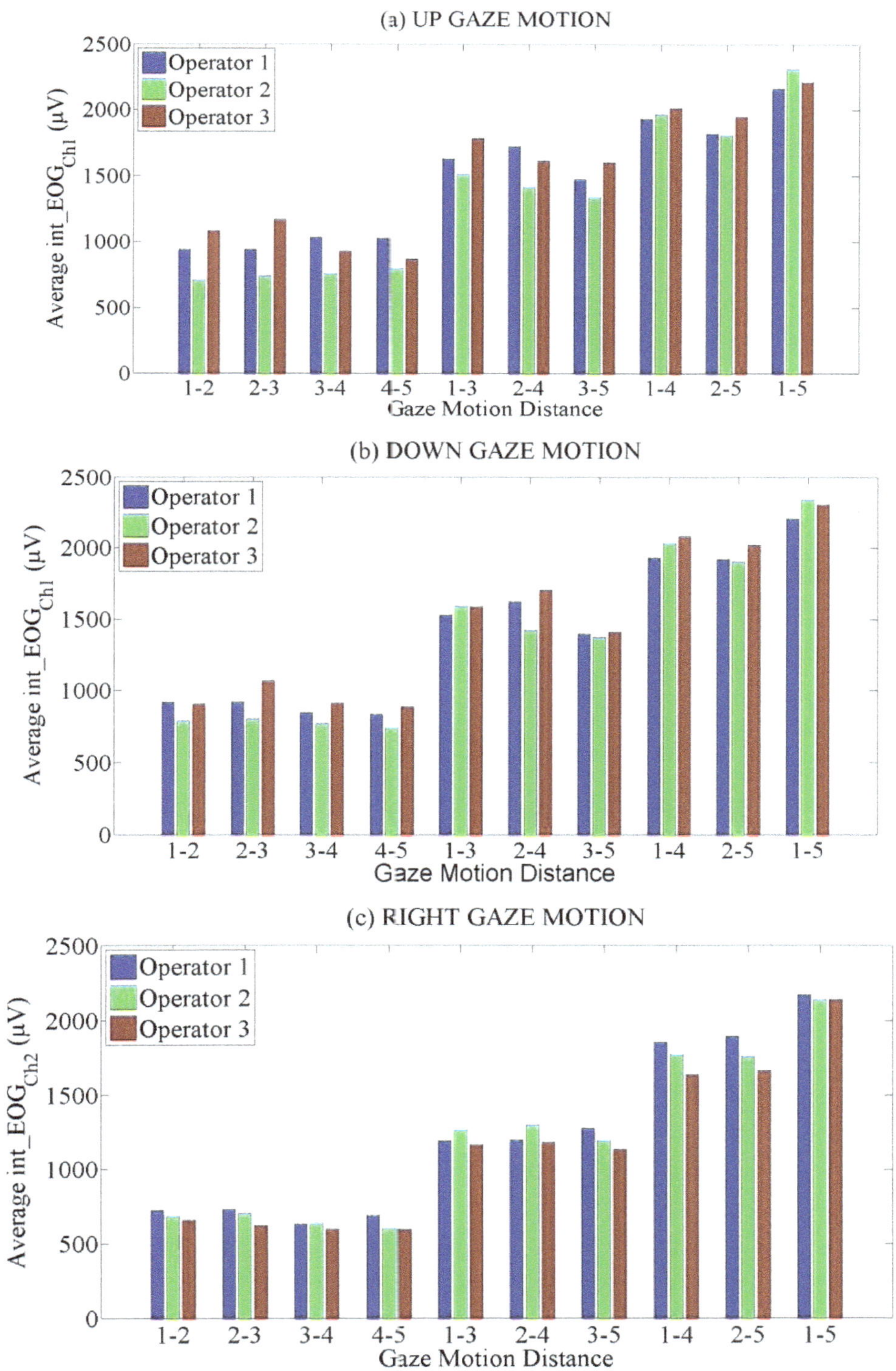

Figure 10. *Cont.*

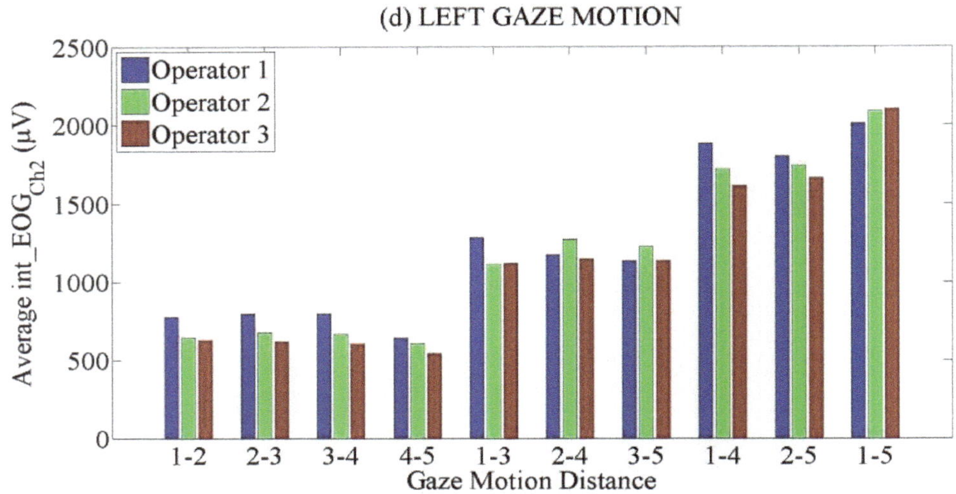

Table 4. ANOVA for the EOG signal for vertical and horizontal gaze motions among four areas in the 1-point gaze motion distance.

	Operator	Up			Down		
		F	*p*-value	*F_crit*	*F*	*p*-value	*F_crit*
Vertical	Operator 1	0.112	0.952	3.239	1.935	0.165	3.239
	Operator 2	0.675	0.577	3.072	0.720	0.554	3.239
	Operator 3	1.364	0.274	2.947	2.79	0.063	3.028
	Operator	**Right**			**Left**		
		F	*p*-value	*F_crit*	*F*	*p*-value	*F_crit*
Horizontal	Operator 1	1.013	0.412	3.239	1.720	0.216	3.490
	Operator 2	1.783	0.174	2.960	1.032	0.397	3.028
	Operator 3	2.483	0.0888	3.0725	1.834	0.172	3.072

After examining the similarity of the signal in all areas, the linear relationship between gaze motion and distance was determined. Although there was no difference between signals in the same distance, the EOG values among the operators were varied. Therefore, the normalization of integral EOG ($Norm\_EOG_{Chi,i=1,2}$), as shown by Equation (13), was used in this system. It could be calculated by dividing the absolute integral of the EOG signal ($\text{int}\_EOG_{Chi,i=1,2}$) by the average of the maximum integral EOG ($\overline{Maximum\_\text{int}\_EOG_{Chi,i=1,2}}$). Table 5 provides the average maximum of the integral EOG for right, left, up and down from three operators. Finally, the linear relationship between the normalized EOG and the pixel are shown by Equations (14–17). The constant variables of those equations (*A*, *B*, *C*, *D*, *E*, *F*, *G*, *H*, *F* and *G*) are provided in Table 6.

$$Norm\_EOG_{Chi,i=1,2} = \frac{\text{int}\_EOG_{Chi,i=1,2}}{\overline{Maximum\_\text{int}\_EOG_{Chi,i=1,2}}} \tag{13}$$

$$\text{v}\_up = A \times Norm\_EOG_{Ch2} + B \tag{14}$$

$$v\_down = C \times Norm\_EOG_{Ch2} + D \tag{15}$$

$$u\_right = E \times Norm\_EOG_{Ch1} + F \qquad (16)$$

$$u\_left = F \times Norm\_EOG_{Ch1} + G \qquad (17)$$

Table 5. $\overline{Maximum\_\mathrm{int}\_EOG_{ch_{i,i=1,2}}}$ from three operators.

Gaze Direction	$\overline{Maximum\_\mathrm{int}\_EOG_{ch_{i,i=1,2}}}$		
	Operator 1	**Operator 2**	**Operator 3**
Up	2,462	2,417	2,288
Down	2,237	2,368	2,368
Right	2,207	2,044	2,393
Left	2,151	1,986	2,402

Table 6. Constant variables from three operators to complete Equations (10–13).

Operator	Vertical (Ch1)				Horizontal (Ch2)			
	Up		Down		Right		Left	
	A	B	C	D	E	F	G	H
1	976	−299	−876	199	1,092	−108	−1,091	104
2	938	−309	−816	175	1,124	−185	−1,142	207
3	956	−292	−878	204	1,241	−121	−1,113	101

5. Rotation Matrix

The rotation matrix had an important role in this study, since it corrected linear pixel differences between desired targets and actual gaze motions. Some tests were needed to get the rotation matrixes. These tests were almost the same as the experiments, which obtained the four linear equations. The experiments were done in one day, and the operators gazed 20 times at the training targets for every single session. A session depended on the directions, areas and distances. The difference was if, in the previous experiment, Ch1 was just involved in the vertical gaze motion and Ch2 was only used for the horizontal gaze motion, now both of them worked together for all gaze motions. Therefore, in the case that the operators gazed at a point, the system must consider the signal combinations as mentioned early in the Electrooculography section.

There were still some undetermined signals when the operators looked at a point. There was a signal from Ch1 that indicated down movement when the operators actually only performed a left gaze motion. Ch1 produced a negative EOG signal that indicated an up gaze motion when the operators only gazed to the right. In the vertical case, the undetermined signals came from Ch2. The offset position in the vertical axis for right and left gaze motion, as well the offset position in the horizontal axis for up and down movement were called errors.

Figure 11. Illustration of the error pixels in the four basic gaze motion directions.

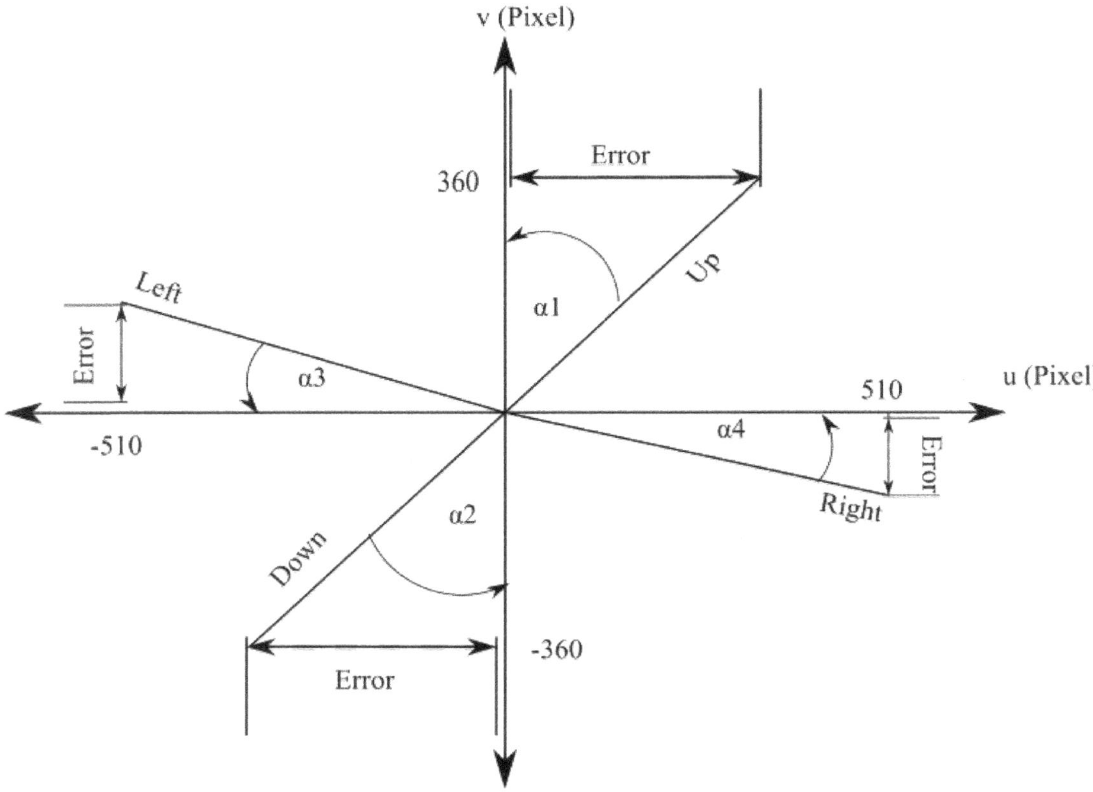

The errors produced slopes between the four lines (up, down, right and left) and the axis, as shown in Figure 11. Table 7 shows the values of $\alpha 1$, $\alpha 2$, $\alpha 3$ and $\alpha 4$ from three operators. Those values varied by subject. Rotating the four lines (up, down, right and left) in a counter-clockwise direction using Equation (18) was investigated in this research. The goal was to bring the up and down line exactly on the v-axis, as well as the left and right line on the u-axis. These methods are illustrated by Figure 12.

Table 7. The angles of rotation matrixes for the up, down, right and left gaze motions.

Operator	$\alpha 1$	$\alpha 2$	$\alpha 3$	$\alpha 4$
1	41.48°	42.25°	18.48°	18.89°
2	42.25°	46.99°	19.12°	18.19°
3	52.78°	51.57°	31.16°	33.16°

$$\begin{pmatrix} u' \\ v' \end{pmatrix} = \begin{pmatrix} \cos\alpha_i & -\sin\alpha_i \\ \sin\alpha_i & \cos\alpha_i \end{pmatrix} \begin{pmatrix} u \\ v \end{pmatrix} i = 1,2,3,4 \tag{18}$$

u = Pixel in u before rotation, v = Pixel in v before rotation

u' = Pixel in u after rotation, v' = Pixel in v after rotation

Figure 12. The positions of pixels. **(a)** Before rotation process. **(b)** After the rotation process.

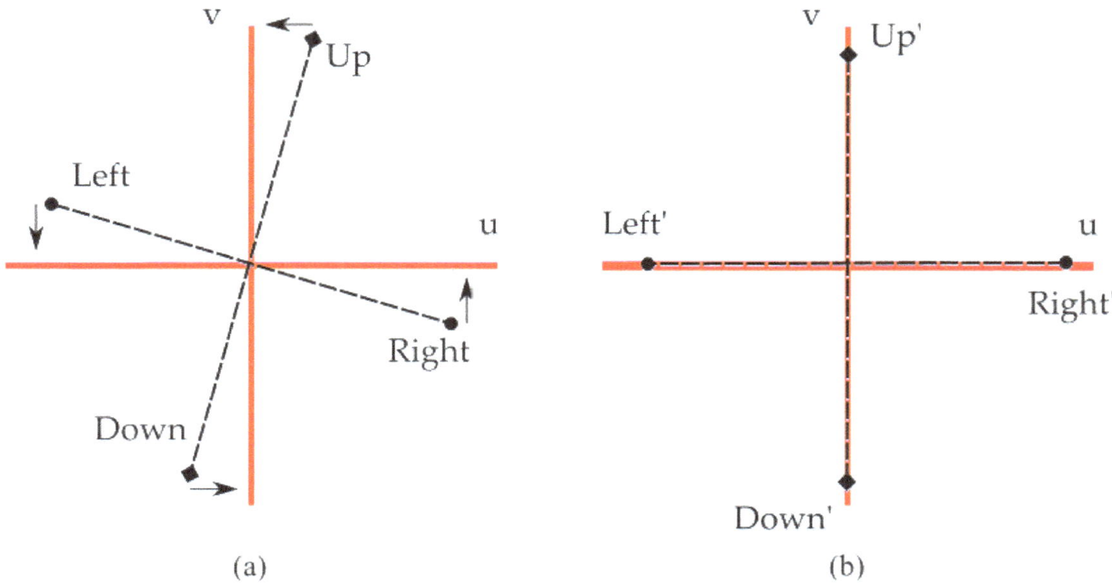

Table 8 shows some examples of average data for gaze motions from the (0, 0) pixel to the four desired positions. For the right or left gaze motion, there were some errors, especially on the v-axis, because of the signal coming from Ch1. These errors were successfully reduced by rotation matrixes from −147 pixels to 30 pixels on the right gaze motion and 255 pixels to −27 pixels on the left gaze motion. On the other hand, the errors for vertical gaze motions coming from Ch2 on the u-axis were reduced from 207 pixels to 43 pixel in the up gaze motion and from −180 pixels to −35 pixels in the down gaze motion. After the rotation process, the errors were approximately ±25 pixels (u) and ±30 pixels (v) for horizontal gaze motion. For the vertical gaze, the errors were approximately ±40 pixels (u) and ±30 pixels (v). Comparing this error to the total pixels in the horizontal (1020 pixels) and the vertical (720 pixels) directions, the errors were around 3% for the u-axis and 4% for the v-axis.

Table 8. Pixel positions before and after rotation for some gaze motions from the (0, 0) pixel to the desired pixels.

	Signal		**Desired Pixel**		**Actual Pixel before Rotation**		**Actual Pixel after Rotation**	
	$Norm\_EOG_{Ch1}$	$Norm\_EOG_{Ch2}$	**ud**	**vd**	**u**	**v**	**u'**	**v'**
Right	0.42	0.52	510°	0°	461°	−147°	483°	30°
Left	0.59	0.53	−510°	0°	−476°	255°	−530°	−27°
Up	0.73	0.36	0°	360°	207°	332°	43°	389°
Down	0.69	0.26	0°	−360°	−180°	−296°	−35°	−340°

6. Results and Discussion

A sequence of tests was already done after all parts of the rotation matrix were established. That means that the EOG part, the robot and the gaze motion distance algorithm were connected well. Figure 13a shows the display on the monitor when the testing process was being started. There was a piece of the robot manipulator in front of the screen, which was Link2. The blue dot was the initial eye position, and the red dot was the target point. Operators started their gaze motion from the blue dot, then moved to the red dot. This movement was tracked by the robot, as shown by Figure 13b. At this time, the target point in the Figure 13a became the eye position or blue dot. The red line connecting the two points illustrates the pathway of the end-effector. In this case, there were no error pixels, either on the u-axis or the v-axis.

Figure 13. The display for testing the tracking of a 2D analog object using EOG. (**a**) Initial position for the end-effector of robot and the position of target. (**b**) The end-effector follows the gaze motion.

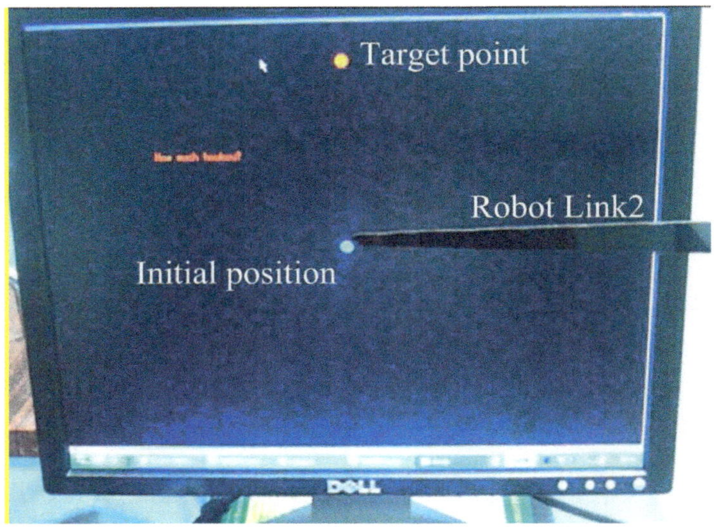

(a)

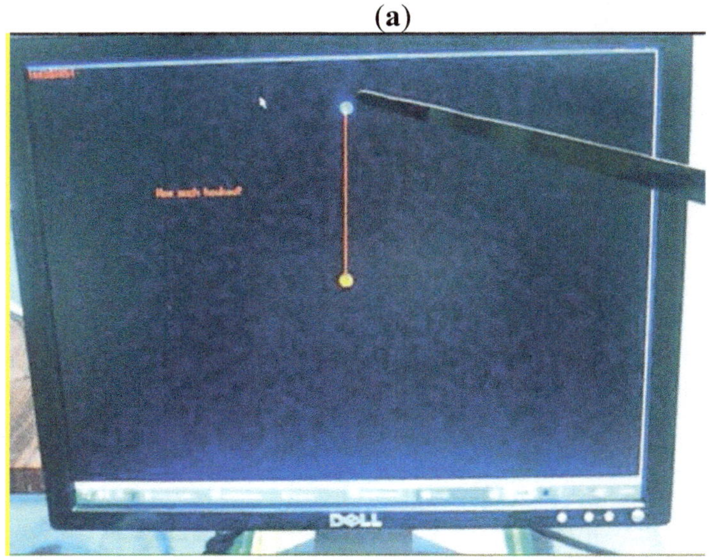

(b)

Another example of tracking points is shown in Figure 14. There were two sequences of gaze motions in this example. The first gaze was diagonal in the left-up direction, and the second gaze motion was diagonal in the right-up direction. After the second gaze motion, the actual eye position (blue dot) was not exactly at the same position as the target point (red dot). This difference was called the Error_Distance. The error pixel in horizontal is calculated by Equation (19) and the error pixel in vertical was shown by Equation (20), The Error_Distance was the Euclidian distance between the desired pixel position and the actual pixel position. It is calculated by Equation (21).

Figure 14. The error pixel between the desired pixel position and the actual pixel position (1 = Error_Distance, 2 = u_Error, 3 = v_Error, 4 = desired position, 5 = actual position).

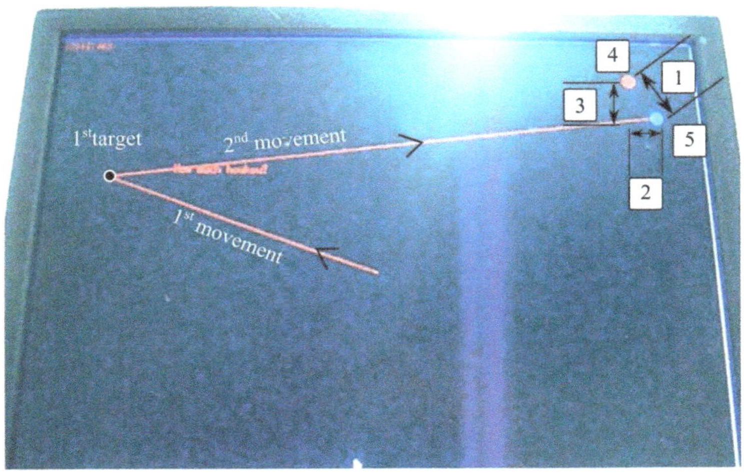

$$u\_error = ud - u' \quad (19)$$

$$v\_error = vd - v' \quad (20)$$

$$Error\_Distance = \sqrt{u\_error^2 + v\_error^2} \quad (21)$$

A sequence of experiments was done by three operators. Three operators gazed in total 20 times to track 20 random points. Based on (1), every pixel in the horizontal direction was equal to 0.0333 cm, since the length of the monitor was 34 cm or 1020 pixels. Calculated by Equation (2), every pixel in the vertical direction was equal to 0.0375 cm, because the monitor height was 27 cm or 720 pixels. The Error_Distance is given in pixels and cm, as shown by Table 9. As long as the distance between operators and the display was 40 cm, the angle errors, as shown in Figure 15, were also available for horizontal and vertical gaze motion. The errors in degrees are given in Table 10.

Figure 15. The error gaze motion in degrees.

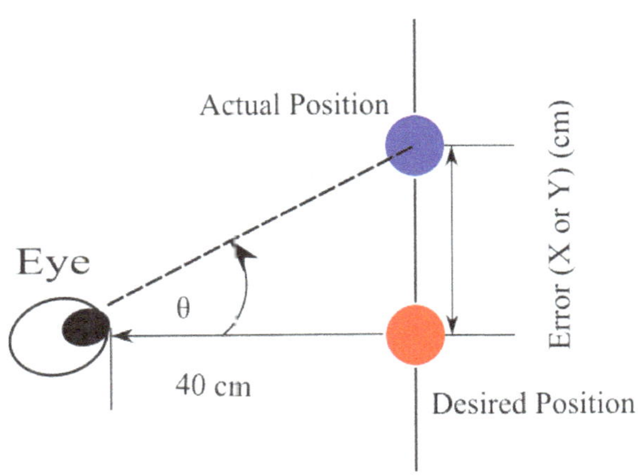

Table 9. Average error (mean ± standard deviation) from three operators in pixels and cm.

	Operator	$\overline{x\_error}$	$\overline{y\_error}$	$\overline{Error\_Distance}$
Pixel	1	54 ± 19	51 ± 14	76 ± 17
	2	48 ± 17	44 ± 23	69 ± 19
	3	50 ± 18	48 ± 21	70 ± 17
cm	1	1.80 ± 0.65	1.94 ± 0.53	2.71 ± 0.60
	2	1.60 ± 0.57	1.67 ± 0.88	2.44 ± 0.69
	3	1.67 ± 0.61	1.68 ± 0.79	2.48 ± 0.64

Table 10. Average error (mean ± standard deviation) from three operators in degrees..

Operator	**Degree**	
	$\theta_{\overline{x\_error}}$	$\theta_{\overline{y\_error}}$
1	2.57° ± 0.94°	2.77° ± 0.76°
2	2.30° ± 0.89°	2.39° ± 1.25°
3	2.40° ± 0.87°	2.40° ± 1.14°

At a glance, the average error in the u-axis was almost the same as that in the *y*-axis. The average error in the u-axis was 50 pixels ± 18 pixels, and the average error in the v-axis was 47 pixels ± 20 pixels. However, there was a conversion coordinate from pixel (u, v) to robot coordinates (*x*, *y*) in cm. In the robot coordinates, the average maximum error pixels in the *x*-axis were 2.31 cm and 2.50 cm in the *y*-axis. The maximum error distances from Operators 1, 2 and 3 were 3.31 cm, 3.14 cm and 3.13 cm. Overall, the maximum average error distance was around 3.20 cm. Table 9 gives the data for the average error in degrees. The average angle error in the horizontal direction was around 2.42° ± 0.87° and 2.52° ± 1.50° in the vertical direction. These results show that the linear EOG equations in the horizontal direction have a slightly better performance than the vertical direction. Horizontal gaze motion has a higher accuracy than vertical gaze motion. Horizontal gaze motion is slightly more stable than vertical gaze motion based on the standard deviation value.

This system successfully gives an alternative method of a tracking system using EOG. In [17], ten random targets were tracked by three operators and the operator's blinked their eye when the robot reached a target. Using rotation matrixes, the robot directly moved to the target when a gaze action was performed. This ability can avoid errors that occur due to the process of blinking detection. The maximum error using rotation matrixes was approximately 3.1 cm. This system was also more stable based on the standard deviation.

7. Conclusions

This test was performed with a fixed head position with a maximum angle to the horizontal axis of 46° and a maximum gaze motion in the vertical axis of 38°. An EOG system that was used to drive the two-DoF planar robot manipulator for the analog tracking of an object was developed not only for direction but also for distance. The direction was not only the vertical and horizontal axes, but also the diagonal gaze motion. There were five targets, each in a horizontal line and a vertical line, to check the similarity of the EOG in each axis. The results showed that there were no differences among the EOG potential values, as long as the gaze motions were performed in the same axis and at an equal distance.

To reduce pixel error, the rotation matrix was attached to the process based on the four basic directions. This system was tested by three operators on 20 random points. The test shows that this system successfully tracked objects indicated by the monitor display and the robot movement. The level of accuracy in the u-axis is slightly better than the v-axis, since the maximum angle error in the u-axis was 3.31° and that in the v-axis was 3.58°. This system, with good accuracy, is acceptable to be further developed. By attaching a gripper at the end-effector, the developed EOG-based robot control system could help disabled people handle materials, such as a glass, books, or to point to an object that they want to access.

Although this system could track freely moving objects with a good level of accuracy, there are still many challenges to improve it. One of them is its robustness, because EOG values are varied not only across different people, but also, each individual can produce a different EOG signal for the same gaze motion distance at two different times. These situations need to be standardized in order to develop EOG for more general applications.

Author Contributions

This is a team work research in Sasaki and Ito Laboratory, Gifu University. Minoru Sasaki has the idea about the alternative way to connect human and machine. Minoru Sasaki and Satoshi Ito have been developing research in human and machine interface. Takeo Okamoto and Muhammad Rusydi analyzed the EOG characteristic and built the relationship between EOG and gaze distance. Muhammad Rusydi is the main author of this paper.

Conflicts of Interest

The authors declare no conflict of interest.

References

1. Harun, H.; Mansor, W. EOG Signal detection for home appliances activation. In Peoceedings of the 5th International Colloquium on Signal Processing and Its Applications (CSPA), Kuala Lumpur, Malaysia, 6–8 March 2009.
2. Murata, A. Eye-gaze input *versus* mouse: Cursor control as a function of age. *Int. J. Hum. Comput. Interact.* **2006**, *21*, 1–14.
3. Pinheiro, C.G.; Naves, E.L.; Pino, P.; Losson, E.; Andrade, A.O.; Bourhis, G. Alternative communication systems for people with severe motor disabilities: A survey. *Biomed. Eng. Online* **2011**, doi:10.1186/1475-925X-10-31.
4. Barreto, A.B.; Scargle, S.D.; Adjouadi, M. A practical EMG-based human-computer interface for users with motor disabilities. *J. Rehabil. Res. Dev.* **2007**, *37*, 53–63.
5. Wolpaw, J.R.; McFarland, D.J.; Neat, G.W.; Forneris, C.A. An EEG-based brain-computer interface for cursor control. *Electroencephalogr. Clin. Neurophysiol.* **1991**, *78*, 252–259.

6. Magosso, E.; Ursino, M.; Zaniboni, A.; Gardella, E. A Wavelet-based energetic approach for the analysis of biomedical signals: Application to the electroencephalogram and electro-oculogram. *Appl. Math. Comput.* **2009**, *207*, 42–62.
7. Dhines, K.; Eric, P. Classification of EOG for Human Computer Interface. In Proceedings of the 2nd Joint EMBS/BMES Conference, Houston, TX, USA, 23–26 October 2002.
8. Zhao, L.V.; Wu, X.-P.; Li, M.; Zhang, C. Implementation of the EOG-Based Human Computer Interface System. In Proceedings of the 2nd International Conference on Bioinformatics and Biomedical Engineering, Shanghai, China, 16–18 May 2008.
9. Choi, K.H.; Minoru, S. Brain-wave bio potentials based mobile robot control: Wavelet-neural network pattern recognition approach. In Proceedings of the IEEE International Conference on Systems, Man, and Cybernetics, Tucson, AZ, USA, 7–10 October 2001.
10. Deng, L.Y.; Hsu, C.-L.; Lin, T.-C.; Tuan, J.-S.; Chang, S.-M. EOG-based human-Computer Interface system development. *Expert Syst. Appl.* **2010**, *37*, 3337–3343.
11. Barea, R.; Boquete, L.; Ortega, S.; López, E.; Rodríguez-Ascariz, J.M. EOG-based eye movements codification for human computer interaction. *Expert Syst. Appl.* **2012**, *39*, 2677–2683.
12. Cristian-Cezar, P.; Florin, G.; Doru, T. EOG-based Visual Navigation Interface Development. *Expert Syst. Appl.* **2012**, *39*, 10857–10866.
13. Michale, F.M.; Mitchell, G.B.; Daphane, L.M.; Caro, A.W.; Michael, B. ISCEV Standard for Clinical Electro-oculograph. *Doc. Ophthalmol.* **2011**, *122*, 1–7.
14. Gu, J.J.; Meng, M.; Cook, A.; Liu, X.P. Design, Sensing and control of a robotic prosthetic eye for natural eye movement. *Appl. Bionics Biomech.* **2006**, *3*, 29–41.
15. Aysegul, G.; Sadik, K. Classification of electro-oculogram signals using artificial neural network. *Expert Syst. Appl.* **2006**, *21*, 199–205.
16. Aungsakun, S.; Phinyomark, A.; Phukpattaranont, P.; Limsakul, C. Robust eye movement recognition using EOG signal for human-computer interface. *Softw. Eng. Comput. Syst.* **2011**, *180*, 714–723.
17. Eduardo, I.; Jose, M.A.; Eduardo, F.; Andre, U. Interface based on Electrooculography for Velocity Control of a Robot Arm. *Appl. Bionics Biomech.* **2010**, *7*, 199–207.
18. Zhao, L.V.; Wu, X.-P.; Li, M.; Zhang, D. A Novel Eye Movement Detection Algorithm for EOG Driven Human Computer Interface. *Pattern Recognit. Lett.* **2010**, *31*, 1031–1047.
19. Barea, R.; Boguete, L.; Mazo, M.; Lopez, E. Wheelchair Guidance Strategies Using EOG. *J. Intell. Robot. Syst.* **2002**, *34*, 279–299.
20. Rusydi, M.I.; Okamoto, T.; Sasaki, M.; Ito, S. Optimizing Electrode Positions for Determination of Gaze Movement Using the NF 5201. In Proceedings of the 7th Asia Pacific Symposium on Applied Electromagnetics and Mechanis, Ho Chi Minh City, Vietnam, 25–27 July 2012; pp. 266–271.
21. Malmivuo, J.; Plonsey, R. *Bioelectromagnetism*; Oxford Unviersity Press: New York, NY, USA, 1995; pp. 538–542.
22. Croft, R.J.; Barry, R.J. EOG correction of blinks with saccade coefficients: A test and revision of the aligned-artefact average solution. *Clin. Neurophysiol.* **2000**, *111*, 444–451.
23. Nöjd, N.; Hannula, M.; Narra, N.; Hyttinen, J. Electrode position optimization for facial EMG measurements for human-computer interface. *Methods Inf. Med.* **2008**, *47*, 192–197.

24. Timmins, N.; Marmor, M.F. Studies on the stability of the clinical electro-oculogram. *Doc. Ophthalmol.* **1992**, *81*, 163–171.

11

A Magnetic Microrobot with *in situ* Force Sensing Capabilities

Wuming Jing and David Cappelleri *

School of Mechanical Engineering, Purdue University, 585 Pudue Mall, West Lafayette, IN 47907, USA; E-Mail: jing6@purdue.edu

* Author to whom correspondence should be addressed; E-Mail: dcappell@purdue.edu

---

Abstract: This paper presents a proof-of-concept prototype of a micro force sensing mobile microrobot. The design consists of a planar, elastic mechanism serving as computer vision-based force sensor module, while the microrobot body is made from a magnetic layer driven by a magnetic field. From observing the deformation of the elastic mechanism, manipulation forces can be determined. The deformation is tracked by a CCD camera attached to an optical microscope. This design is validated through experimental tests with a micromachined prototype. The preliminary results verify this first microrobot prototype is indeed capable of in situ force sensing. This concept can be scaled down further for next generation designs and can be designed for real biomedical applications on microscale.

Keywords: vision-based force sensing; PDMS compliant mechanism; magnetic mobile microrobot

---

1. Introduction

Untethered mobile robots on microscale have emerged as one of the new frontiers in intelligent automation systems only since the last decade. They have shown potential in biomedical applications, such as in biological processes at cellular level and *in vivo* medical tasks. The microscale robot agents provide attractive working modes due to their small size [1]. At the same time, the limited footprint also makes it challenging to carry function modules on board, such as the power for actuation and in situ sensing. However, the real advantages of microrobot system can only be utilized if it possesses high resolution sensing functions [2]. Therefore, it is desired to incorporate force sensing functionality into a microscale robotic system for biomedical applications. Additionally, micro-force interactions on real biomedical single cell manipulations are still underexplored. When manipulating biological cells, the typical forces ranges are at the micro-Newton (μN) level or lower. Therefore, the goal of our work in this paper is to come up with a wireless mobile micro device that incorporates a micro force sensing module on board (Figure 1), which is able to send force feedback information in situ.

Figure 1. The concept system of the **micro-F**orce **S**ensing **M**obile **M**icrorobot ($\mu FSMM$). The vision-based micro force sensor end-effector is connected with a magnetic base. The base is driven by a series of electromagnetic coils. Deformation of vision-based force sensor is observed with the vision system.

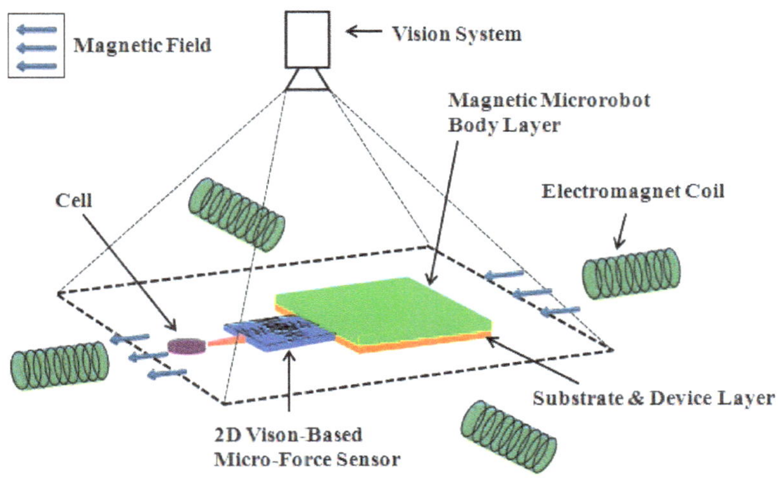

2. Related Work

For wireless mobile microrobots, actuation itself is challenging due to the constraint space for power storage. Electrical [3] and thermal [4] energy have been applied to power microrobots, which are based electrostatic attraction and laser heating, respectively. These designs require an engineered working surface that is not always available in practical biomedical applications. Thereafter, magnetic actuation has routinely been applied in microrobotic systems.

There are various magnetic microrobot systems that have been developed successfully. In fluid media, a micro spiral magnetic structure can swirl forward with a rotating magnetic field [5]. On the dry surface, devices made of asymmetric magnetic hammers, named "MagMite" [6] and "PolyMite" [7], are able to translate with an oscillating magnetic field. The micro agent made of pure neodymium-iron-boron magnetic block [8] can translate in a rocking (stick-slip) mode with pulsing magnetic field signals. The ball shape [9] and rod shape ("RodBot") [10] prototypes have also been fabricated to roll on surfaces. Tumbling locomotion has also been realized through a microscale magnetic microrobot [11,12]. Driving the magnetic microrobot directly by magnetic field gradients has also been conducted successfully [13,14]. The magnetic assembly named "OctoMag" has been manipulated in five dimensions of freedom. On the smaller scale, Martel *et al.* [15] have used an MRI scanner to actuate a hydrogel-based shrinkable micro agent mixed with magnetic particles.

Various micro force sensors based on microelectromechanical systems (MEMS) have been developed to derive information used for cellular force measurements [16]. Indirect force measurement methods have been applied such as piezoresistive principle [17,18], strain gauge [19,20] and capacitance-based force measurement [21,22]. In [17], the micro force is sensed by the silicon cantilever beam deposited with resistors based on the piezoresistive effect. In [19], an orthotweezers system has been developed for automated microassembly. Beyeler *et al.* [22] have developed six-axis force-torque sensor consisting of microfabricated comb capacitors with micro-Newton (μN) and nano-Newton-meter ($nN \cdot m$) resolution. Koch *et al.* [23] have fabricated and tested a compliant surface-micromachined spring which is able to measure forces as small as 1 pico-Newton (pN). The common feature of these devices are complicated structures and fabrication process. They are also usually not compatible with working in fluidic environments. Moreover, the auxiliary drive electronics will clutter the workspace, when being used for microscale biomedical applications. However, a vision based force sensor is suitable for the applications in limited workspace. A vision based computational force sensor for elastically deformable objects is presented in [24]. Greminger *et al.* [25] present a method to visually measure the force distribution the contour data in the image of linearly elastic object. On the microscale, optics are used to sense contact forces in [26,27]. For single cell study, polydimethylsiloxane (PDMS) microbeams have been used to sense one dimensional micro-scale forces [28,29].

This paper presents the **micro-F**orce **S**ensing **M**obile **M**icrorobot ($\mu FSMM$, Figure 1), which couples a two-dimensional (2D) vision based micro force sensor with a magnetic mobile microrobot. This wireless micro agent owns mobility and in situ force sensing function at the same time, that is preferable to the biomedical applications on microscale. It is the first truly microscale mobile robot with onboard sensing capabilities. The design, modeling, prototyping, preliminary calibration and testing of this first $\mu FSMM$ are presented accordingly.

3. Micro-Force Sensing Mobile Microrobot ($\mu FSMM$) Design

The target application for the $\mu FSMM$ is force controlled biological cell manipulation (Figure 1). Therefore, a 2D elastic end-effector mounting on a magnetic microrobot body is desirable. The end-effector, that serves as the vision based micro force sensor, is based on our previous work in [30]. The tip of the end-effector is designed with a point contact in an effort to reduce surface contact, and therefore stiction, with the cell being manipulated. Furthermore, the micro force sensor needs to have a resolution which ranges from μN down to nN level for this application. Therefore, the vision based micro force sensor has a maximum threshold of stiffness, corresponding to the vision system being utilized.

3.1. Design Specifications

In order to meet the function requirements, the design parameters are evaluated by the constraints imposed by our present vision system and electromagnetic microrobot testbed. The observing vision system consists of a CCD camera (FL3-FW-14S3C, $www.ptgrey.com$) and zoom imaging lens (VZM 330i, $www.edmundoptics.com$). The imaging parameters are summarized in Table 1. Therefore, assuming a pixel tracking accuracy of ± 1 for the sensor deformation, the micro force sensor needs to deflect more than $\Delta_{min} \approx 4\ \mu m$ at maximum magnification or $15\ \mu m$ at minimum magnification. (Note that the accuracy of vision based sensing not only relies on the image resolution, but also on the robustness and accuracy of the feature tracking algorithm utilized.)

Table 1. Parameter of vision system.

Camera sensor size (mm)	$7.60\ (H) \times 6.20\ (V)$
Pixels	$1,360\ (H) \times 1,024\ (V)$
Primary magnification (PMAG)	$0.75\ X - 3\ X$
Field of view (FOV, mm)	$10.13\ (H) \times 8.27\ (V)$ at $0.75\ X$ $2.53\ (H) \times 2.07\ (V)$ at $3\ X$
Measurement accuracy (μm)	$14.9\ (H) \times 16.2\ (V)$ at $0.75\ X$ $3.7\ (H) \times 4.0\ (V)$ at $3\ X$

Note: (1) H and V stands for horizontal and vertical directions of imaging view, respectively. (2) The pixel error is assumed as ± 1 for a high contrast image.

As the first generation of this $\mu FSMM$ prototype design, the agent is designed at a large scale for easy prototyping, but the feature sizes are still at the microscale with the total footprint no larger than the smallest field of view (FOV, $2.53\ mm \times 2.07\ mm$) of vision system. Thereafter, the magnetic microrobot has an area of $1.5\ mm \times 1.0\ mm$ for evaluation of its magnetic force exerting capabilities. The first generation agent magnetic base is prototyped with nickel foil whose thickness is about $25\ \mu m$. The electromagnetic coil testbed [11] utilized is able to produce a magnetic field with flux intensity (B) greater than $50\ mT$ within the workspace and a field gradient (∇B) greater than $1.5\ T \cdot m^{-1}$. Thus, the typical magnetic force F_m is evaluated at 50% power as:

$$F_m = V_m \cdot (M \cdot \nabla)B = 1.5 \times 1.0 \times 25(mm \cdot mm \cdot \mu m) \times 2 \times 10^3 (A \cdot m^{-1}) \times 0.75 (T \cdot m^{-1}) \approx 56\ (nN) \tag{1}$$

where V_m is the magnetic volume and M is the magnetization of the nickel body. Thereafter, the required maximum stiffness K_{max} of the proposed 2D elastic micro force sensor corresponds to the condition where this magnetic force value F_m results in the minimum deflection $\Delta_{min} = 4\ \mu m$ detected by the vision system. That is calculated based on:

$$F = K \cdot \Delta \tag{2}$$

where the acting force $F = F_m$, the deflection of the micro force sensor $\Delta = \Delta_{min} = 4\ \mu m$. Therefore, the maximum stiffness K_{max} equals to $0.014\ N/m$.

3.2. Micro Force Sensor Design

The micro force sensor end-effector is essentially a two-dimensional low stiffness spring [30]. The general topology for the design is shown in Figure 2. The suspended structure of multiple spring beam geometries are designed for low stiffness in both lateral (X) and transverse (Y) directions. The geometric parameters of fourteen candidate designs ($P1 - P14$) in [30] range as following: $n: 1 - 4$; $L_1 = 30 - 300\ \mu m$; $L_2 = 125 - 250\ \mu m$; $L_3 = 125 - 250\ \mu m$; $L_4 = 125 - 250\ \mu m$; $d_1 = 250 - 450\ \mu m$; $d_2 = 50 - 100\ \mu m$; $d_3 = 200 - 590\ \mu m$, where n is the number of springs on each symmetric side relative to Y axis.

Figure 2. Diagram of the micro force sensor end-effector of the $\mu FSMM$. The probe length and spring length are variable for tuning the device stiffness [30].

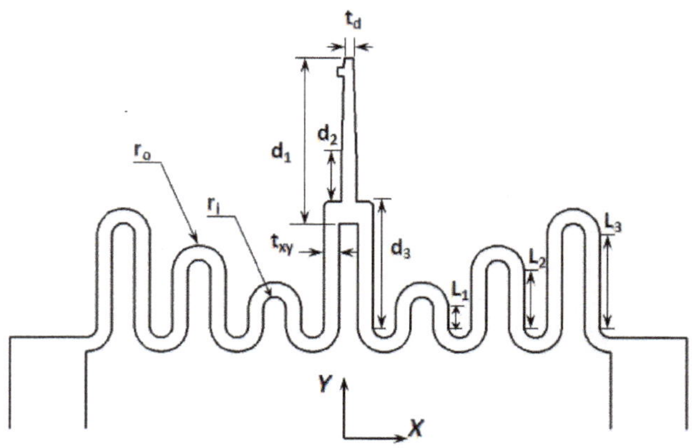

The material of the micro force sensor is selected as polydimethylsiloxane (PDMS) due to its small Young's modulus, as low as hundreds of KPa. In addition, PDMS is compatible with biological applications [31]. After the geometry is chosen, the stiffness of each design is numerically evaluated through finite element analysis (FEA) models in COMSOL software ($www.comsol.com$). Design candidates that meet the stiffness design requirement ($K \leq 0.014\ N/m$) for the $\mu FSMM$ system are selected for further investigation.

3.3. Mobile Microrobot Design

In fluidic environments, the mobile agent needs to overcome the resistance which is generated by pressure gradients and friction drag on the object surface. When the agent is small compared to the environment, the pressure gradient can be treated as zero and the fluid drag F_d only depends on the friction on the agent surfaces due to the motion in fluid. Therefore, the requirement of mobility for the $\mu FSMM$ can be expressed as:

$$F_m > F_d \tag{3}$$

where F_m is the magnetic driving force evaluated as Equation (1). Considering the $\mu FSMM$ as a 2D flat plate, the fluid drag force F_d can be evaluated through the drag coefficient C_d as [32]:

$$C_d = \frac{F_d}{\frac{1}{2}\rho v^2 S} = \frac{1.328}{\sqrt{Re}} \tag{4}$$

where ρ is mass density of the fluid media, v is flow velocity, S is the total surface area in contact with the fluid and Re is the Reynolds number as $Re = vL/\nu$. ν is the kinematic viscosity of the fluid media and L can be treated as the length of the microrobot. Equation (4) is valid only if the microrobot is translating in a laminar flow. Due to the small inertia, the microscale robot has a low Reynolds number, which indicates the laminar flow domain. Thereafter, the mobility requirement for the $\mu FSMM$ agent can be assessed as:

$$F_m > F_d = 0.664\frac{\rho v^2 S}{\sqrt{Re}} \tag{5}$$

where the acting magnetic force can be evaluated in Equation (1). For quantitative estimation, assume the magnetic volume is $L_m \times L_m \times t_m$, the dimension to the total device is $L \times L \times t$ and $t/L = 1/10$, $L_m/L = 2/3$, where L is the length and t is the thickness. In addition, assume the microrobot works in water environment, where $\rho = 1 \times 10^3\ kg/m^3$, $\nu = 1.0\ cSt = 1 \times 10^{-6}\ m^2/s$, and the set translation velocity is αL, which means the microrobot travels α times of body length per second. Let $\alpha = 2$ and plug Equation (1) and Re into Equation (5), we can evaluate the required thickness of magnetic layer as:

$$t_m > \frac{1.59 \cdot \rho \cdot \alpha^{\frac{3}{2}} \cdot \nu^{\frac{1}{2}}}{(M \cdot \nabla)B} \cdot (\frac{L}{L_m})^2 \cdot L = 0.0067\ L \tag{6}$$

which means in the above working environment, the minimum thickness of the nickel magnetic layer for a $1\ mm$ device is about $7\ \mu m$. Equation (6) indicates that this requirement is related to the fluid properties, the available magnetic field, the desired mobility and the body length of the $\mu FSMM$ microrobot.

4. Prototype Fabrication

The $\mu FSMM$ prototype is fabricated through multiple step surface micromachining process and subsequent assembly (Figure 3). As shown in Figure 3a, in order to manufacture the PDMS micro force sensor, a mold made of negative photoresist (KMPR 1050, MicroChem, $www.microchem.com$) is patterned at first in step (1) and (2). The thickness of the mold is customized in the spin coat process in step (1), which determines the resultant thickness of the micro force sensor. For easy releasing of the force sensor part, the patterned photoresist mold is then exposed to CF_4 plasma for 1 min. PDMS (Sylgard 184, $www.dowcorning.com$) mixed with base:cure ratio of 12.5:1 (by mass) is then filled in the mold by a spin coating process (step (3)). After curing, the overflowed PDMS layer is planed down (step (4)) either by dry etching with CO_2 laser or wet etching with a solution of tetrabutylammonium fluoride (TBAF) in N-methylpyrrolidinone (NMP) (3:1; v/v; NMP/75% TBAF in water). Discussion of these two techniques are presented in [30]. Afterwards, the photoresist mold is dissolved in Remover PG bath (step (5)) and the PDMS part is released in step (6).

After the individual PDMS part is released, a beryllium copper base is epoxied to the base section of the PDMS force sensor. The base layer is used to strengthen the device and helps the attachment with magnetic body layer. The overall body length of the fabricated prototype is around $2.5\ mm$. Based on Equation (6), the required thickness of the magnetic layer is about $16\ \mu m$. Therefore, a customized $25\ \mu m$ thick nickel foil ($www.alfa.com$) is attached to the previous subassembly. This nickel layer serves as the magnetic microrobot body, which makes it possible to drive the whole assembly by the magnetic field gradients.

One example prototype is shown in Figure 4. This prototype corresponds to the $P12$ design configuration in [30] and is depicted in Figure 4a. The fabrication result shows clean cut of the geometry. The device was fabricated with a larger thickness for better mechanical strength, because the soft workpiece is easy to bend and crash during the handling and transportation. In addition, thicker part is also easier for the releasing process after the fabrication. The stiffness of the micro force sensor will be analyzed and calibrated in the following section.

Figure 3. The fabrication process of $\mu FSMM$ prototype. (**a**) The microfabrication process of the micro force sensor. (**b**) Assembly of integrated prototype.

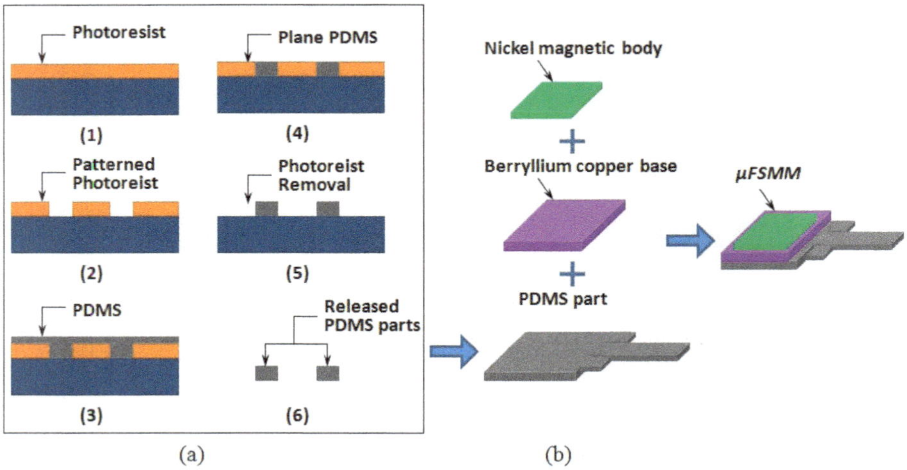

Figure 4. Example of $\mu FSMM$ prototyping result. (**a**) The design diagram of corresponding micro force sensor. (**b**) Photograph of final resulting assembly.

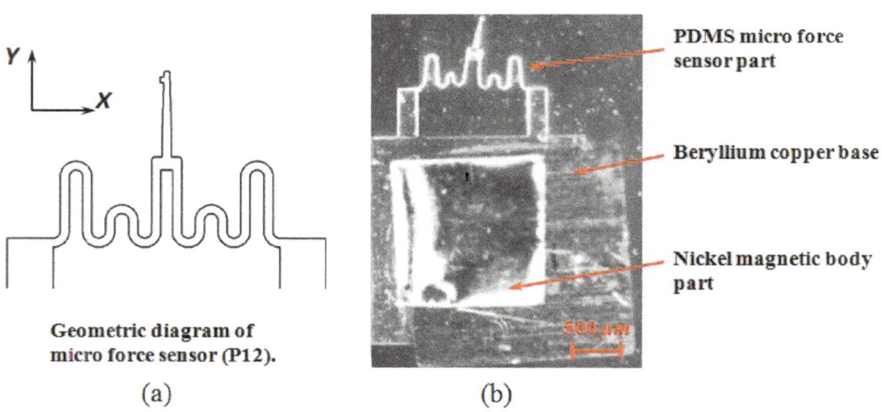

5. Calibration

The prototype is experimentally calibrated with a micro force sensing probe (FT-S1000, FemtoTools, Zurich, Switzerland) as shown in Figure 5. The micro force sensing probe is able to sense the force in the range of $\pm 1,000\ \mu N$ and with resolution of $0.05\ \mu N$ at $10\ Hz$ sampling frequency. The $\mu FSMM$ prototype is set along X and Y directions with tip right against the force sensing probe. Therefore, the traveled distance of the micro force sensing probe is also the deflection of the micro force sensor on $\mu FSMM$. Both tests along X and Y direction are repeated three times, and the average of the results are plotted in Figure 6. The standard deviations of the stiffness are $3.0 \times 10^{-4}\ N/m$ and $3.7 \times 10^{-4}\ N/m$ on the X and Y directions, respectively. For both results of simulation and measurement, the stiffness in

X direction is lower than that in Y direction. From the summarized stiffness results of the micro force sensor (Table 2), it is noted that the experimentally tested stiffness is higher than the simulation result of original design. The result shows that the thicker fabricated device layer will increase the stiffness as expected, but does meet the design requirement.

Figure 5. The experimental setup for the calibration of micro force sensor. The position of micro force sensor and force sensing probe is adjusted by linear motion stages with resolution under 20 μm.

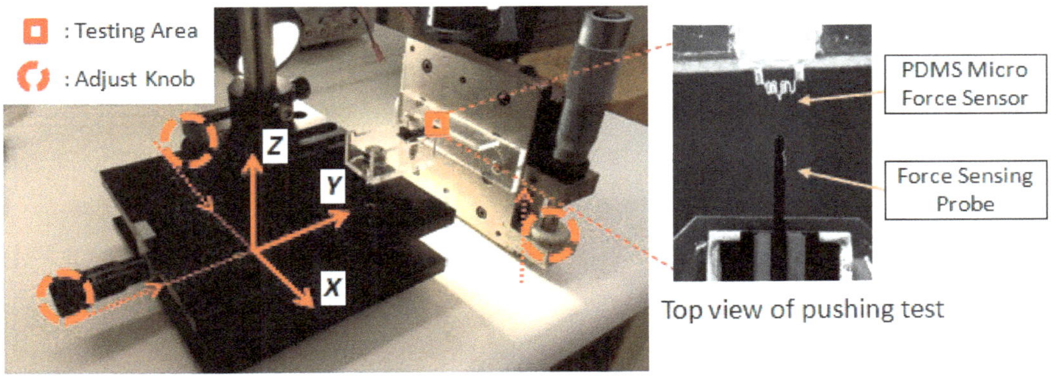

Figure 6. The experimental calibration of the micro force sensor part on $\mu FSMM$. The micro force sensing probe pushes the tip of PDMS part at low speed (0.01 mm/s). The stiffness of the PDMS force sensor is fitted with the test results as $K_x \approx 0.00534\ N/m,\ K_y \approx 0.0105\ N/m$.

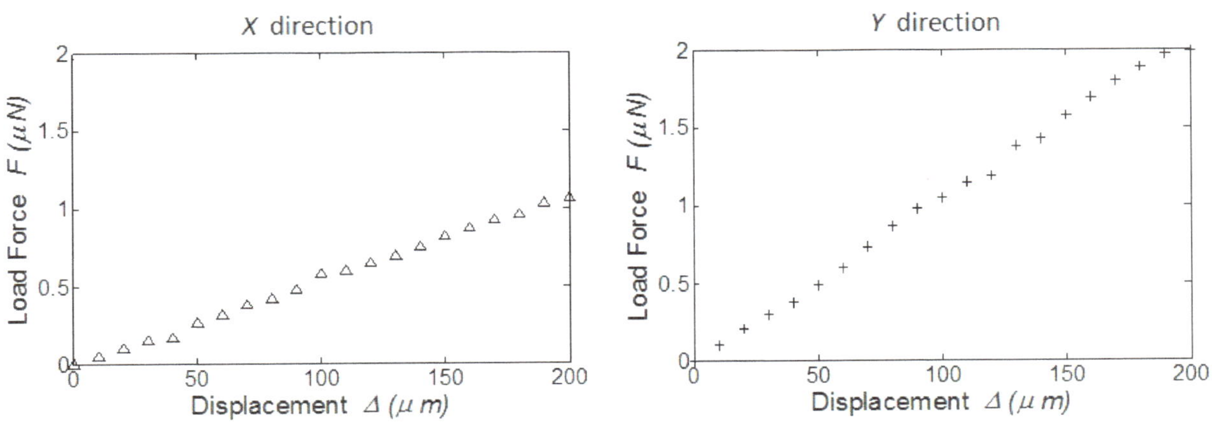

Table 2. Stiffness of design, modeling and measuring results.

	K_x (N/m)	K_y (N/m)
Design criteria	0.014	0.014
FEA-design parameters	0.0012	0.0070
Experimental measurements	0.0053	0.0105

6. Characterization

6.1. Mobility

The fundamental purpose of this $\mu FSMM$ design is to serve as a mobile microrobot. Therefore, the mobility of this magnetic assembly is tested with the electromagnetic coil system which is described in [11]. Before inducing the 2D motion of $\mu FSMM$ agent by the side coils of the coil system, the produced magnetic field on the horizontal plane is analyzed through FEA model in COMSOL software (Figure 7a). The coil is able to produce straight magnetic field in the central area of workspace. The magnetic field is also verified with measurements using Gauss meter (GM1-ST, AlphaLab Inc., West Salt Lake City, UT, USA). The field strengths are measured at the grid points of 5×5 isometric mesh of the workspace and the field is interpolated based on the measurements, as shown in Figure 7b.

Figure 7. The magnetic field actuating the $\mu FSMM$ microrobot. (**a**) The FEA model using COMSOL software for a single coil. (**b**) The magnetic field measurements at discrete points in workspace using Gauss meter. Both input currents are $1.0\ A$ in the coil.

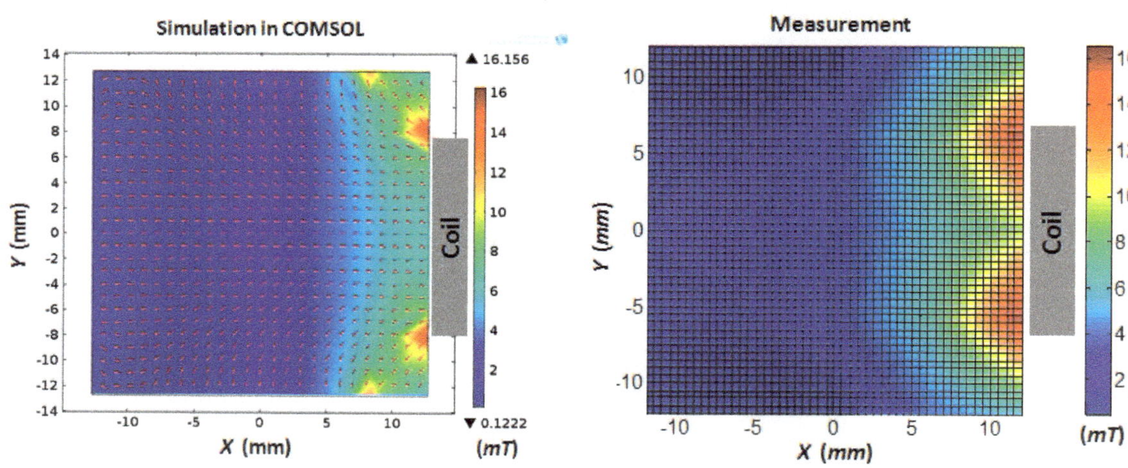

(**a**) (**b**)

With the electromagnetic coil drive system, this $P12$ $\mu FSMM$ prototype (Figure 4) is able to translate in DI water with current input as low as $0.05\ A$. A series of translation tests are conducted under increasing magnetic field flux intensities, where the average velocities during translations are shown in Figure 8a. The microrobot demonstrates good mobility with low magnetic field intensity levels. The agent has also been rotated under a rotating magnetic field signal. The deviation of the geometric center point from its original position has been recorded in Figure 8b. The small deviation shows stable controllability of the orientation of the $\mu FSMM$ agent.

Moreover, the eventual $\mu FSMM$ is targeting biomedical tasks on the microscale, where the robot is required to own a footprint at the μm level. The mobility of such a microscale magnetic structure has been tested using the microscale I-bar shaped prototype (Figure 9). The outline of this planar structure is patterned through an SU-8 photoresist layer and magnetic layer made of nickel by physical vapor deposition. The rectangular geometry of this agent is varied to investigate its affects on mobility in a fluid environment, where the linear motion results in DI water are summarized in Table 3. It shows that thicker magnetic layer leads to faster speed, which indicates the magnetic volume plays a significant role for the mobility. For the same magnetic volume, the larger aspect ratio of the rectangular shape has better mobility. This is indicated by increasing velocity results on the main diagonal in Table 3. These rules can serve as reference for the future design of a fully microscale $\mu FSMM$ agent.

Figure 8. The results of the mobility tests for the $\mu FSMM$ prototype. (**a**) Linear translation velocity *vs.* magnetic field flux intensity. (**b**) The deviation of geometric center point when the prototype is rotated with 360°.

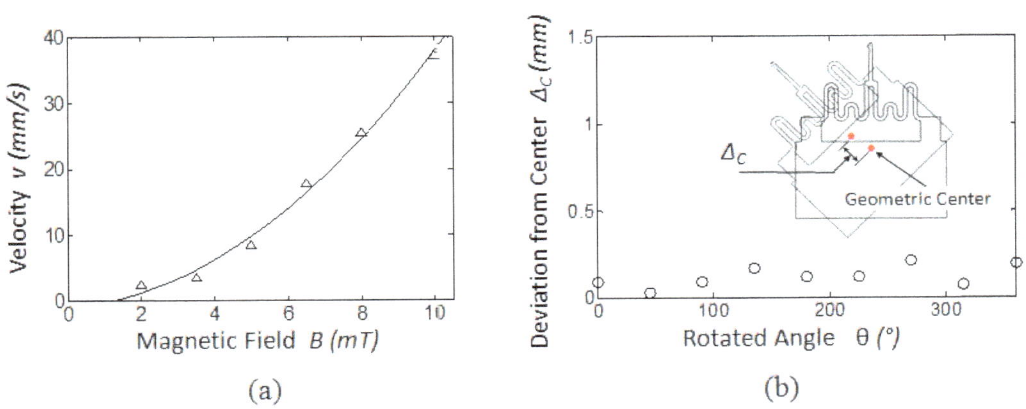

Figure 9. The design of I-bar shape micro magnetic agent. The inset photograph shows the fabricated prototypes.

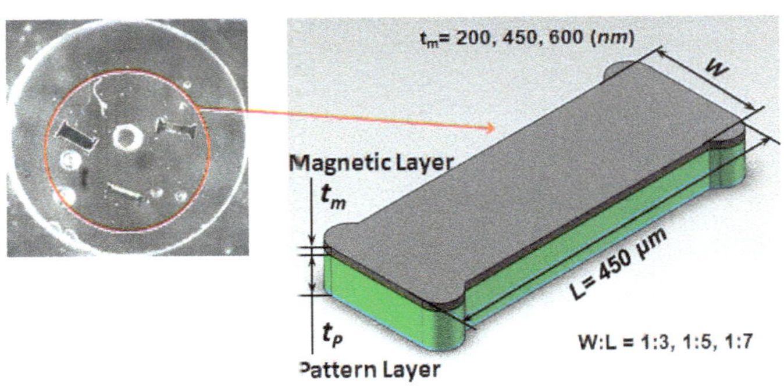

Table 3. The linear motion results of micro planar I-bar shaped magnetic agent

	Velocity	(mm/s)	
t_m (nm)	AR **1:3**	AR **1:5**	AR **1:7**
200	1.55	1.29	1.10
400	4.70	4.27	3.64
600	10.24	9.21	6.02

Note: (1) AR means the aspect ratio for $W : L$. (2) The motion tests are conducted in DI water environment. The input current is $0.5\ A$ and the magnetic field is $2\ mT$.

6.2. Calibration of Input Current to Force Relationship

The purpose of this $\mu FSMM$ design is to couple a micro force sensing end-effector with a mobile microrobot. In practical microrobot applications, the blocking force acting on the robot body is due to the motion/momentum of the mobile microrobot. In the scenario of our magnetic mobile microrobot system, the sensed micro force corresponds to the current input into the electromagnetic drive coil. The sensed blocking force depends on the motion of $\mu FSMM$ agent, which relies on the magnetic field generated by the drive coil. Therefore, with a calibrated micro force sensor, the in situ force can be derived through the current input into the drive coil. This relation is examined here with the prototype calibrated in previous section ($K_y = 0.0105\ N/m$).

The microrobot assembly is powered with a known current input and driven against an obstacle (wall), which blocks the agent in the Y direction. The blocking force can be evaluated by the exerting magnetic force, which depends on the derived magnetic field due to the current input. The resultant deflection is recorded and the corresponding blocking force is then backed out based on the known stiffness of the micro force sensor. Therefore, the acting force on the magnetic mobile microrobot can be calibrated with the vision based micro force sensor.

A series of calibration tests have been conducted with increasing current input I: $0.5A$, $1.0A, 2.0A, 3.0A, 4.0A, 5.0A, 6.0A$. After the producing magnetic field is assessed with simulation and measurement, the exerting magnetic force is calculated with Equation (1). At the same time, the corresponding force is also calculated by the resultant deflection of PDMS micro force sensor with known stiffness. The series of resulting forces are compared in Figure 10. The maximum differential is $0.021\ \mu N$ ($\approx 5\%$ error), which shows good consistency between driving the magnetic microrobot and the micro force sensing. Because the acting force is proportional to the current input, this linear relation can be fitted according to the calibration result with force sensor (Figure 10) as $F\ (\mu N) = 0.0634 \times I\ (A)$, where F is the acting force and I is the input current.

Figure 10. The force calibration according to the input current. F_m indicates the force derived through magnetic principle (Equation (1)), while F represents the evaluated force with micro force sensor (Equation (2)).

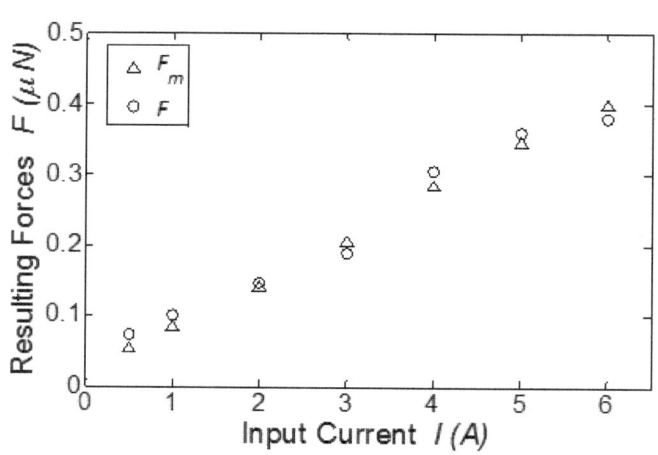

7. Conclusions and Future Work

This paper describes a micro force sensing mobile microrobot ($\mu FSMM$). It is the first mobile microrobot with in situ force sensing capabilities. The micro force sensing module is realized through a 2D vision based compliant mechanism end-effector made of PDMS, which is able to detect the forces at low μN down to nN levels. This micro force sensor module is experimentally tested with a micro force sensing probe. A magnetic layer is attached to the micro force sensor which serves as drive part of the microrobot assembly. The sensed force is also calibrated against the acting magnetic force. This builds up the relation between the micro force and the current input to the electromagnetic coil for the magnetic microrobot drive system.

The work presented here only includes the preliminary analysis and prototyping as a proof-of-concept. Comprehensive experimentation is required before real biomedical applications for this $\mu FSMM$ design can be realized. The calibration is needed mainly due to the errors generated during the fabrication process. Although the manufactured prototype still meets the design requirement, the deviation of geometric parameters will increase the stiffness of the micro force sensor. Moreover, the eventual $\mu FSMM$ agent is intended to own a footprint at μm level. Therefore, it is critical to control the accuracy of the fabrication process. In addition to the PDMS micro force sensor end-effector, the magnetic drive part also needs precise fabrication methods at the microscale, such as the photolithography process proposed in [11], or physical and chemical etching process to outline the magnetic volume. This is also important because magnetic volume determines the acting force on the microrobot, which requires delicate calibration. Nevertheless, this proof-of-concept work presented here proves the possibility of incorporating a force sensing function in the magnetic microrobot system that is desirable for biomedical tasks at the microscale.

Acknowledgements

The authors gratefully acknowledge the support of NSF IIS Award # 1149827.

Authors Contributions

David Cappelleri came up with the idea for creating a magnetic microrobot with a vision-based micro-force sensor. The magnetic microrobot body design is based on the work of Wuming Jing. The vision-based micro-force sensor design is based on the work of Cappelleri. Both authors equally contributed to the fabrication process design and fabrication of the prototypes. Jing performed the calibration and characterization analysis and wrote the paper.

Conflicts of Interest

The authors declare no conflict of interest.

References

1. Dong, L.; Nelson, B.J. Robotics in the small, part 2: Nanorobotics. *IEEE Robot. Autom. Mag.* **2007**, doi:10.1109/MRA.2007.905335.

2. Fatikow, S.; Rembold, U. *Microsystem Technology and Microrobotics*; Springer: New York, NY, USA, 1997.
3. Donald, B.R.; Levey, C.G.; McGray, C.D.; Paprotny, I.; Rus, D. An Untethered, Electrostatic, Globally Controllable MEMS Micro-Robot. *J. Microelectromech. Syst.* **2006**, *15*, 1–15.
4. Sul, O.J.; Falvo, M.R.; Taylor, R.M.; Washburn, S.; Superfine, R. Thermally actuated untethered impact-driven locomotive microdevices. *Appl. Phys. Lett.* **2006**, doi:10.1063/1.2388135.
5. Zhang, L.; Abbott, J.; Dong, L.; Kratochvil, B.E.; Bell, D.; Nelson, B.J. Artificial bacterial flagella: Fabrication and magnetic control. *Appl. Phys. Lett.* **2009**, *94*, 064107:1– 064107:3.
6. Frutiger, D.R.; Vollmers, K.; Kratochvil, B.E.; Nelson, B.J. Small, Fast, and Under Control: Wireless Resonant Magnetic Micro-agents. *Int. J. Robot. Res.* **2010**, *29*, 613–636.
7. Tung, H.W.; Frutiger, D.R.; Pan, S.; Nelson, B.J. Polymer-Based Wireless Resonant Magnetic Microrobots. In Proceedings of the IEEE International Conference on Robotics and Automation (ICRA), St. Paul, MN, USA, 14–18 May 2012.
8. Pawashe, C.; Floyd, S.; Sitti, M. Modeling and experimental characterization of an untethered magnetic micro-robot. *Int. J. Robot. Res.* **2009**, *28*, 1077–1094.
9. Jiang, G.L.; Guu, Y.H.; Lu, C.N.; Li, P.K.; Shen, H.M.; Lee, L.S.; Yeh, J.A.; Hou, M.T.-K. Development of rolling magnetic microrobots. *J. Micromech. Microeng.* **2010**, *20*, 085042.
10. Tung, H.W.; Peyer, K.E.; Sargent, D.F.; Nelson, B.J. Noncontact manipulation using a transversely magnetized rolling robot. *Appl. Phys. Lett.* **2013**, *103*, 114101.
11. Jing, W.; Pagano, N.; Cappelleri, D.J. A novel micro-scale magnetic tumbling microrobot. *J. Micro-Bio Robot.* **2013**, *8*, 1–12.
12. Jing, W.; Pagano, N.; Cappelleri, D. A Tumbling Magnetic Microrobot with Flexible Operating Modes. In Proceedings of the IEEE International Conference on Robotics and Automation (ICRA), Karlsruhe, Germany, 6–10 May 2013.
13. Abbott, J.J.; Ergeneman, O.; Kummer, M.; Hirt, A.M.; Nelson, B.J. Modeling Magnetic Torque and Force for Controlled Manipulation of Soft-Magnetic Bodies. *IEEE Trans. Robot.* **2007**, *23*, 1247–1252.
14. Kummer, M.; Abbott, J.J.; Kratochvil, B.E.; Borer, R.; Sengul, A.; Nelson, B.J. OctoMag: An Electromagnetic System for 5-DOF Wireless Micromanipulation. *IEEE Trans. Robot.* **2010**, *26*, 1006–1017.
15. Tabatabaei, S.N.; Lapointe, J.; Martel, S. Shrinkable Hydrogel-Based Magnetic Microrobots for Interventions in the Vascular Network. *Adv. Robot.* **2011**, *25*, 1049–1067.
16. Sun, Y.; Nelson, B. MEMS for cellular force measurements and molecular detection. *Int. J. Inf. Acquis.* **2004**, *1*, 23–32.
17. Li, X.; Su, D.; Zhang, Z. A novel technique of microforce sensing and loading. *Sens. Actuators A: Phys.* **2009**, *153*, 13–23.
18. Gnerlich, M.; Perry, S.F.; Tatic-Lucic, S. A submersible piezoresistive MEMS lateral force sensor for a diagnostic biomechanics platform. *Sens. Actuators A: Phys.* **2012**, *188*, 111–119.
19. Hoover, A.; Fearing, R. Rapidly prototyped orthotweezers for automated microassembly. In Proceedings of the IEEE International Conference on Robotics and Automation (ICRA), Roma, Italy, 10–14 April 2007.

20. Perez, R.; Chaillet, N.; et al.. Fabrication, modeling and integration of a silicon technology force sensor in a piezoelectric micro-manipulator. *Sens. Actuators A: Phys.* **2006**, *128*, 367–375.
21. Beyeler, F.; Nelid, A.; Oberti, S.; Bell, D.J.; Sun, Y.; Dual, J.; Nelson, B.J. Monolithically Fabricated Microgripper With Integrated Force Sensor for Manipulating Microobjects and Biological Cells Aligned in an Ultrasonic Field. *J. Microelectromech. Syst.* **2007**, *16*, 7–15.
22. Beyeler, F.; Muntwyler, S.; Nelson, B.J. A Six-Axis MEMS Force? Torque Sensor With Micro-Newton and Nano-Newtonmeter Resolution. *J. Microelectromech. Syst.* **2009**, *18*, 433–441.
23. Koch, S.J.; Thayer, G.E.; Corwin, A.D.; de Boer, M.P. Micromachined piconewton force sensor for biophysics investigations. *Appl. Phys. Lett.* **2006**, *89*, 173901.
24. Wang, X.; Ananthasuresh, G.; Ostrowski, J. Vision-based sensing of forces in elastic objects. *Sens. Actuators A: Phys.* **2001**, *94*, 142–156.
25. Greminger, M.; Nelson, B. Vision-based force measurement. *IEEE Trans. Pattern Anal. Mach. Intell.* **2004**, *26*, 290–298.
26. Jeong, K.H.; Keller, C.; Lee, L. Direct force measurements of biomolecular interactions by nanomechanical force gauge. *Appl. Phys. Lett.* **2005**, *86*, 193901.
27. Ioppolo, T.; Kozhevnikov, M.; Stepaniuk, V.; Otugen, M.V.; Sheverev, V. Micro-optical force sensor concept based on whispering gallery mode resonators. *Appl. Opt.* **2008**, *47*, 3009–3014.
28. Sasoglu, F.; Bohl, A.; Layton, B. Design and microfabrication of a high-aspectratio pdms microbeam array for parallel nanonewton force measurement and protein printing. *J. Micromech. Microeng.* **2007**, *17*, 623–632.
29. Liu, X.; Kim, K.; Zhang, Y.; Sun, Y. Nanonewton force sensing and control in microrobotic cell manipulation. *Proc. Robot.: Sci. Syst. IV* **2009**, *28*, 1065–1076.
30. Cappelleri, D.J.; Piazza, G.; Kumar, V. A two dimensional vision-based force sensor for microrobotic applications. *Sens. Actuators A: Phys.* **2011**, *171*, 340–351.
31. Sia, S.K.; Whitesides, G.M. Microfluidic devices fabricated in Poly(dimethylsiloxane) for biological studies. *Electrophoresis* **2003**, *24*, 3563–3576.
32. Fox, R.W.; McDonald, A.T. *Introduction to Fluid Mechanics*; John Wiley & Sons, Inc.: Hoboken, NJ, USA, 1992.

12

Vision-Based Cooperative Pose Estimation for Localization in Multi-Robot Systems Equipped with RGB-D Cameras

Xiaoqin Wang *, Y. Ahmet Şekercioğlu and Tom Drummond

Department of Electrical and Computer Systems Engineering, Monash University, Melbourne 3800, Australia; E-Mails: ahmet.sekercioglu@monash.edu (Y.A.S.); tom.drummond@monash.edu (T.D.)

* Author to whom correspondence should be addressed; E-Mail: xiaoqin.wang@monash.edu

Academic Editor: Wenjie Dong

Abstract: We present a new vision based cooperative pose estimation scheme for systems of mobile robots equipped with RGB-D cameras. We first model a multi-robot system as an edge-weighted graph. Then, based on this model, and by using the real-time color and depth data, the robots with shared field-of-views estimate their relative poses in pairwise. The system does not need the existence of a single common view shared by all robots, and it works in 3D scenes without any specific calibration pattern or landmark. The proposed scheme distributes working loads evenly in the system, hence it is scalable and the computing power of the participating robots is efficiently used. The performance and robustness were analyzed both on synthetic and experimental data in different environments over a range of system configurations with varying number of robots and poses.

Keywords: pose estimation; RGB-D camera; self-calibration; cooperative localization; multi-robot coordination

---

1. Introduction

Multi-Robot Systems (MRSs), which were first proposed in early 1980s, are becoming increasingly popular mobile sensor platforms to measure and estimate quantities of interest at spatially distributed locations. Compared to a single-robot operation, MRSs have the advantages on faster task completion, more extensive coverage, increased reliability to sensor failures, and higher estimation accuracy through sensor fusion. MRSs have been widely used in a variety of tasks such as data collection [1], surveillance [2–4], target tracking [5–7], formation tracking and control [8–10], and visual SLAM [11–13].

Precise knowledge on locations and orientations (poses) of the robots is a prerequisite for the successful accomplishment of the collaborative tasks. One approach to localize the robots is to equip them with Global Positioning System (GPS) receivers and use the GPS system. However, GPS signals are not available indoors and they cannot directly provide the orientation information. An alternative approach is cooperative localization in which robots work together and use the robot-to-robot measurements to construct a map of their network. Cooperative localization operates in two broad stages: The *initialization process*, which involves *relative pose estimation* (RPE), provides initial location and orientation estimates of the robots. Then the *refinement process* updates the initial estimates iteratively to enhance the accuracy.

In this study, we focus on the initialization process of the cooperative localization in MRSs. The MRSs of interest here are equipped with cameras or visual sensors. In this kind of MRSs, robots' locations and orientations can be estimated through careful calibration of the rigid body transform operations between the cameras mounted on the robots. Generally there are two kinds of vision-based approaches to achieve this goal: (1) manual calibration; or (2) self-calibration. Manual calibration approaches require a special calibration pattern to be visible in all images [14] or the precise pose information of calibration patterns/objects have to be known [15]. Some easily detectable single features which require human interaction, such as moving a LED in a dark room, can also be used to manually calibrate multiple cameras [16–18]. Manual calibration methods, even though provide good results, require special equipment or time consuming manual measurements. Self-calibration algorithms simultaneously process several images captured by different cameras and find the correspondences across images. Correspondences are established through extracting 2D features from images automatically and matching them between different images. Then, based on the established correspondences, cameras' relative poses can be estimated from the essential matrix. Besides the algorithm which uses static features [19–22], Aslan *et al.* [23] detect people walking in the room and use the point on the top of each person's head as the calibration feature The accuracy of the self-calibration highly depends on the reliability of the relative pose estimates. This problem was first discussed in [24] with the concept of the vision graph. Kurillo *et al.* [25], Cheng *et al.* [26], and Vergés-Llahí *et al.* [27] later used and refined it for this purpose as well. It is becoming a useful general tool for describing the directionality of networked visual sensors. It has been more recently addressed by Bajramovic *et al.* [28–30]. They proposed a graph based calibration method which measures the uncertainty of the relative pose estimation between each camera pair. All of the self-calibration algorithms measure the epipolar structure of the system and suffer from scale ambiguity. If there is not any object or pattern with known geometry in the scene, the orientations and locations between robots are determined up to a scale.

In this paper, we consider the initialization process for localization in a multi-robot system with $N \geq 3$ robots operating in GPS-denied indoor environments (see Figure 1). A RGB-D camera, which provides both color images and per-pixel depth information, is mounted at the top of each robot. A central node with high performance processor is also implemented in the system which can operate computationally expensive computer vision algorithms. We present a novel self-calibration algorithm to determine the locations and orientations of the robots in this RGB-D camera equipped multi-robot system. The propose scheme can be arranged over an indoor scenarios without imposed constraints for all robots to share a common field-of-view (FoV). Our approaches assume that at least any two given robots have overlapping FoVs and that the cameras on robots have been internally calibrated prior to deployment. Our proposed algorithms consist of the following steps: (1) each robot extracts color features locally and sends the descriptors of these features to the central node; (2) the central node performs feature matching to determine neighboring robots and generates an Initial Pose Matrix (IPM); (3) the central node constructs a robot dependency graph and selects a number of relative poses to connect robots as a calibration tree; (4) after the central node broadcasts the information of the calibration tree, robots work collaboratively to determine the relative poses according to the calibration tree; (5) the determined relative poses are then transmitted to the central node to compute the poses of all the robots in the system. We formulate the selection of relative poses as a shortest path problem, which consists of finding shortest path from a vertex to the other vertices in an edge-weighted graph. The graph represents FoVs of robots as vertices and overlapping FoVs as edges respectively.

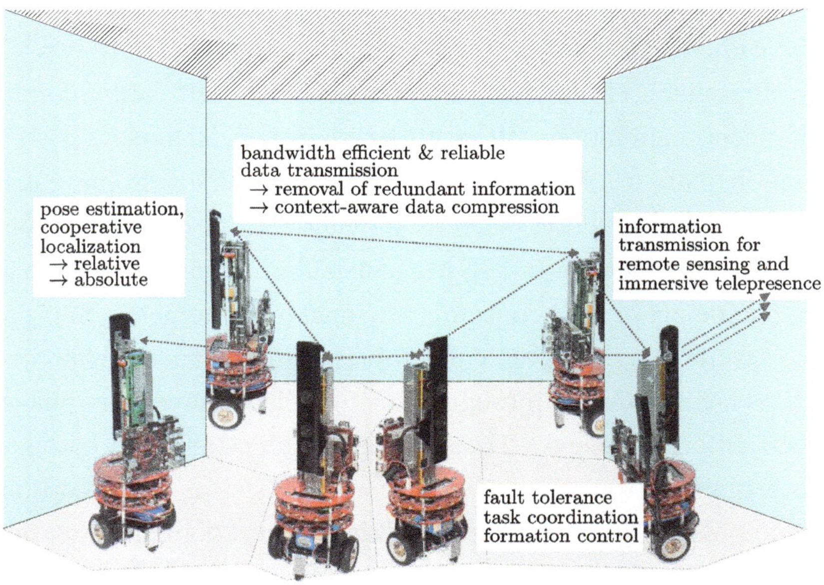

Figure 1. An indoor mapping and exploration scenario showing the Monash University's RGB-D equipped experimental mobile robots "eyeBugs". A typical application would be mapping indoors after a disaster such as Fukushima nuclear reactor accident. As shown in the diagram, there are numerous challenges that need to be addressed. In this paper, our focus is initialization problem in cooperative localization.

The main contributions of this paper are:

- Construction of a robot dependency graph based on the overlapping ratio between neighboring robots.
- Development of a procedure to determine the relative pose of multiple RGB-D camera equipped robots.
- By contrast to the conventional approaches that only utilize color information, our approach takes the advantages of the combination of RGB and depth information.
- The locations and orientations of robots are determined up to the real world scale directly without involving scale ambiguity problem.

- Extensive experiments using synthetic and real world data were conducted to evaluate the performance of our algorithms in various environments.

The rest of the paper is organized as follows. In Section II, the characteristics of the RGB-D camera and the multi-robot system used in this paper are introduced. In Section III, we formulate the robot localization problem and propose our solutions. In Section IV, we present the experiments and results, and our concluding remarks can be found in Section V.

2. A Multi-Robot System Using RGB-D Cameras As Visual Sensors

2.1. eyeBug: A Robot Equipped with RGB-D Camera

At the Wireless Sensor and Robot Networks Laboratory (WSRNLab) [31] we have created a multi-robot system consisting of experimental mobile robots called eyeBugs (Figure 2) for computer vision, formation control, visual sensing research activities. A single-board computer, BeagleBoard-xM [32] is the main computational resource of an eyeBug. Each BeagleBoard-xM has an ARM37x 1 GHz processor, a USB hub, and an HDMI video output. Communication service between robots is provided through WiFi links. We run an ARM processor optimized GNU/Linux operating system (Ubuntu Version 11.10) [33]. OpenKinect [34], OpenCV [35] and libCVD [36] libraries were installed to capture and process image information. Microsoft Kinect, which produces color and disparity-based depth images, is mounted vertically at the center of the top plate of each robot. The default RGB video stream provided by Kinect uses 8 bits for each color at VGA resolution (640 $\times$ 480 pixels, 24 bits/pixel). The depth video stream is also in VGA resolution.

Figure 2. eyeBug-the robot developed for the Monash WSRNLab's [31] experimental multi-robot platform. The RGB-D data generated by the Kinect RGB-D sensor is processed on BeagleBoard-xM running GNU/Linux operating system.

2.2. Characteristics of RGB-D Camera

Kinect has an infrared (IR) projector-camera-pair and a RGB camera. The depth sensing of Kinect is based on a fixed structured light source positioned at a known baseline from the IR camera. The depth information is measured through a triangulation process which is based on the detection of transverse shifts of local dot patterns in the IR speckle with respect to its reference patterns at a known distance to the device [37]. This process runs repeatedly on all regions in the IR speckle and generates a disparity-based depth image. It should be noted that the depth signal inevitably degrades when multiple RGB-D cameras are pointing at the same scene. It is because the camera projects a structured light dot pattern onto the scene continuously without modulation and devices interfere with one another [38]. This interference can be eliminated by a "Shake 'n' Sense" approach [38].

The normalized disparity values returned by the Kinect are inversely proportional to their depth [39]. Furthermore, [40,41] show that there is a non-linear relationship between the normalized disparity values and their depth value in Euclidean space. Therefore, it is more suitable to represent the data in inverse depth coordinates. Consider the vector $\mathbf{p}_e = [x\ y\ z\ 1]^T$ which represents a real world point in Euclidean space by using homogeneous coordinates. The relation between a real world point in inverse depth coordinates and its corresponding pixel in the depth image can be established as follows,

$$\mathbf{p}_e \equiv \frac{1}{z}\begin{bmatrix} x & y & z & 1 \end{bmatrix}^T \equiv \begin{bmatrix} u & v & 1 & q \end{bmatrix}^T \equiv \begin{bmatrix} \frac{i-i_c}{f_x} & \frac{j-j_c}{f_y} & 1 & \frac{1}{z} \end{bmatrix}^T \tag{1}$$

where (i, j) denotes the pixel coordinates of this real world point projection in the depth image, and z is the corresponding depth value returned by the camera.

3. Self-Calibration Cooperative Pose Estimation

3.1. Overview

Given $N(N \geq 3)$ robots equipped with intrinsically calibrated RGB-D cameras, the goal is to automatically determine the initial pose of each robot in a common coordinate system using only the color and depth data. A central node with a high performance processor is also included in the system which runs the computationally expensive algorithms. The function of this node is explained in Section 3.3.

When two robots a and b have sufficiently overlapping FoVs, the relative pose between two robots can be represented by a transformation matrix, $\mathbf{M}_{ab}$, in SE(3) as follow,

$$\mathbf{M}_{ab} = \begin{bmatrix} & \mathbf{R} & & \mathbf{t} \\ 0 & 0 & 0 & 1 \end{bmatrix} \tag{2}$$

where $\mathbf{R}$ is a 3×3 rotation matrix and $\mathbf{t}$ is a 3×1 translation vector. $\mathbf{M}_{ab}$ denotes relative pose of robot b with respect to robot a and is the rigid transformation from the coordinate system of robot b to that of robot a. If there is a robot c and the relative pose between robots c and b is $\mathbf{M}_{bc}$, then the relative pose between robots a and c can be derived via composition as,

$$\mathbf{M}_{ac} = \mathbf{M}_{bc}\mathbf{M}_{ab} \tag{3}$$

This transformation provides a mapping from the coordinate system of c to that of b, then from that of b to that of a. Robot b is the *intermediate node* in this process. This operation is transitive, therefore one robot's pose relative to another can be determined indirectly over an arbitrary number of intermediate poses if they exist.

Thus, the system's topology can be built up from the pairwise relative poses between robots that have common FoVs. In order to achieve this, we first need to determine the robots with sufficiently overlapping FoVs. Secondly, robots are grouped in pairs to determine rough estimations of the relative poses, and a number of relative poses are selected based on the reliability of the pose information. In the final step, we calibrate the overall system based on the selected pairwise relative poses. A general description of the scheme we propose is shown in Figure 3. Each step is described in details in the following sections. The list of symbols used through the paper is given in Table 1.

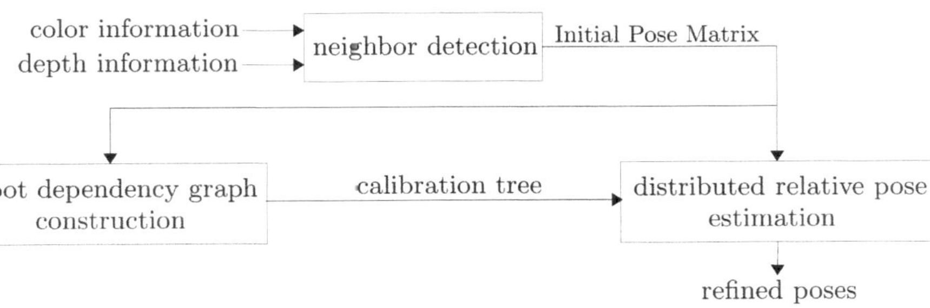

Figure 3. Operational overview of the proposed self-calibration scheme for cooperative pose estimation.

Table 1. Mathematical Notation.

Notation	**Description**
$\mathbf{Z}_a$	Depth image captured by robot a.
$\mathbf{p}_e$	Vector representing a real world point in Euclidean space.
$(i_{c,a}, j_{c,a})$	Principal point coordinates of the pinhole camera model.
$(f_{x,a}, f_{y,a})$	Focal length of the camera in horizontal and vertical axes.
$\mathbf{M}_{ab}$	Transformation matrix describing the relative pose between robots a and b.
$\mathbf{p}_a^l$	Sampled points on the depth image captured by robot a.
$\mathbf{p}_b^{l*}$	Corresponding points of $\mathbf{p}_a^l$ on the depth image captured by robot b.
N_a	Number of sampled points on $\mathbf{Z}_a$.
P_a	Set of sample points on $\mathbf{Z}_a$.
P_b^*	Set of corresponding points of P_a on $\mathbf{Z}_a$.
$\vec{n}_{l*,b}$	Surface normal at point $\mathbf{p}_b^{l*}$.
$w_{l,a}$	Weight parameter for correspondence established between $\mathbf{p}_a^l$ and $\mathbf{p}_b^{l*}$.
$\mathbf{E}$	Update transformation matrix in each iteration.
α_j	An element of a 6D motion vector.
$\mathbf{G}_j$	6D motion generator matrices.

3.2. Assumptions

We make the following assumptions about the multi-robot system:

- Intrinsic parameters of the RGB-D camera on each robot are calibrated prior to deployment,
- At least two robots in the system have overlapping FoVs;
- The scene is static and the robots do not move during the localization process, and

- The robots can form an ad-hoc network and directly communicate with each other.

3.3. Neighbor Detection and Initial Relative Pose Estimation

We define robots with overlapping FoVs as neighbors. One robot's neighbors can be detected through searching for image pairs sharing a common FoVs. This search can be viewed as a matching of point correspondences that considers the local environment of each feature point set. There are three steps in neighbor detection process: feature detection, feature description, and feature matching. The first two steps are performed on each robot locally. Taking processing speed and accuracy into consideration, we implement FAST [42,43] for feature detection and ORG [44] for feature description on each robot. Instead of transmitting the complete images, each robot sends the feature descriptions to the central node to minimize the transmission load. The corresponding depth information of each feature is also transmitted in conjunction with the feature descriptors.

Associating the feature descriptors with their corresponding depth values, central node can generate feature points in 3D. Central node performs feature matching between every two sets of the feature descriptors. In order to increase the matching reliability and robustness against outliers, we adopt both the symmetrical matching scheme and the geometric constraint to refine the matched points. In the symmetrical matching scheme, the correspondences between two sets of feature descriptors are established bidirectionally. One group of correspondences is generated from matching the first feature set to the second feature set. The other group is produced from matching the second feature set to the first feature set. For a pair of matched features to be accepted, two features must be the best matching candidate of each other in both directions.

Then, we use RANSAC to find a coarse registration, $\mathbf{M}_{ij}^*$, between every two matched feature sets. The error metric used to find the best alignment is

$$\mathbf{M}_{ij}^* = \arg\max_{\mathbf{M}_{ij}} \left(\sum_{l=1}^{n} |\mathbf{M}_{ij}\mathbf{p}_i^l - \mathbf{p}_j^l|^2 \right) \tag{4}$$

Here, $\mathbf{p}_i^l$ and $\mathbf{p}_j^l$ contain the depth information of two matched feature points as described in Equation (1). Each term in the summation indicates the squared distance between the transformed pose of a feature point $\mathbf{p}_i^l$ in robot i's feature set and the matched feature point $\mathbf{p}_j^l$ in the robot j's feature set. Between every two matched feature sets, the central node samples a number of matched feature point pairs and determine the transformation matrix repeated. The determined transformation in each iteration is evaluated based on the number of inliers in the remaining 3D feature points. Ultimately, only the matched feature points which agree with the optimal transformation matrix are kept as the good matches. The determined coarse registration between every two matched feature sets are stored as the *initial relative poses*. Initial relative poses are not accurate which require further refinements.

After operating the above process on every two feature sets, an Initial Pose Matrix (IPM) can be constructed. As shown in Table 2, each element, $\mathbf{M}_{ij}^*$, represents the initial relative pose between robot i and robot j. The diagonal elements represent the relative pose with itself, thus they are negligible.

Table 2. Initial Pose Matrix (IPM) and Uncertainty Matrix (UM) of a multi-robot system with four robots.

FMM				
No.	**1**	**2**	**3**	**4**
1	×	$\mathbf{M}^*_{12}$	$\mathbf{M}^*_{13}$	$\mathbf{M}^*_{14}$
2	$\mathbf{M}^*_{21}$	×	$\mathbf{M}^*_{23}$	$\mathbf{M}^*_{24}$
3	$\mathbf{M}^*_{31}$	$\mathbf{M}^*_{32}$	×	$\mathbf{M}^*_{34}$
4	$\mathbf{M}^*_{41}$	$\mathbf{M}^*_{42}$	$\mathbf{M}^*_{43}$	×

UM				
No.	**1**	**2**	**3**	**4**
1	×	w_{12}	w_{13}	w_{14}
2	w_{21}	×	w_{23}	w_{24}
3	w_{31}	w_{32}	×	w_{34}
4	w_{41}	w_{42}	w_{43}	×

3.4. Selection of Relative Pose

After determining the neighboring robots and initial relative poses, we will show the problem on estimating all robots' poses can be transformed to the all-pair shortest path problem.

The relative pose between two neighboring robots can be estimated by RPE algorithm. In order to calibrate the whole system, we require to select a number of relative poses to link all robots together. Since different overlapping areas in FoVs lead to various uncertainty values in the RPE. This process should choose the relative poses with the minimum overall uncertainty between two robots. Furthermore, it is known that the accuracy of the estimation of the relative pose between two robots may significantly degrade when an increasing number of intermediate nodes are added into the computations. This is mainly due to uncertainty accumulated in each time RPE algorithm operates between two robots. In order to ensure each robot has the reliable knowledge of the other robots' locations and orientations, we require to select the relative poses which introduce the smallest overall amount of uncertainty value to calibration the system.

3.4.1. Robot Dependency Graph Construction

To efficiently consider all possible combinations of robot poses, we suggest the usage of the graph structure *robot dependency graph*. Robot dependency graph consists of a set of vertices representing each view of the scene observed by a robot. The weight on each edge indicates to the degree of uncertainty of the pair of views being connected. Thus, estimating all robots' poses can be transformed to finding the shortest path between every two vertices in the robot dependency graph.

In order to determine the weight on each edge, we need to first derive the uncertainty degree of relative pose estimation between every two neighbors. The relative pose between two neighboring robots can be estimated through aligning the 3D point clouds extracted from the depth images captured by different robots. The motion between two depth images can be estimated by various approaches, such as ICP variants [45–47], feature-based registrations [48,49], and combinational methods [50,51]. We choose to use our previous proposed algorithm [47] among the existing approaches, since it reports more accurate and robust results in environments with various amount of occlusions than the state-of-the-art works. The performance of the RPE algorithms depends on the overlapping area between two FoVs. In the same circumstance, a larger overlapping area leads to a more accurate estimate. The overlapping area between two neighbors can be estimated by initial relative pose determined in Section 3.3.

Let $\mathbf{M}_{ab}$ denote the relative pose between robots a and b. The pixels of the depth image captured by robot a can establish a relation with their projections on the depth image captured by robot b as

$$\begin{bmatrix} u_b & v_b & 1 & q_b \end{bmatrix}^T = \mathbf{M}_{ab} \begin{bmatrix} u_a & v_a & 1 & q_a \end{bmatrix}^T \tag{5}$$

$$\begin{bmatrix} \frac{i_b - i_{c,b}}{f_{x,b}} & \frac{j_b - j_{c,b}}{f_{y,b}} & 1 & \frac{1}{z_b} \end{bmatrix}^T = \mathbf{M}_{ab} \begin{bmatrix} \frac{i_a - i_{c,a}}{f_{x,a}} & \frac{j_a - j_{c,a}}{f_{y,a}} & 1 & \frac{1}{z_a} \end{bmatrix}^T \tag{6}$$

Here, (i_a, j_a) and (i_b, j_b) are the pixel coordinates on different depth images, $(i_{c,a}, j_{c,a})$ and $(i_{c,b}, j_{c,b})$ are the principal points of cameras on two robots, $(f_{x,a}, f_{y,a})$ and $(f_{x,b}, f_{y,b})$ are the focal lengths, and z_a and z_b are the depth values of the same real world point's projections in different depth images. Applying Equation (6) on the depth image observed by robot a, we can generate a synthetic view which is virtually taken at robot b's viewpoint. The overlapping area between two robots' FoVs can be determined through comparing the real and synthetic depth images. We define the *overlapping ratio* between two neighbors as the proportion of overlapping area in the observed image. However, in this approach the central node requires the knowledge of complete depth image of robot a. If all the robots have to transmit their observed depth images to the central node, the considerable transmission load will be generated. In order to efficiently estimate the overlapping ratio, robot a can only send the values and coordinates of four pixels in its observed depth image. These four pixels are the nearest pixels with valid depth values to the four corners (top left, top right, bottom left, and bottom right) of a image. After applying Equation (6) on these four pixels, the quadrangle constructed by the reprojections of these four pixels indicates the region observed by robot a in robot b's view. Though the points in the scene lay on different planes and have various range values, this approach can still provide a rough estimate of the overlapping ratio. An example is shown in Figure 4.

(a) Robot a's view (b) Robot b's view (c) Synthetic view of robot b (d) Estimated overlapping area

Figure 4. Overlapping area estimation. (**a**) Depth image captured by robot a, 4 corner pixels are highlighted in red; (**b**) Depth image captured by robot b; (**c**) Depth image virtually captured at robot b's position. It is generated from (**a**); (**d**) The rough estimate of overlapping area in robot b's view. White region indicates overlapping region.

We conducted numerical simulations on the RPE algorithm [47] to explore the relation between the overlapping ratio and the uncertainty in estimation. The equation

$$w_{ij} = \begin{cases} 1 & \text{if } \phi_{ij} \geq 0.7 \\ 1.5 & \text{if } 0.7 > \phi_{ij} \geq 0.6 \\ 2.4 & \text{if } 0.6 > \phi_{ij} \geq 0.5 \\ \infty & \text{if } 0.5 > \phi_{ij} \end{cases} \tag{7}$$

is adopted to quantize the overlapping ratio and uncertainty degree. Here, ϕ_{ij} represents the overlap ratio between two robots i and j, and w_{ij} indicates the uncertainty degree in RPE between two neighboring robots. A larger w_{ij} indicates a larger uncertainty value in the RPE. Based on IPM and criteria in Equation (7), the Uncertainty Matrix (UM) can be generated.

According to the UM, we can generate the robot dependency graph, $G = (V, A)$. There is an edge between any two neighboring robots iff the overlapping ratio is within the range in Equation (7). The weight of the edge linking robots i and j is w_{ij}, which indicates the uncertainty degree. A lower w_{ij} indicate a smaller uncertainty value in relative pose estimation. Then, the problem of relative pose selection is transformed to the all-pairs shortest path problem in the robot dependency graph, which minimizes the uncertainty in the RPE between every two robots.

The shortest path between every two vertices can be determined using Floyd–Warshall algorithm. The central node first generates $Dist$ as a $|V| \times |V|$ array of minimum distance and initializes $Dist$ according to UM. Then, Floyd–Warshall algorithm is used to determine the shortest paths between every pair of robots and update $Dist$. Then, the central node needs to select one robot as the *primary robot* and make all the other robots calibrate their poses according to the primary robot's coordinate system. In order to minimize the uncertainty, the central node selects the robot which has the smallest overall weight on the shortest paths to all the other robots as the primary robot. At last, the robots can be connected as a *calibration tree* with the primary robot as the root. In this method, the RPE algorithm only requires to operate $|V| - 1$ times to connect all the robots. The time complexity of the overall scheme is $O(V)$. Thus, this scheme is scalable to initialize multi-robot systems with a large number of robots.

3.5. Distributed Relative Pose Estimation Algorithm

Though the initial relative poses on the edges of the calibration tree have already be obtained in neighbor detection process, these estimations are too inaccurate to calibrate the overall system. Therefore, after the calibration tree is built, the central node will broadcast the information of the calibration tree and the related initial relative poses to all the robots. Then our earlier work [47] implemented on every robot will operate to refine the initial relative poses. In our earlier work [47], an Iterative Closest Points (ICP) variant was proposed to estimate the relative pose between two RGB-D camera equipped robots. Different from the methods for conventional RGB camera which use feature correspondences to determine the rotation and translation up to a scale, this distributed, peer-to-peer algorithm determines the relative pose in consistent real world scale through explicit registration of surface geometries extracted from two depth images. The registration problem is approached by iteratively minimizing a cost function in which error metrics are defined based on the bidirectional point-to-plane geometrical relationship.

$\mathbf{Z}_a$ and $\mathbf{Z}_b$ are two depth images captured by robots a and b. The correspondences between $N = N_a + N_b$ pairs of points extracted from $\mathbf{Z}_a$ and $\mathbf{Z}_b$ are established. We can then estimate the transformation matrix $\mathbf{M}_{ab}$ by minimizing the bidirectional point-to-plane error metric, $\mathcal{C}$, expressed in normal least squares form as follows,

$$\mathcal{C} = \sum_{l=1}^{N_a} \left[w_{l,a}(\mathbf{M}_{ab}\mathbf{p}_a^l - \mathbf{p}_b^{l^*}) \cdot \vec{n}_{l^*,b} \right]^2 + \sum_{k=1}^{N_b} \left[w_{k,b}(\mathbf{M}_{ab}{}^{-1}\mathbf{p}_a^{k^*} - \mathbf{p}_b^k) \cdot \vec{n}_{k,b} \right]^2 \tag{8}$$

where $\mathbf{p}_a^l$ and $\mathbf{p}_b^k$ are the sampled points in depth frames $\mathbf{Z}_a$ and $\mathbf{Z}_b$, $\mathbf{p}_b^{l*}$ and $\mathbf{p}_a^{k*}$ are their corresponding points on the other depth image respectively. The variables $w_{l,a}$ and $w_{k,b}$ are weight parameters for correspondences established in opposite directions between pairs. The variables $w_{l,a}$ and $w_{k,b}$ are weight parameters for correspondences established in opposite directions between pairs. Also, $\vec{n}_{l^*,b}$ and $\vec{n}_{k,b}$ are the surface normals at the corresponding points $\mathbf{p}_b^{l^*}$ and $\mathbf{p}_b^k$ in real world coordinates, and

$$\vec{n}_{l^*,b} = \begin{bmatrix} \alpha_{l^*,b} & \beta_{l^*,b} & \gamma_{l^*,b} & 0 \end{bmatrix}^T \tag{9}$$

$$\vec{n}_{k,b} = \begin{bmatrix} \alpha_{k,b} & \beta_{k,b} & \gamma_{k,b} & 0 \end{bmatrix}^T \tag{10}$$

The cost function presented in Equation (8) consists of two parts:

1. the sum of squared distances in the forward direction from depth images $\mathbf{Z}_a$ to $\mathbf{Z}_b$, and
2. the sum of square distances in the backward direction from $\mathbf{Z}_b$ to $\mathbf{Z}_a$.

Algorithm 1 Relative pose refinement procedure

1: Capture a depth image, $\mathbf{Z}_a$, on robot a, and capture a depth image, $\mathbf{Z}_b$, on robot b.
2: Initialize the transformation matrix, $\mathbf{M}_{ab}$, by the initial relative pose.
3: **procedure** REPEAT UNTIL CONVERGENCE
4: Update depth frame $\mathbf{Z}_a$ according to transformation matrix.
5: Randomly sample N_a points from $\mathbf{Z}_a$ to form set P_a,
$P_a = \{\mathbf{p}_a^k \in \mathbf{Z}_a, k = 1, \ldots, N_a\}$,
6: Randomly sample N_b points from $\mathbf{Z}_b$ to form set P_b,
$P_b = \{\mathbf{p}_b^k \in \mathbf{Z}_b, k = 1, \ldots, N_b\}$.
7: Find the corresponding point set, P_b^*, of P_a in $\mathbf{Z}_b$,
$P_b^* = \{\mathbf{p}_b^{k*} \in \mathbf{Z}_b, k = 1, \ldots, N_a\}$;
Find the corresponding point set, P_a^*, of P_b in $\mathbf{Z}_a$,
$P_a^* = \{\mathbf{p}_a^{k*} \in \mathbf{Z}_a, k = 1, \ldots, N_b\}$.
▷ The correspondences are established using the project and walk method with a neighborhood size of 3x3 based on the nearest neighbor criteria
8: Apply the weight function bidirectionally,
$P_a \mapsto P_b^*$, $P_b \mapsto P_a^*$
9: Compute and update transformation matrix based on current bidirectionally weighted correspondences
10: **end procedure**

We can then estimate $\mathbf{M}_{ab}$ by re-weighting the least squares operation in an ICP framework. Details of the criteria for selecting the weight function can be found in [47,52]. An overview of the entire process is presented in Algorithm 1. In the first iteration, $\mathbf{M}_{ab}$ is initialized by the initial relative pose determined in neighbor detection process. Afterwards, in this coarse-to-fine algorithm, each iteration generates an update $\mathbf{E}$ to the robot's pose which modifies the transformation matrix $\mathbf{M}_{ab}$. $\mathbf{E}$ takes the same form as $\mathbf{M}_{ab}$ which may be parameterized by a 6-dimensional motion vector having the elements $\alpha_1, \alpha_2, \ldots, \alpha_6$ via the exponential map and their corresponding group generator matrices $\mathbf{G}_1, \mathbf{G}_1, \ldots, \mathbf{G}_6$ as

$$\mathbf{E} = \exp\left(\sum_{j=1}^{6} \alpha_j \mathbf{G}_j\right) \tag{11}$$

where

$$\mathbf{G}_1 = \begin{bmatrix} 0 & 0 & 0 & 1 \\ 0 & 0 & 0 & 0 \\ 0 & 0 & 0 & 0 \\ 0 & 0 & 0 & 0 \end{bmatrix} \mathbf{G}_2 = \begin{bmatrix} 0 & 0 & 0 & 0 \\ 0 & 0 & 0 & 1 \\ 0 & 0 & 0 & 0 \\ 0 & 0 & 0 & 0 \end{bmatrix} \mathbf{G}_3 = \begin{bmatrix} 0 & 0 & 0 & 0 \\ 0 & 0 & 0 & 0 \\ 0 & 0 & 0 & 1 \\ 0 & 0 & 0 & 0 \end{bmatrix}$$

$$\mathbf{G}_4 = \begin{bmatrix} 0 & 0 & 0 & 0 \\ 0 & 0 & -1 & 0 \\ 0 & 1 & 0 & 0 \\ 0 & 0 & 0 & 0 \end{bmatrix} \mathbf{G}_5 = \begin{bmatrix} 0 & 0 & 1 & 0 \\ 0 & 0 & 0 & 0 \\ -1 & 0 & 0 & 0 \\ 0 & 0 & 0 & 0 \end{bmatrix} \mathbf{G}_6 = \begin{bmatrix} 0 & -1 & 0 & 0 \\ 1 & 0 & 0 & 0 \\ 0 & 0 & 0 & 0 \\ 0 & 0 & 0 & 0 \end{bmatrix}$$

Here $\mathbf{G}_1$, $\mathbf{G}_2$ and $\mathbf{G}_3$ are the generators of translations in x, y and z directions, while $\mathbf{G}_4$, $\mathbf{G}_5$ and $\mathbf{G}_6$ are rotations about x, y and z axes respectively.

Afterwards, the task becomes finding the $\alpha_1, \ldots, \alpha_6$ that describe the relative pose. Through determining the partial derivatives of u_b, v_b and q_b with respect to the unknown elements of the motion vector $\alpha_1, \ldots, \alpha_6$, the Jacobian matrix for each established corresponding point pair can be obtained from

$$\mathbf{J} = \begin{bmatrix} q_a & 0 & -u_a q_a & -u_a v_a & 1 + u_a^2 & -v_a \\ 0 & q_a & -v_a q_a & -1 - v_a^2 & v_a u_a & u_a \\ 0 & 0 & -q_a^2 & -v_a q_a & u_a q_a & 0 \end{bmatrix} \tag{12}$$

The six-dimensional motion vector, which minimizes Equation (8), is then determined iteratively by the least squares solution

$$\mathbf{B} = (\mathbf{K}^T \mathbf{W} \mathbf{K})^{-1} \mathbf{K}^T \mathbf{W} \mathbf{Y} \tag{13}$$

in which $\mathbf{W}$ is a diagonal matrix weighting the bidirectional point-to-plane correspondences. $\mathbf{B}$, $\mathbf{Y}$, and $\mathbf{K}$ are matrices

$$\mathbf{B} = \begin{bmatrix} \alpha_1 \\ \alpha_2 \\ \alpha_3 \\ \alpha_4 \\ \alpha_5 \\ \alpha_6 \end{bmatrix}, \mathbf{Y} = \begin{bmatrix} -(\mathbf{p}_a^1 - \mathbf{p}_b^{1^*}) \cdot \vec{n}_{1^*,b} \\ \vdots \\ -(\mathbf{p}_a^{N_a} - \mathbf{p}_b^{N_a^*}) \cdot \vec{n}_{N_a^*,b} \\ -(\mathbf{p}_a^{1^*} - \mathbf{p}_b^1) \cdot \vec{n}_{1,b} \\ \vdots \\ -(\mathbf{p}_a^{N_b^*} - \mathbf{p}_b^{N_k}) \cdot \vec{n}_{N_b,b} \end{bmatrix} \tag{14}$$

$$\mathbf{K} = \left[\vec{n}'_{1^*,b}\mathbf{J}_1 \ldots \vec{n}'_{N_a^*,b}\mathbf{J}_{N_a}\ \vec{n}'_{1,b}\mathbf{J}_{1^*} \ldots \vec{n}'_{N_b,b}\mathbf{J}_{N_b^*}\right]^T \tag{15}$$

Here, $\vec{n}'_{l,b} = \begin{bmatrix}\alpha_{l,b} & \beta_{l,b} & \gamma_{l,b}\end{bmatrix}^T$ is the surface normal expressed in a slightly different form than the one shown in Equations (9) and (10). To detect the convergence of our algorithm, we use the thresholds for the ICP framework presented in [46]. Once the algorithm converges, the registration is considered as completed and the $\mathbf{M}_{ab}$ is refined based on the initial relative pose.

In reality each robot only has its own captured depth image. In order to efficiently accomplish the centralized working principle of the algorithm described above, we distribute the tasks to two robots. Instead of transmitting a complete depth image from one robot to another, each robot only transmits a number of sampled points on its captured depth image to the other robot. The distributed process is illustrated in Figure 5.

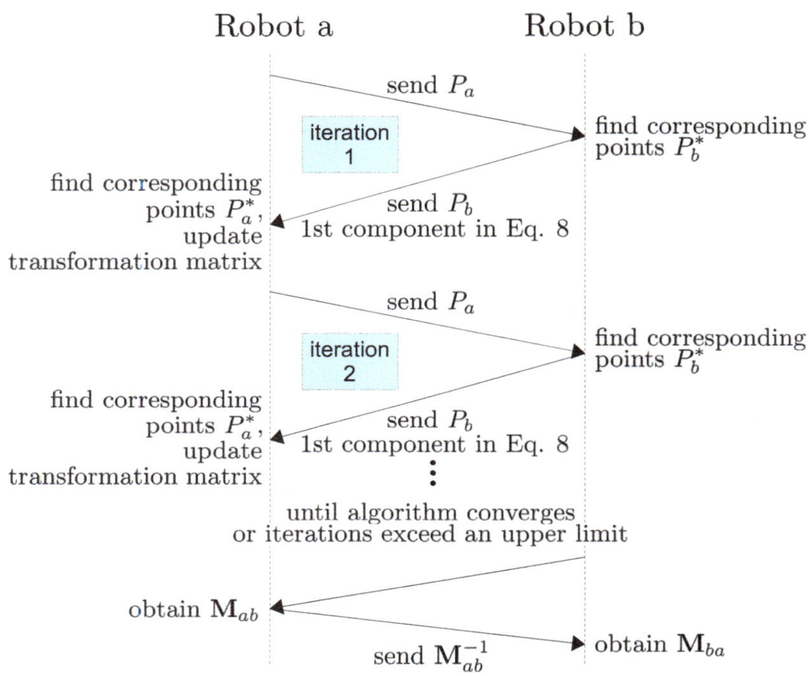

Figure 5. Distributing the tasks to two robots.

At each iteration, after robot b receives the sampled point set, P_a, from robot a, robot b will find the corresponding point set, P_b^*, on its captured depth frame $\mathbf{Z}_b$. The first component in Equation (8) will be derived. The information representing the first component will be sent along with the sampled point set, P_b, from robot b to robot a. At robot a, P_b's corresponding point set, P_a^*, will be determined. And the second component in Equation (8) will be derived. Thereby, robot a will acquire the information of both first and second component in Equation (8), and the motion parameters can be determined. These procedures will be performed in each iteration. The transformation matrix describing the relative pose between two robots will be obtained by robot a until the algorithm converges. More details and performance evaluations of this approach can be found in [47].

The relative poses that construct the calibration tree are transmitted to the central node after being determined by robots. Then all robots' locations and orientations can be calibrated according to primary robot's coordinate system. A simple example of the working process is shown below.

Figure 6 depicts a calibration tree for initializing a multi-robot system with 4 robots. Robots operate RPE algorithm to derive the relative poses $\mathbf{M}_{ab}$, $\mathbf{M}_{ac}$, and $\mathbf{M}_{cd}$ according to the tree. By using these three pose matrices, the relative pose between every two robots in the network can be derived. For instance, robot d's location and orientation in robot b's coordinate system can be derived as

$$\mathbf{M}_{bd} = \mathbf{M}_{ad}\mathbf{M}_{ba} = \mathbf{M}_{cd}\mathbf{M}_{ac}\mathbf{M}_{ab}^{-1} \quad (16)$$

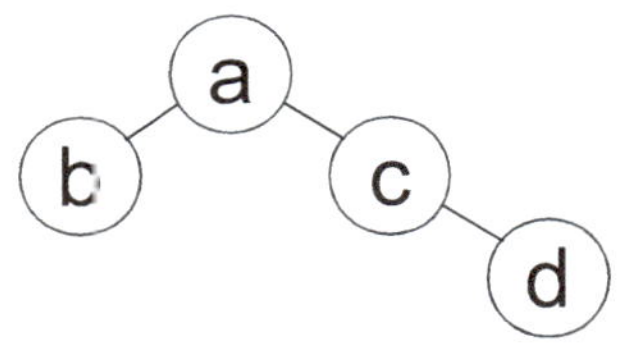

Figure 6. Example of a calibration tree.

4. Experimental Results and Discussion

In order to quantitatively evaluate the performance of the proposed method, we conducted a series of experiments both in simulation and with our custom built multi-robot system. Section 4.1 describes the localization experiments that were carried out with our experimental multi-robot system in indoor environments. Section 4.2 presents the results of a set of simulations that were designed to further evaluate the behavior of the method.

4.1. Indoor Experiments

We used the color and depth images captured by our experimental multi-robot system consisting of seven RGB-D camera equipped robots. The images were taken from various locations in and around WSRNLab facility. Color images of collected five scenes are shown in Figure 7. As the robots are deployed on the same plane in this set of experiments, the ground truth locations and orientations can be easily measured manually. The estimated robots' poses are shown in Figure 8, in which the estimated poses are depicted in red circles and the ground truth are represented in blue stars. We derived the average absolute errors accordingly and presented in Table 3. The calibration trees of 5 scenes are illustrated in Figure 9. We can see the pose estimations of Scene 1 have the smallest absolute error, while the estimates in Scene 5 have the largest absolute error. We also measure the average relative error for localization. The relative errors are computed based on the absolute errors and systems' dimensions. It is clear that the pose estimation results in Scene 4 and Scene 5 are the most and least accurate among five scenes respectively. Through analyzing the robots' sensing ranges in different scenes, we find that sensing range is a main factor that affects the performance of our proposed scheme. As reported in [39], the errors in the depth measurements of Kinect increase quadratically from a few millimeters at 0.5 m distance to about 4 cm at the maximum range of the sensor. The inaccurate depth information obtained by the Kinect on each robot influences the performance of the distributed relative pose estimation algorithm, thereby decreasing the accuracy of the overall scheme. Due to this limitation of the RGB-D camera, the robots' sensing ranges should be controlled between around 0.5 m to 3.5 m.

Figure 7. Color images captured by the multi-robot system in 5 scenes.

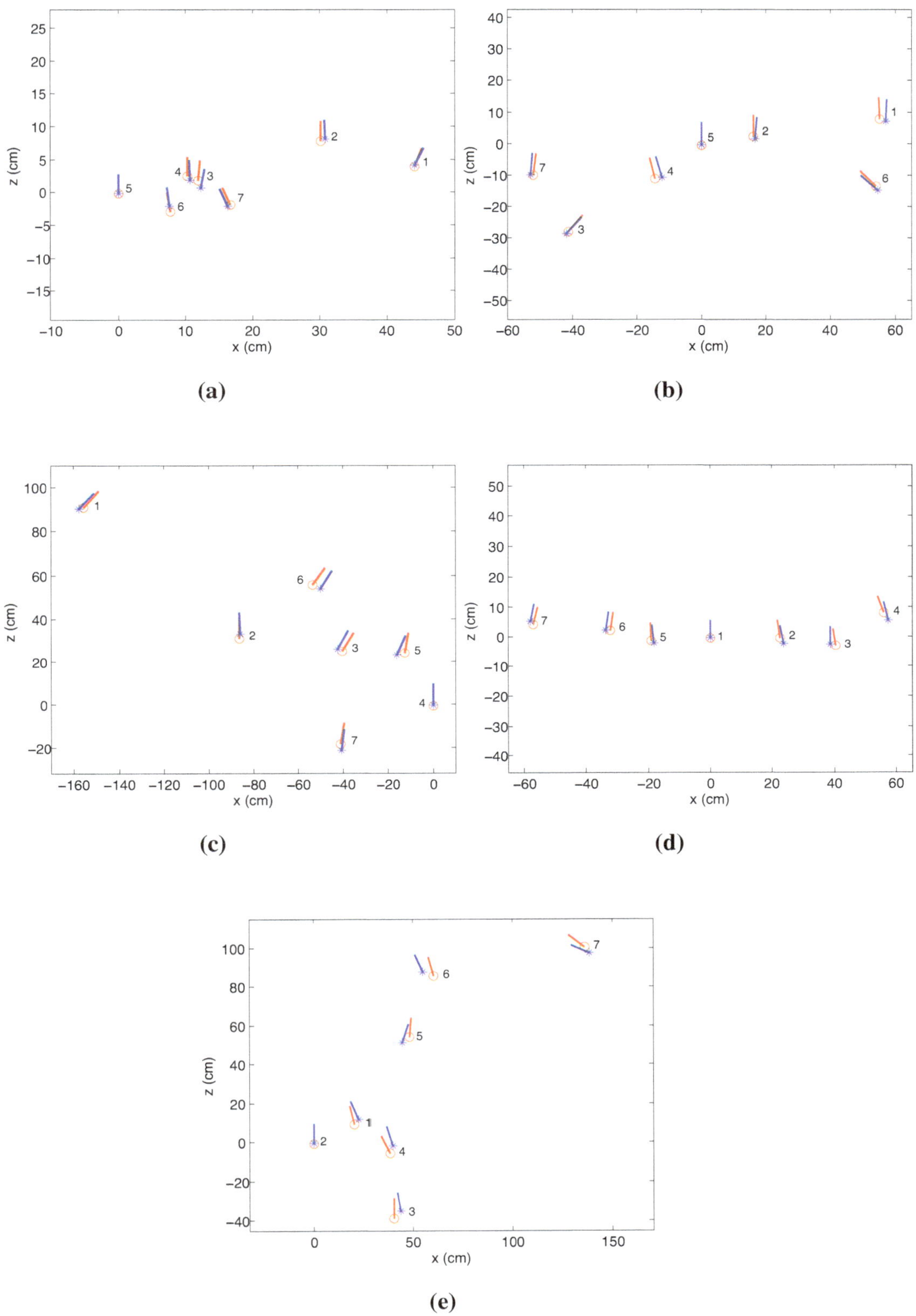

Figure 8. Estimated and ground truth robots' poses. Estimated locations are depicted in red circles and the ground truth are represented in blue stars. The line segments on different markers indicate orientations. (**a**) Scene 1; (**b**) Scene 2; (**c**) Scene 3; (**d**) Scene 4; (**e**) Scene 5.

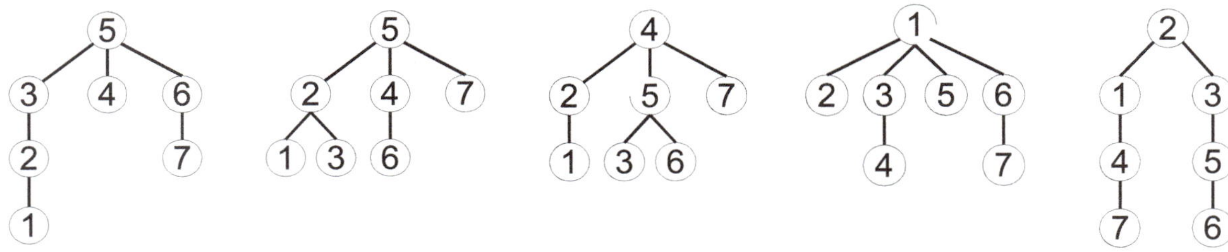

(a) Scene 1 (b) Scene 2 (c) Scene 3 (d) Scene 4 (e) Scene 5

Figure 9. Calibration trees of indoor experiments.

Table 3. Average error between the estimated poses and the ground truth.

Data Set	Sensing Range		Average Absolute Error		Localization Average Relative Error
	Max (m)	Average (m)	Location (mm)	Orientation (°)	
1	1.92	1.47	10.0	1.6	2.26%
2	6.23	1.72	14.8	2.3	1.36%
3	3.95	1.86	25.1	2.7	1.39%
4	1.79	1.41	12.6	2.1	1.12%
5	6.02	4.14	64.7	6.2	3.81%

4.2. Simulation Experiments

A series of simulation experiments were conducted to investigate the accuracy and bandwidth consumption of the proposed scheme implemented on systems that were larger and have more complicated topologies than the ones we could construct with our available hardware. When robots are deployed on different planes, the ground truth poses can hardly be measured precisely by the manual methods. Therefore, in this set of simulations we applied 3D image warping technique [53] on the color and corresponding depth images captured in Scene 4 to generate synthetic image sets with known transformation matrices. Image sets are generated for systems consisting of 10, 15 , 20, 25, 30, 35, and 40 robots. In this process, we make sure that each robot has sufficiently overlapping FoV with at least one another robot and can be connected in calibration tree. The results presented in Figure 10 are averaged over 10 runs of the simulations with vertical bars indicating the variance.

Figure 10a,b, indicate the absolute errors in both location and orientation increase as the number of robots in the system grows. It is because when the number of robots raises, the calibration tree becomes larger and the primary robot requires more intermediate nodes to establish the connection with another robot in the system. Though this effect accumulates small errors, the average absolute errors are still quite small and the average relative error is within 1.3%. The average bandwidth consumption of each robot is presented in Figure 10c. As expected, the bandwidth usage per robot remains consistent approximately in relation to the number of robots in the system. There are two processes in our proposed scheme require communication between robots: (1) Neighbor detection; and (2) distributed relative pose estimation. As the number of times that distributed relative pose estimation runs increase linearly with number of robots, the each robot's transmission load in this process will not be influenced by the number of robots in the system. The only variable that affects the transmission load is the number of the feature points in the neighbor detection process. However, the number of feature points, which depends on the structure of the captured scene, is irrelative with the number of robots. Therefore, the evenly distributed communication load throughout the system indicates the good scalability of our proposed scheme.

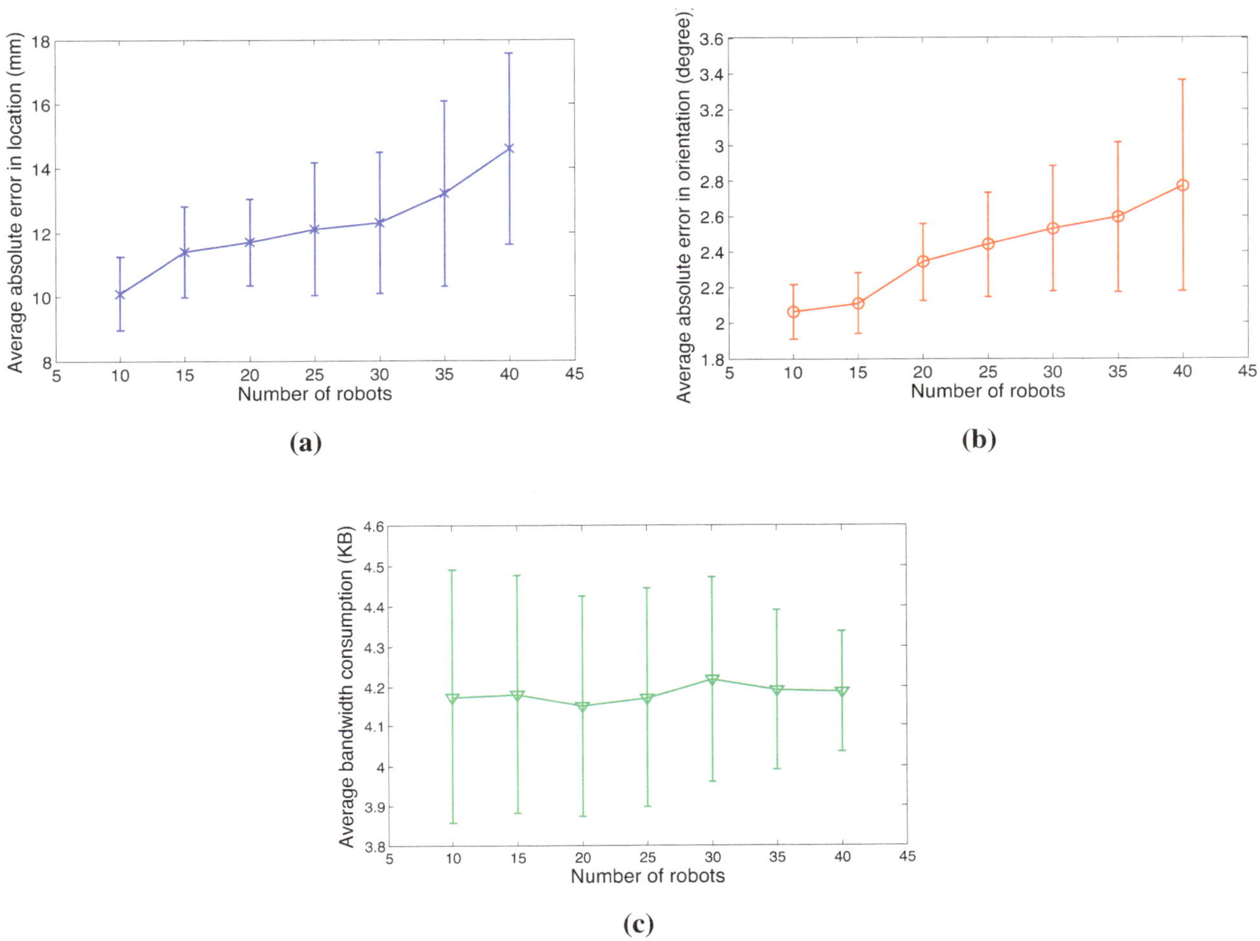

Figure 10. Simulation results. (**a**) Average absolute error in location estimation with increasing number of robots in the system; (**b**) Average absolute error in estimated orientation with increasing number of robots in the system; (**c**) Bandwidth consumption per robot with increasing number of robots in the system.

5. Conclusions

This paper describes the first approach which uses color and depth information to initialize robots' poses in a RGB-D camera equipped multi-robot system. Our scheme first detects each robot's neighbors using robust feature matching. Then the overlapping area between FoVs of neighboring robots are determined to establish the robot dependency graph. A calibration tree is generated through finding the shortest path between the primary robot to all the other robots in the system. At last, a distributed relative pose estimation algorithm is performed to precisely compute relative pose between every two connected robots in the calibration tree. Extensive real world experiments and synthetic dataset simulations have been conducted. The results show that our scheme is robust and accurate in various environments and densities of robots. Importantly, the proposed scheme operates distributively and allows the robots to use the limited wireless bandwidth more efficiently. This calibration algorithm, which provides initial location and orientation information, has great potentials to be used in a wide range of applications, such as visual SLAM and 3D reconstruction.

In future, we plan to present a distributed solution to enhance the performance of the current semi-centralized framework. In order to achieve this goal, we require to upgrade the hardware of eyeBug. Therefore, eyeBug obtains better computational ability and can operate more complicated computer vision algorithms. Further, we would like to develop an efficient algorithm which can keep updating robots' relative poses in real time. This algorithm can be built on our framework to realize the refinement process in cooperative localization.

Author Contributions

Xiaoqin Wang designed the overall scheme, created the datasets using the multi-robot system developed by WSRNLab researchers, tested the scheme and analyzed the results, and prepared the first draft of the manuscript. Y. Ahmet Şekercioğlu proposed the idea about using depth data in conjunction with color information, created the robot system, provided the experimental facilities and valuable input to the manuscript. Tom Drummond contributed the mathematical deduction and formulas in distributed relative pose estimation algorithm.

Conflicts of Interest

The authors declare no conflict of interest.

References

1. Oyekan, J.; Hu, H. Ant Robotic Swarm for Visualizing Invisible Hazardous Substances. *Robotics* **2013**, *2*, 1–18.
2. Parker, L.E. Distributed Algorithms for Multi-Robot Observation of Multiple Moving Targets. *Auton. Robots* **2002**, *12*, 231–255.
3. Delle Fave, F.; Canu, S.; Iocchi, L.; Nardi, D.; Ziparo, V. Multi-Objective Multi-Robot Surveillance. In Proceedings of the 4th International Conference on Autonomous Robots and Agents, Wellington, New Zealand, 10–12 February 2009; pp. 68–73.
4. Karakaya, M.; Qi, H. Collaborative Localization in Visual Sensor Networks. *ACM Trans. Sens. Netw.* **2014**, *10*, 18:1–18:24.
5. Stroupe, A.W.; Martin, M.C.; Balch, T. Distributed Sensor Fusion for Object Position Estimation by Multi-Robot Systems. In Proceedings of the IEEE International Conference on Robotics and Automation, Seoul, Korea, 21-26 May 2001; Volume 2, pp. 1092–1098.
6. Soto, C.; Song, B.; Roy-Chowdhury, A.K. Distributed Multi-Target Tracking in a Self-Configuring Camera Network. In Proceedings of the IEEE Conference on Computer Vision and Pattern Recognition, Miami, FL, USA, 20–25 June 2009; pp. 1486–1493.
7. Xu, Y.; Qi, H. Mobile Agent Migration Modeling and Design for Target Tracking in Wireless Sensor Networks. *Ad Hoc Netw.* **2008**, *6*, 1–16.
8. Dong, W. Tracking Control of Multiple-Wheeled Mobile Robots With Limited Information of a Desired Trajectory. *IEEE Trans. Robot.* **2012**, *28*, 262–268.

9. Dong, W.; Chen, C.; Xing, Y. Distributed estimation-based tracking control of multiple uncertain non-linear systems. *Int. J. Syst. Sci.* **2014**, *45*, 2088–2099.
10. Dong, W.; Djapic, V. Leader-following control of multiple nonholonomic systems over directed communication graphs. *Int. J. Syst. Sci* **2014**, doi: 10.1080/00207721.2014.955553.
11. Samperio, R.; Hu, H. Real-Time Landmark Modelling for Visual-Guided Walking Robots. *Int. J. Comput. Appl. Technol.* **2011**, *41*, 253–261.
12. Gil, A.; Reinoso, Ó.; Ballesta, M.; Juliá, M. Multi-Robot Visual SLAM Using a Rao-Blackwellized Particle Filter. *Robot. Auton. Syst.* **2010**, *58*, 68–80.
13. Chow, J.C.; Lichti, D.D.; Hol, J.D.; Bellusci, G.; Luinge, H. IMU and Multiple RGB-D Camera Fusion for Assisting Indoor Stop-and-Go 3D Terrestrial Laser Scanning. *Robotics* **2014**, *3*, 247–280.
14. Zhang, Z. A Flexible New Technique for Camera Calibration. *IEEE Trans. Pattern Anal. Mach. Intell.* **2000**, *22*, 1330–1334.
15. Kitahara, I.; Saito, H.; Akimichi, S.; Ono, T.; Ohta, Y.; Kanade, T. Large-scale virtualized reality. In Proceedings of the International Conference on Computer Vision and Pattern Recognition, Kauai, HI, USA, 8–14 December 2001.
16. Chen, X.; Davis, J.; Slusallek, P. Wide Area Camera Calibration Using Virtual Calibration Objects. In Proceedings of the IEEE Conference on Computer Vision and Pattern Recognition, Hilton Head, SC, USA, 13–15 June 2000.; Volume 2, pp. 520–527.
17. Svoboda, T.; Hug, H.; Gool, L.J.V. ViRoom-Low Cost Synchronized Multicamera System and Its Self-calibration. In Proceedings of the 24th DAGM Symposium on Pattern Recognition, Zurich, Switzerland, 16–18 September 2002; pp. 515–522.
18. Svoboda, T.; Martinec, D.; Pajdla, T. A Convenient Multi-Camera Self-Calibration for Virtual Environments. *Teleoperators Virtual Environ.* **2005**, *14*, 407–422.
19. Hörster, E.; Lienhart, R. Calibrating and Optimizing Poses of Visual Sensors in Distributed Platforms. *Multimed. Syst.* **2006**, *12*, 195–210.
20. Läbe, T.; Förstner, W. Automatic Relative Orientation of Images. In Proceedings of the 5th Turkish-German Joint Geodetic Days, Berlin, Germany, 28–31 March 2006; Volume 29, p. 31.
21. Rodehorst, V.; Heinrichs, M.; Hellwich, O. Evaluation of Relative Pose Estimation Methods for Multi-Camera Setups. *Int. Arch. Photogram. Remote Sens.* **2008**, pp. 135–140.
22. Jaspers, H.; Schauerte, B.; Fink, G.A. Sift-Based Camera Localization Using Reference Objects for Application in Multi-camera Environments and Robotics. In Proceedings of the International Conference on Pattern Recognition Applications and Methods, vilamoura, portugal, 6–8 February 2012; pp. 330–336.
23. Aslan, C.T.; Bernardin, K.; Stiefelhagen, R.; others. Automatic Calibration of Camera Networks Based on Local Motion Features. In Proceedings of the Workshop on Multi-Camera and Multi-Modal Sensor Fusion Algorithms and Applications, Marseille, France, October 2008.
24. Devarajan, D.; Radke, R.J. Distributed Metric Calibration of Large Camera Networks. In Proceedings of the First Workshop on Broadband Advanced Sensor Networks (BASENETS), San Jose, CA, USA, 25 October 2004; Volume 3, pp. 5–24.

25. Kurillo, G.; Li, Z.; Bajcsy, R. Wide-Area External Multi-Camera Calibration Using Vision Graphs and Virtual Calibration Object. In Proceedings of the Second ACM/IEEE International Conference on Distributed Smart Cameras, Stanford, CA, USA, 7–11 September 2008; pp. 1–9.
26. Cheng, Z.; Devarajan, D.; Radke, R.J. Determining Vision Graphs for Distributed Camera Networks Using Feature Digests. *EURASIP J. Appl. Signal Process.* **2007**, *2007*, 220–220.
27. Vergés-Llahı, J.; Moldovan, D.; Wada, T. A New Reliability Measure for Essential Matrices Suitable in Multiple View Calibration. In Proceedings of the International Joint Conference on Computer Vision, Imaging and Computer Graphics Theory and Applications, Funchal, Portugal, 22–25 January 2008; pp. 114–121.
28. Bajramovic, F.; Denzler, J. Global Uncertainty-based Selection of Relative Poses for Multi Camera Calibration. In Proceedings of the British Machine Vision Conference, 1–4 September 2008, Leeds, UK; pp. 1–10.
29. Bajramovic, F.; Brückner, M.; Denzler, J. An Efficient Shortest Triangle Paths Algorithm Applied to Multi-Camera Self-Calibration. *J. Math. Imaging Vis.* **2012**, *43*, 89–102.
30. Brückner, M.; Bajramovic, F.; Denzler, J. Intrinsic and Extrinsic Active Self-Calibration of Multi-Camera Systems. *Mach. Vis. Appl.* **2014**, *25*, 389–403.
31. Wireless Sensor and Robot Networks Laboratory (WSRNLab). Available online: http://wsrnlab.ecse.monash.edu.au (accessed on 1 December 2014).
32. Beagleboard-xM System Reference Manual. Available online: http://beagleboard.org/static/ (accessed on 1 December 2014).
33. Ubuntu Server for ARM Processor Family. Available online: http://www.ubuntu.com/download/server/arm (accessed on 1 December 2014).
34. OpenKinect Library. Available online: http://openkinect.org (accessed on 1 December 2014).
35. OpenCV: Open Source Computer Vision Library. Available online: http://opencv.org (accessed on 1 December 2014).
36. libCVD-Computer Vision Library. Available online: http://www.edwardrosten.com/cvd/ (accessed on 1 December 2014).
37. Arieli, Y.; Freedman, B.; Machline, M.; Shpunt, A. Depth Mapping Using Projected Patterns. U.S. Patent 8,150,142, 2012.
38. Butler, D.A.; Izadi, S.; Hilliges, O.; Molyneaux, D.; Hodges, S.; Kim, D. Shake'n'Sense: Reducing Interference for Overlapping Structured Light Depth Cameras. In Proceedings of the SIGCHI Conference on Human Factors in Computing Systems, ACM, Austin, Texas, 5–10 May 2012; pp. 1933–1936.
39. Khoshelham, K. Accuracy Analysis of Kinect Depth Data. In Proceedings of the ISPRS Workshop Laser Scanning, Calgary, Canada, 29–31 August 2011; Volume 38, p. W12.
40. Stowers, J.; Hayes, M.; Bainbridge-Smith, A. Altitude Control of a Quadrotor Helicopter Using Depth Map from Microsoft Kinect Sensor. In Proceedings of the 2011 IEEE International Conference on Mechatronics, Istanbul, Turkey, 13–15 April 2011; pp. 358–362.
41. Khoshelham, K.; Elberink, S.O. Accuracy and Resolution of Kinect Depth Data for Indoor Mapping Applications. *Sensors* **2012**, *12*, 1437–1454.

42. Rosten, E.; Drummond, T. Machine Learning for High-Speed Corner Detection. In *Computer Vision–ECCV 2006*; Springer: Berlin/Heidelberg, Germany, 2006; pp. 430–443.

43. Rosten, E.; Porter, R.; Drummond, T. Faster and Better: A Machine Learning Approach to Corner Detection. *IEEE Trans. Pattern Anal. Mach. Intell.* **2010**, *32*, 105–119.

44. Rublee, E.; Rabaud, V.; Konolige, K.; Bradski, G. ORB: An Efficient Alternative to SIFT or SURF. In Proceedings of the 2011 IEEE International Conference on Computer Vision, Barcelona, Spain, 6–13 November 2011; pp. 2564–2571.

45. Izadi, S.; Kim, D.; Hilliges, O.; Molyneaux, D.; Newcombe, R.; Kohli, P.; Shotton, J.; Hodges, S.; Freeman, D.; Davison, A.; Fitzgibbon, A. Kinectfusion: Real-Time 3D Reconstruction and Interaction Using a Moving Depth Camera. In Proceedings of the UIST, Santa Barbara, CA, USA, 16–19 October 2011; pp. 559–568.

46. Lui, W.; Tang, T.; Drummond, T.; Li, W.H. Robust Egomotion Estimation Using ICP in Inverse Depth Coordinates. In Proceedings of the 2012 IEEE International Conference on Robotics and Automation (ICRA), Saint Paul, MN, USA, 14–18 May 2012; pp. 1671–1678.

47. Wang, X.; Şekercioğlu, Y.A.; Drummond, T. A Real-Time Distributed Relative Pose Estimation Algorithm for RGB-D Camera Equipped Visual Sensor Networks. In Proceedings of the 7th ACM/IEEE International Conference on Distributed Smart Cameras (ICDSC 2013), Palm Springs, CA, USA, 29 October–1 November 2013.

48. Zou, Y.; Chen, W.; Wu, X.; Liu, Z. Indoor Localization and 3D Scene Reconstruction for Mobile Robots Using the Microsoft Kinect Sensor. In Proceedings of the 10th IEEE International Conference on Industrial Informatics, Beijing, China, 25–27 July 2012; pp. 1182–1187.

49. Wang, H.; Mou, W.; Ly, M.H.; Lau, M.; Seet, G.; Wang, D. Mobile Robot Egomotion Estimation Using RANSAC-Based Ceiling Vision. In Proceedings of the 24th Chinese Control and Decision Conference, Taiyuan, China, 23–25 May 2012; pp. 1939–1943.

50. Henry, P.; Krainin, M.; Herbst, E.; Ren, X.; Fox, D. RGB-D Mapping: Using Kinect-Style Depth Cameras for Dense 3D Modeling of Indoor Environments. *Int. J. Robot. Res.* **2012**, *31*, 647–663.

51. Endres, F.; Hess, J.; Engelhard, N.; Sturm, J.; Cremers, D.; Burgard, W. An Evaluation of the RGB-D SLAM System. In Proceedings of the IEEE International Conference on Robotics and Automation (ICRA 2012), Saint Paul, MN, USA, 14–18 May 2012; pp. 1691–1696.

52. Holland, P.W.; Welsch, R.E. Robust regression using iteratively reweighted least-squares. *Commun. Stat.-Theory Meth.* **1977**, *6*, 813–827.

53. Mori, Y.; Fukushima, N.; Fujii, T.; Tanimoto, M. View Generation with 3D Warping Using Depth Information for FTV. In Proceedings of the 3DTV Conference: The True Vision-Capture, Transmission and Display of 3D Video, Potsdam, Germany, 4–6 May 2008; pp. 229–232.

13

An Artificial Neural Network Based Robot Controller that Uses Rat's Brain Signals

Marsel Mano [1,*], Genci Capi [2], Norifumi Tanaka [3] and Shigenori Kawahara [4]

[1] Graduate School of Science and Engineering for Education, University of Toyama, Gofuku Campus, 3190 Gofuku, Toyama 930-8555, Japan

[2] Faculty of Engineering, University of Toyama, Gofuku Campus, 3190 Gofuku, Toyama 930-8555, Japan; E-Mail: capi@eng.u-toyama.ac.jp

[3] Department of Computational Intelligence and Systems Science, Tokyo Institute of Technology, 4259-G5-17, Yokohama-shi, Kanagawa 226-8503, Japan; E-Mail: norifumi0816@softbank.ne.jp

[4] Department of Life Sciences and Bioengineering, University of Toyama, Gofuku Campus, 3190 Gofuku, Toyama 930-8555, Japan; E-Mail: kawahara@eng.u-toyama.ac.jp

* Author to whom correspondence should be addressed; E-Mail: marsel.mano@gmail.com

---

Abstract: Brain machine interface (BMI) has been proposed as a novel technique to control prosthetic devices aimed at restoring motor functions in paralyzed patients. In this paper, we propose a neural network based controller that maps rat's brain signals and transforms them into robot movement. First, the rat is trained to move the robot by pressing the right and left lever in order to get food. Next, we collect brain signals with four implanted electrodes, two in the motor cortex and two in the somatosensory cortex area. The collected data are used to train and evaluate different artificial neural controllers. Trained neural controllers are employed online to map brain signals and transform them into robot motion. Offline and online classification results of rat's brain signals show that the Radial Basis Function Neural Networks (RBFNN) outperforms other neural networks. In addition, online robot control results show that even with a limited number of electrodes, the robot motion generated by RBFNN matched the motion generated by the left and right lever position.

Keywords: brain machine interface; learning and adaptive systems; radial basis function neural controllers

---

1. Introduction

During the last two decades, brain machine interface (BMI) has received increasing attention. The reason being, that brain signals recorded with multiple electrodes pasted on scalp or implanted inside the brain, are regarded as a communication channel with high potential for the control of prosthetic devices [1,2]. This could be very useful to patients with disabilities, severe motor dysfunctions, locked-in syndrome, *etc.* Various BMI works, have investigated the control of robotic devices using animal or human brain activity. Trained animals have been shown to accomplish basic tasks, like getting food or water with a robotic arm by using their brain activity, recorded with brain implanted electrode arrays [3–7]. Human BMI applications include robotic arm movements, wheelchair navigation, or gaming [2,8–10]. Due to high risk and complicated surgical procedures, human invasive BMI is exploited less than animal BMI [11].

Depending on the application and the specific neural activity, the brain signals used for BMI include the Electro-Myographic (EMG), the Electro-Oculographic (EOG) and the Electro-Encephalographic (EEG). Typically, any of the above brain signals are recorded while humans (animals) do a specific mental or physical task. The signals, together with the mental activity event time logs, are used to build classifier models. These classifiers are later used to determine the mental activity online.

A common problem of all BMI applications is the acquired signal quality. The signal quality depends on the electrode location, type of signal, type of neural activity, *etc.*, but generally the recorded signal is noisy. Thus, determining specific mental activity of interest is a difficult task. Usually, in non-invasive BMI applications, this is addressed by increasing the number of electrodes and applying spatial and temporal filtering to the acquired multichannel signal [12,13]. In invasive BMI, although the signal has less noise, a high number of electrodes (generally from eight to more than 100) are required to achieve a good classification rate. In contrast to non-invasive BMI, adding electrodes through brain surgery is very complicated, risky and expensive.

In this paper, we employ a radial basis function neural network (RBFNN) in order to achieve a good classification rate with only four brain implanted electrodes. The rat's cortical activity is used to train the RBFNN, mapping the brain signal to the rat's voluntary musculoskeletal movement. In our experiment, we tested several multilayer perceptron neural network architectures and we used different training methods. The RBFNN outperformed the classification accuracy of other networks. The superiority of RBFNNs in dealing with noisy input data and selecting specific important input features makes them more tolerant to brain signals which vary from day to day and also less prone to over fitting.

For rat training, we developed a lever based robot control system. The rat's brain activity is recorded (offline) during lever based robot operation and then is used to train various artificial neural networks. After this, the network with the best classification results is used to classify the brain activity (online) and actuate the robot, while the lever functions are disabled. The RBFNN showed a higher performance and thus it was used online.

Experimental results show that the RBFNN is able to classify the rat's brain signals in real time with a high level of confidence during right and left lever press events. During two online sessions, the rat was able to get control of the robot and obtain food, only by using its brain activity. The classification accuracy achieved by the RBFNN is higher than that of similar works shown in [14].

This paper has the following structure. Section 2 explains the rat training. Section 3 presents the method used for signal acquisition, processing and the RBFNN developed for classification. The experimental setup is described in Section 4. In Section 5 the experimental results are summarized. Section 6 concludes the paper.

2. Rat Training

The rat used in our experiments was a five month old male Wistar/ST with electrodes pre-implanted in its brain. The surgery and all of the experimental procedures were performed in accordance with the Guidelines for the Care and Use of Laboratory Animals in the University of Toyama, and the NIH Guide for the Care and Use of Laboratory Animals.

In a typical experiment, food-rewarding is used to train the rats to perform an instrumental action (*i.e.*, lever pressing) [14]. At the beginning, the rat is trained to learn to press the right or left lever as follows:

(1) Press any lever to get food supplied manually by a human.
(2) Press levers as above, except with their head restricted.

After the rat has learned the above tasks, the food is placed on a small robot (e-puck). The robot has a food handler (Figure 1) mounted on the left side and a green LED light under the food handler. By using the LED, the rat is able to visually detect the robot. The rat learns to direct the robot by pressing the levers in scenarios that change from simple to more complicated ones, as follows:

(1) The robot is placed in front of the rat and it moves straight forward when any lever is pressed.
(2) The robot is placed on the right (left) side of the rat and it follows half of a U-shape trajectory when only the right (left) lever is pressed.
(3) The robot is placed initially on the right or left side of the rat and it moves to the right or left of a U-shape when the respective lever is pressed.

Figure 1. Developed system.

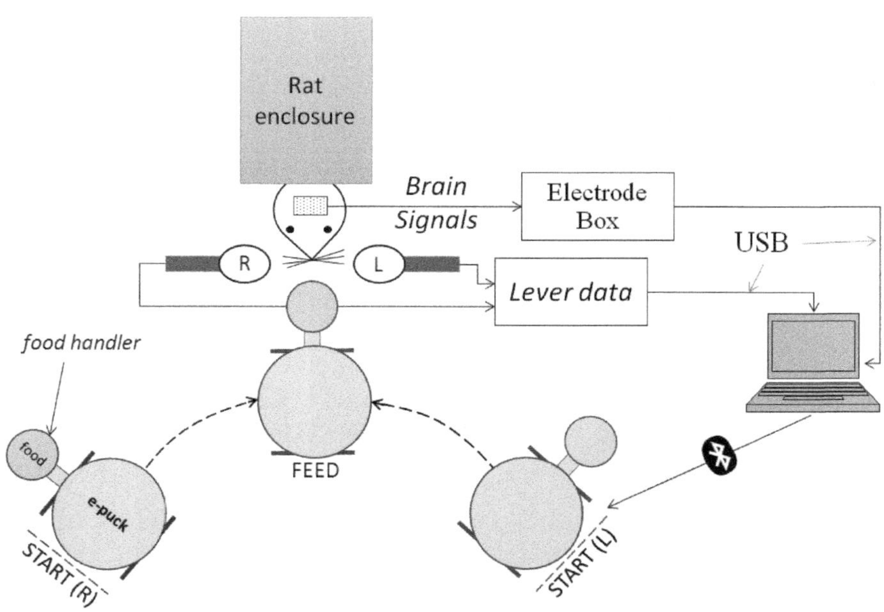

Training is complete when the rat learns to press the lever on the side of the LED in order to move the robot and get the food.

In our experiments, the rat was carefully placed in the rat enclosure shown on Figure 1, where it can easily access both levers. The levers are connected through a serial interface with the computer and the robot is controlled via a Bluetooth interface. Brain signals are acquired through electrodes implanted in its brain and sent to a computer.

The robot movement is in a U-shape trajectory and depending on the pressed lever it can be in either a clockwise or a counterclockwise direction. The robot initially is placed at one of the start positions (Figure 1). If the correct lever is pressed the robot moves towards the food position; if the wrong lever is pressed it moves further away. For instance, when the robot is placed on "START (R)", the LED is on the right side of the rat. If the rat presses the right lever the robot will move clockwise to the feed position, however, if the rat presses the left lever the robot will move counterclockwise and it will go further away from the feed position. The lever control functions were enabled only during offline data recording. During online experiments the levers were used for event recording and the robot was controlled by brain signal mapped into lever actions by RBFNN.

3. Method

3.1. Signal and Event Acquisition

Four electrodes were implanted on the rat's brain (Figure 2).

(1) Electrodes, L1 and R1, were implanted in the motor cortex area, respectively in the left and right hemisphere.
(2) Electrodes, L2 and R2, were implanted in the somatosensory cortex area, respectively in the left and right hemisphere.

Electrodes were connected to the electrode box/amplifier. Next, signals were digitized at 250 Hz and collected on a personal computer via USB. All electrodes were software referenced to earth. Lever press events were simultaneously recorded.

Figure 2. Electrode locations: (**a**) in motor cortex (Interaural); (**b**) in somatosensory area (Interaural); (**c**) dorsal view.

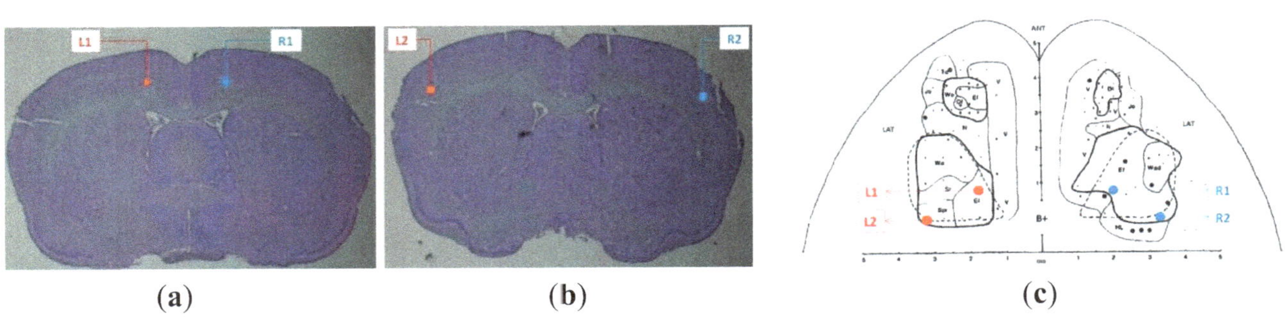

3.2. Feature Extraction and Neural Network

The offline recorded data are represented by matrix $S\epsilon R^{C\times T}$ where C is the number of channels ($L1$, $R1$, $L2$, $R2$) and T the number of sample time points. Epochs $X^{(j)}$ with duration E are extracted in the interval $\left[t_s^{(j)}, t_f^{(j)}\right]$ in the vicinity of the lever press events, as shown in Figure 3.

Figure 3. Epoch extraction interval.

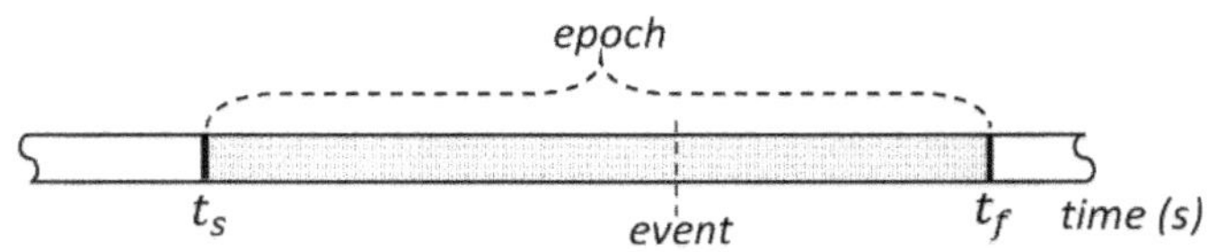

Every epoch is then divided into N ((j) is omitted) time windows with duration $(t_f - t_s)/N$. For every window and for every channel, the sample data mean values are calculated as $m_{c,n} = mean\langle X_c(t)\rangle_{t\in t_n;c\in C}$, where $n \in \{1, ..., N\}$. The data are stored in a feature vector $F(f_i)_{(N\times C)\times 1}$, where $f_i = m_{c,n}\ \forall\ i = N \times c + n$.

With all the feature vectors extracted from offline sessions, we trained feed forward neural networks with $I = N \times C$ input layer neurons, a variable number of hidden layer neurons H and a single output neuron O (Figure 4).

Figure 4. Neural network topology.

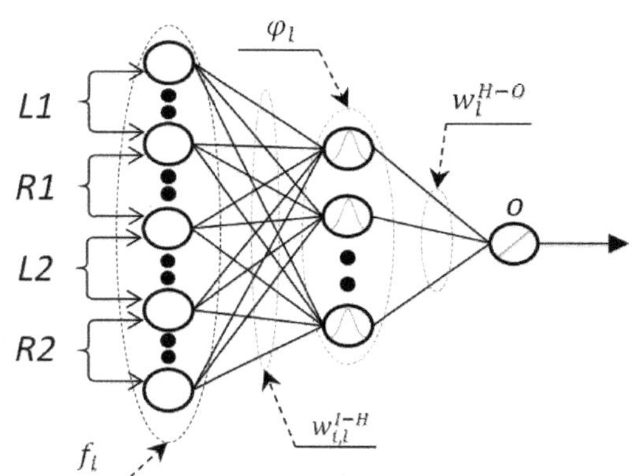

Several multilayer feed-forward neural networks with different topologies and learning functions were tested but the RBFNN performed the best. RBFNN classification clusters are separated by hyper spheres, thus they are better local approximation networks. RBFNN outputs are determined by specific hidden units in certain local receptive fields [15]. This property, allows RBFNN to be more selective of the input features and more tolerant to error. Furthermore, during training the neurons of the hidden layer are added incrementally as the training error decreases. This provides a good initial set up for network connections with shorter training duration.

In this paper, we consider only two neural networks with similar architecture (Figure 4): an RBFNN and multi-layer perceptron neural network trained with Bayesian regulation back propagation (BPNN) [16].

During online experiments, every time a lever was pressed the brain signals were collected in the interval $[t_s, t_f]$. The data were similarly processed to extract the feature vector, which was classified by the RBFNN. In online sessions, only the RBFNN was used to control the robot motion. In any case, BPNN performance was tested with the online acquired datasets, after the online sessions were finished, and the results are shown in Section 5.

3.2.1. RBFNN

The RBFNN's hidden layer neurons have a radial basis transfer function φ^1:

$$\varphi^1(v_l) = \exp\left(-\frac{\|v_l - c_h\|}{\sigma_h}\right) \tag{1}$$

And the output neuron has a linear transfer function:

$$o^1 = \sum_{l=1}^{H} \varphi_l^1(v_l) w_l^{H-O} + w_0^{h-o} \tag{2}$$

where: $v_l = \left[f_1 w_{1,l}^{I-H}, f_2 w_{2,l}^{I-H} \cdots f_I w_{I,l}^{I-H}\right]$, w^{I-H} and w^{H-O} are the input-to-hidden layer weights and hidden-to-output layer weights, respectively. The activation function of the hidden layer neurons φ^1, is a Gaussian function with center c_h and spread σ_h. In our RBFNN, σ_h is set to 1.

In a supervised learning method, neurons are added to the hidden layer of the RBFNN until it meets a specified mean squared error goal. RBFNN performance function is a mean squared error (MSE).

3.2.2. BPNN

In contrast with RBFNN, BPNNs have a fixed number of hidden layer neurons (30). Hidden layer neurons have a hyperbolic tangent transfer function φ^2:

$$\varphi^2(v_h) = tanh(v_h) = \frac{2}{1 - e^{-2v_h}} - 1 \tag{3}$$

The output layer neuron has also a hyperbolic tangent transfer function o^2:

$$o^2 = tanh(\sum_{l=1}^{H} \varphi_l^2(v_l) w_l^{h-o} + w_0^{h-o}) \tag{4}$$

Bayesian regularization is a supervised back propagation training method that updates the network weights and bias values according to the Levenberg-Marquardt optimization [17]. BPNN performance function is a summed square error (SSE).

In our implementation, the output of any neural network is classified as "*left*" if its value is greater than 0.5 and "*right*" if it is equal or less than 0.5.

4. Experimental Setup

In total, three offline recording sessions were done with a trained rat. All experiments were conducted in a dark room in order to make it easier for the rat to spot the LED light. The duration of the sessions varied from 20 to 30 min. The first two sessions were used to train the neural network, with random trial selection. Then the third offline session was used to test network performances and select the best network. Offline sessions were conducted on three different days over the span of a week.

During online sessions the lever information was collected for event marking purposes. However, the robot was controlled online by the RBFNN that mapped the rat's brain signals to the lever's position.

5. Results

5.1. Neural Network

Table 1 summarizes the offline sessions that were used to train and test the neural networks. The dataset of the first two sessions contains 73 events, where 35 are left lever press events and 38 are right lever press events. For each event, epoch data were collected and processed for feature extraction. Samples of the epochs of rat's brain signals for right and left lever press events are shown in Figure 5. The channels on the left hemisphere, L1 and L2 show similar activity to one another. The channels on the right hemisphere, R1 and R2 show less similarity but still have common spikes and dips. On the other hand, left and right hemispheres show different brain activities, but still it is very hard to linearly separate one from the other. The signals show that the brain activity started around 1.5 s before and ended around 0.5 s after the events. Different epoch duration and window lengths were tested and the segment [−1.4, 0.4] with window length 200 ms gave the best performance.

Table 1. Offline sessions event summary.

Session	Duration (min)	Purpose	Lever press events		
			Left	Right	Total
1	20	*train*	16	26	42
2	30	*train*	19	12	31
3	27	*test*	25	13	38
TOTAL			60	51	111

Figure 5. Brain signals during Left and Right lever press: (**a**) Single event epoch; (**b**) Mean of all event epochs recorded in the first offline session.

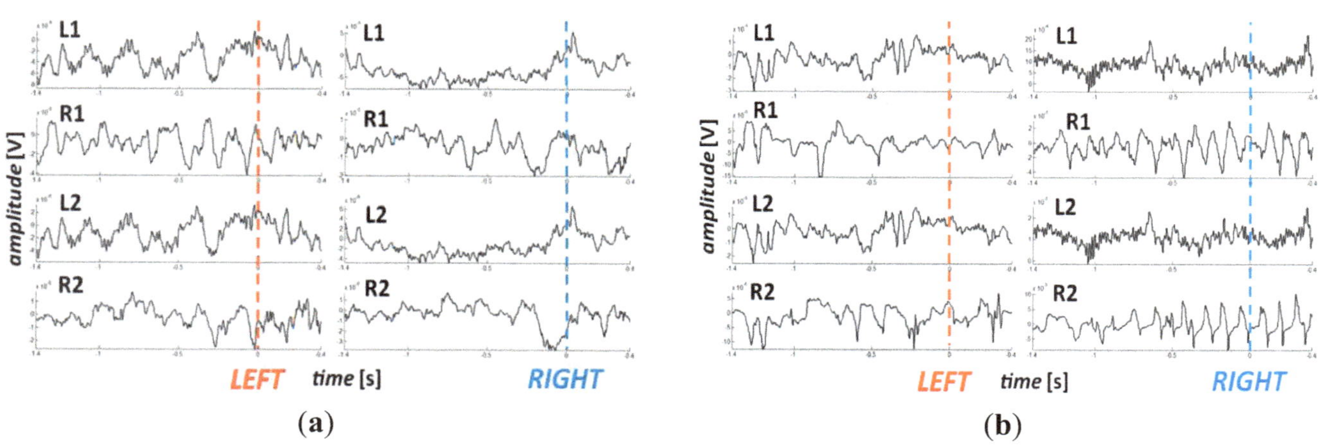

From the epoched data, input matrixes $P_{1(32\times73)}$ and $P_{2(32\times38)}$ together with their respective targets $T_{1(1\times73)}$ and $T_{2(1\times38)}$, were created. P_1 and T_1 were used for supervised training of the neural networks and P_2 and T_2 for testing.

Figure 6 shows the training performances of the RBFNN and BPNN. In order to avoid over fitting, the RBFNN training was finished when the MSE reached the predefined goal value of 0.07. The RBFNN completed learning after 31 epochs. At the end of the training it had 31 neurons in the hidden layer. The training of the BPNN was completed after 75 epochs, when the SSE reached the predefined goal value of 0.01. The RBFNN's selected goal value was established higher than BPNN's since experimental results showed a decrease in performance when the RBFNN training was stressed to achieve lower goal values. Furthermore, both network goal values were determined experimentally to achieve the highest performances.

Figure 6. Neural network training performances: (**a**) Mean squared error (MSE) of Radial Basis Function Neural Networks (RBFNN); (**b**) Sum squared error (SSE) or Bayesian regulation back propagation (BPNN).

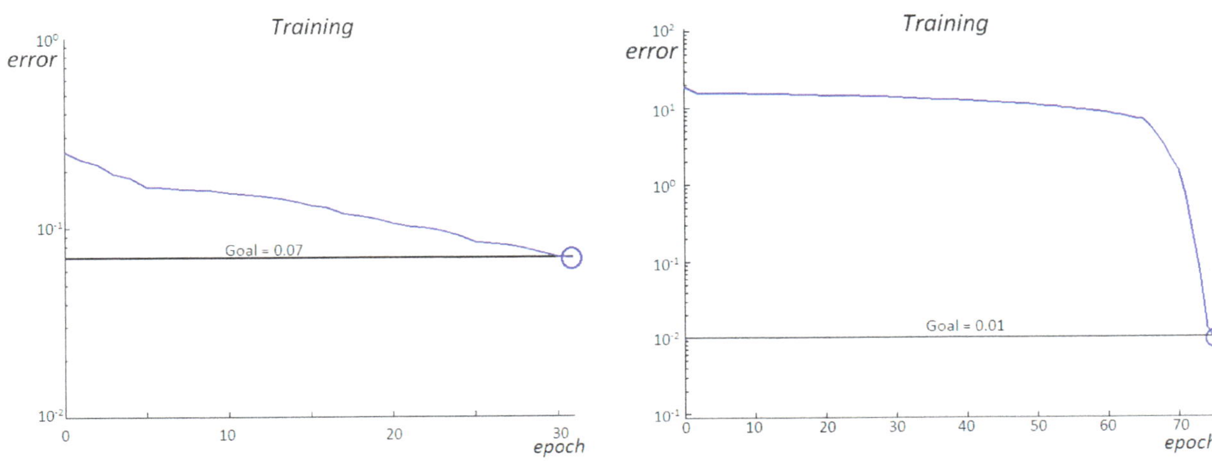

(**a**) (**b**)

Both neural networks were trained and tested with the same offline sessions. Their classification accuracies are shown on Table 2. The first two sessions that were used for training, have very high classification accuracy for both neural networks. The third session, used only to test the networks, shows higher classification accuracy in the case of RBFNN compared to BPNN. Based on the results of the test session, we selected the RBFNN for the online control of the robot.

Table 2. Radial Basis Function Neural Networks (RBFNN) and Bayesian regulation back propagation (BPNN) classification accuracy with offline sessions.

Offline	RBFNN			BPNN		
Session	Left	Right	Total	Left	Right	Total
1 (train)	100.0	96.2	97.6	100.0	100.0	100.0
2 (train)	94.7	100.0	96.8	100.0	100.0	100.0
3 (test)	92.0	61.5	81.6	60.0	69.2	63.2

5.2. Online Robot Control

The offline trained RBFNN, was used to classify the rat's brain signals during online robot control. The results of the online classification are summarized in Table 3.

Table 3. Summary of online sessions.

Online Session	Duration	All Events			Misclassified events		
		Left	Right	Total	Left	Right	Total
1	36.8	18	16	34	3	3	6
2	32.7	20	15	35	4	2	6
TOTAL		38	29	69	7	5	12

The RBFNN classification accuracy during online sessions is graphically illustrated in Figure 7. After the online sets were finished, the data acquired online was used to test the classification performance of the BPNN. As shown in Figure 7, the RBFNN performance is higher than that of the BPNN in both sessions and for each event type individually.

Figure 8(a) shows the robot motion during an experiment where the robot reached the feed position after two RBFNN outputs. The RBFNN accurately classified the brain signals and no extra steps were required. Figure 8(b) shows the robot motion during an experiment where the robot reached the feed position after four RBFNN outputs. The second RBFNN output was misclassified and the robot, instead of going to the feed position, went back to the start position. Two extra steps were needed to move the robot into the feed position.

Figure 7. Online session classification accuracy.

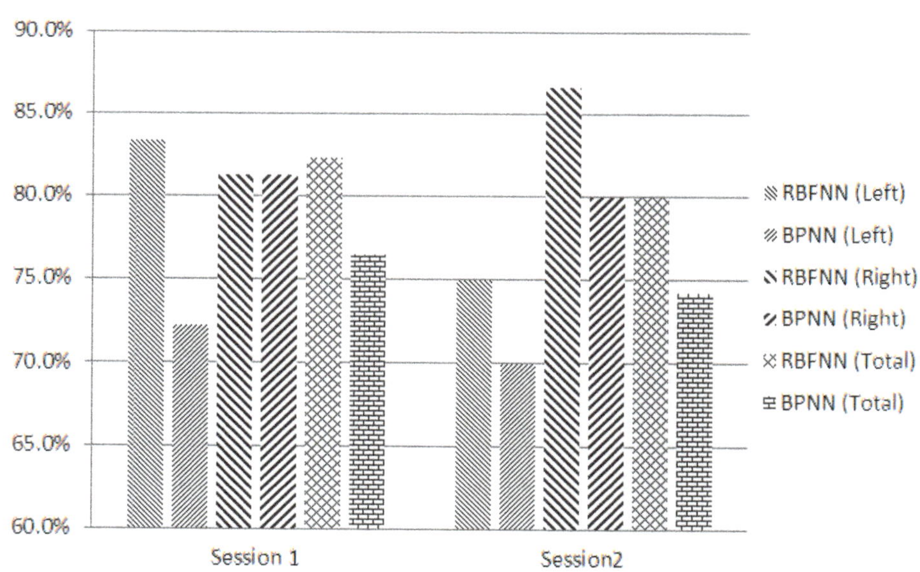

Figure 8. Rat's brain controlled robot motion during: (**a**) a trial on the left side; (**b**) a trial on the right side.

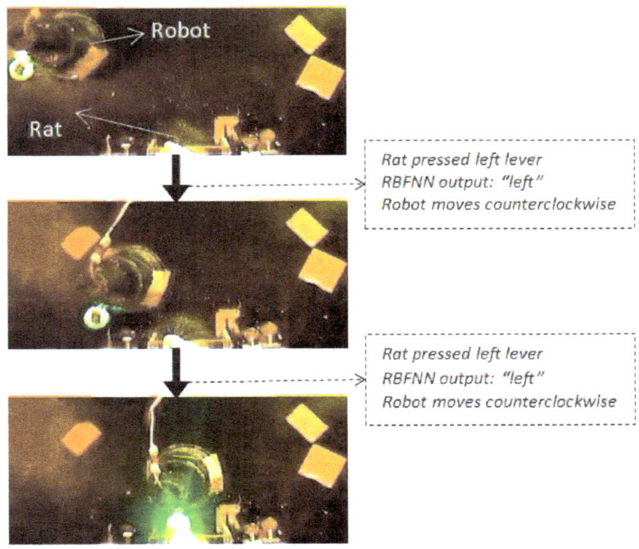

(**a**)

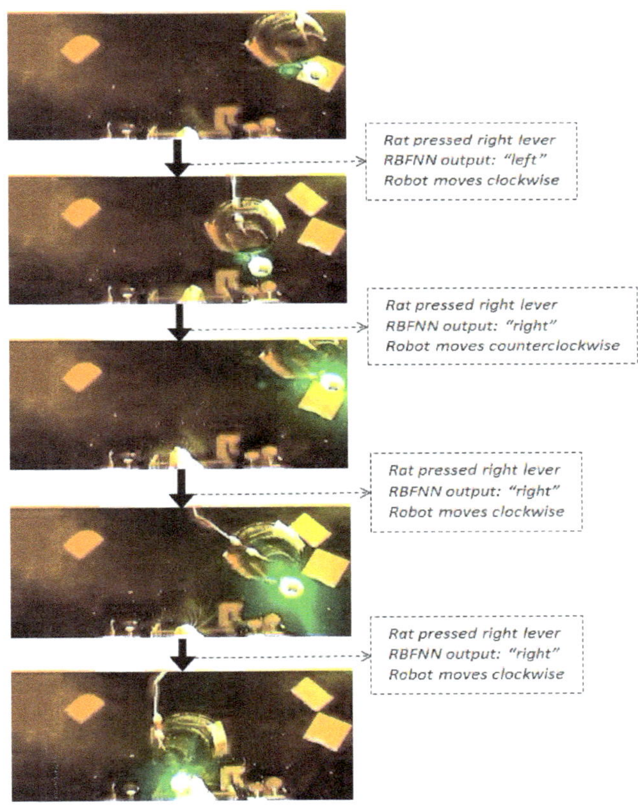

(**b**)

6. Conclusions

In this paper, we presented a neural network based robot controller activated by a rat's brain signals. In our method, we utilized the rat's brain signals, collected by implanted electrodes, to train neural controllers. The trained neural controllers were also tested in online experiments, where the RBFNN was found to give the best performance. The results showed that even with a limited number of implanted electrodes, the RBFNN mapped the brain signals with the robot motion (lever position) with high accuracy.

A future extension of this work would be to investigate how the electrode location influences the neural network classification accuracy. The benefits of the BMI based robot control investigation in animals presented in this paper could be used in BMI applications to restore motor function in humans.

Acknowledgements

The authors would like to express their gratitude to the reviewers for checking this manuscript and giving their valuable suggestions.

References

1. Wolpaw, J.R.; Birbaumer, N.; Heetderks, W.J.; McFarland, D.J.; Peckham, P.H.; Schalk, G.; Donchin, E.; Quatrano, L.A.; Robinson, C.J.; Vaughan, T.M. Brain-computer interface technology: A review of the first international meeting. *IEEE Trans. Rehabil. Eng.* **2000**, *8*, 164–173.
2. Millán, J.D.R.; Rupp, R.; Müller-Putz, G.R.; Murray-Smith, R.; Giugliemma, C.; Tangermann, M.; Vidaurre, C.; Cincotti, F.; Kübler, A; Leeb, R.; *et al.* Combining brain-computer interfaces and assistive technologies: State-of-the-art and challenges. *Front. Neurosci.* **2010**, *4*, 1–15.
3. Taylor, D.M.; Tillery, S.I.H.; Schwartz, A.B. Direct cortical control of 3D neuroprosthetic devices. *Science* **2002**, *296*, 1829–1832.
4. Carmena, J.M.; Lebedev, M.A.; Crist, R.E.; O'Doherty, J.E.; Santucci, D.M.; Dimitrov, D.F.; Patil, P.G.; Henriquez, C.S.; Nicolelis, M.A.L. Learning to control a brain-machine interface for reaching and grasping by primates. *PLoS Biol.* **2003**, *1*, e42, doi: 10.1371/journal.pbio.0000042.
5. Velliste, M.; Perel, S.; Spalding, M.C.; Whitford, A.S.; Schwartz, A.B. Cortical control of a prosthetic arm for self-feeding. *Nature* **2008**, *453*, 1098–1101.
6. Chapin, J.K.; Moxon, K.A.; Markowitz, R.S.; Nicolelis, M.A. Real-time control of a robot arm using simultaneously recorded neurons in the motor cortex. *Nat. Neurosci.* **1999**, *2*, 664–670.
7. Wessberg, J.; Stambaugh, C.R.; Kralik, J.D.; Beck, P.D.; Laubach, M.; Chapin, J.K.; Kim, J.; Biggs, S.J.; Srinivasan, M.A.; Nicolelis, M.A. Real-time prediction of hand trajectory by ensembles of cortical neurons in primates. *Nature* **2000**, *408*, 361–365.
8. Yanagisawa, T.; Hirata, M.; Saitoh, Y.; Kishima, H.; Matsushita, K.; Goto, T.; Fukuma, R.; Yokoi, H.; Kamitani, Y.; Yoshimine, T. Electrocorticographic control of a prosthetic arm in paralyzed patients. *Ann. Neurol.* **2011**, *71*, 353–361.
9. Hinterberger, T.; Veit, R.; Wilhelm, B.; Weiskopf, N.; Vatine, J.-J.; Birbaumer, N. Neuronal mechanisms underlying control of a brain-computer interface. *Eur. J. Neurosci.* **2005**, *21*, 3169–3181.
10. Ang, K.K.; Guan, C.; Chua, K.S.G.; Ang, B.T.; Kuah, C.; Wang, C.; Phua, K.S.; Chin, Z.Y.; Zhang, H. A Clinical Study of Motor Imagery-Based Brain-Computer Interface for Upper Limb Robotic Rehabilitation. In Proceedings of Annual International Conference of the IEEE Engineering in Medicine and Biology Society (EMBC 2009), Minneapolis, MN, USA, 3–6 September 2009; pp. 5981–5984.

11. Majima, K.; Kamitani, Y. An outlook on the present and future of brain-machine interface research. *Brain Nerve.* **2011**, *63*, 241–246.
12. Blankertz, B.; Tomioka, R.; Lemm, S.; Kawanabe, M.; Muller, K. Optimizing spatial filters for robust EEG single-trial analysis. *IEEE Signal Process. Mag.* **2008**, *25*, 41–56.
13. Wu, W.; Chen, Z.; Gao, S.; Brown, E.N. A hierarchical Bayesian approach for learning sparse spatio-temporal decompositions of multichannel EEG. *NeuroImage* **2011**, *56*, 1929–1945.
14. Capi, G. Real robots controlled by brain signals—A BMI approach. *Int. J. Adv. Intell.* **2010**, *2*, 25–35.
15. Xie, T.; Yu, H.; Wilamowski, B. Comparison between Traditional Neural Networks and Radial Basis Function Networks. In Proceedings of the 2011 IEEE International Symposium on Industrial Electronics (ISIE), Gdansk, Poland, 27–30 June 2011; pp. 1194–1199.
16. Haykin, S. *Neural Networks: A Comprehensive Foundation*; Griffin, J., Ed.; Prentice Hall: Upper Saddle River, NJ, US, 1999; Volume 13, pp. 409–412.
17. MacKay, D.J.C. Bayesian interpolation. *Neural Comput.* **1992**, *4*, 415–447.

Permissions

All chapters in this book were first published in Robotics, by MDPI; hereby published with permission under the Creative Commons Attribution License or equivalent. Every chapter published in this book has been scrutinized by our experts. Their significance has been extensively debated. The topics covered herein carry significant findings which will fuel the growth of the discipline. They may even be implemented as practical applications or may be referred to as a beginning point for another development.

The contributors of this book come from diverse backgrounds, making this book a truly international effort. This book will bring forth new frontiers with its revolutionizing research information and detailed analysis of the nascent developments around the world.

We would like to thank all the contributing authors for lending their expertise to make the book truly unique. They have played a crucial role in the development of this book. Without their invaluable contributions this book wouldn't have been possible. They have made vital efforts to compile up to date information on the varied aspects of this subject to make this book a valuable addition to the collection of many professionals and students.

This book was conceptualized with the vision of imparting up-to-date information and advanced data in this field. To ensure the same, a matchless editorial board was set up. Every individual on the board went through rigorous rounds of assessment to prove their worth. After which they invested a large part of their time researching and compiling the most relevant data for our readers.

The editorial board has been involved in producing this book since its inception. They have spent rigorous hours researching and exploring the diverse topics which have resulted in the successful publishing of this book. They have passed on their knowledge of decades through this book. To expedite this challenging task, the publisher supported the team at every step. A small team of assistant editors was also appointed to further simplify the editing procedure and attain best results for the readers.

Apart from the editorial board, the designing team has also invested a significant amount of their time in understanding the subject and creating the most relevant covers. They scrutinized every image to scout for the most suitable representation of the subject and create an appropriate cover for the book.

The publishing team has been an ardent support to the editorial, designing and production team. Their endless efforts to recruit the best for this project, has resulted in the accomplishment of this book. They are a veteran in the field of academics and their pool of knowledge is as vast as their experience in printing. Their expertise and guidance has proved useful at every step. Their uncompromising quality standards have made this book an exceptional effort. Their encouragement from time to time has been an inspiration for everyone.

The publisher and the editorial board hope that this book will prove to be a valuable piece of knowledge for researchers, students, practitioners and scholars across the globe.

List of Contributors

Christopher James Shackleton
School of Systems Engineering, University of Reading, Whiteknights, Reading, Berkshire, RG6 6AY, UK

Rahul Kala
School of Systems Engineering, University of Reading, Whiteknights, Reading, Berkshire, RG6 6AY, UK

Kevin Warwick
School of Systems Engineering, University of Reading, Whiteknights, Reading, Berkshire, RG6 6AY, UK

Takashi Kuremoto
Graduate School of Science and Engineering, Yamaguchi University, Tokiwadai 2-16-1, Ube, Yamaguchi 755-8611, Japan

Tetsuya Tsurusaki
Graduate School of Science and Engineering, Yamaguchi University, Tokiwadai 2-16-1, Ube, Yamaguchi 755-8611, Japan

Kunikazu Kobayashi
School of Information Science & Technology, Aichi Prefectural University, Ibaragabasama 152203, Ngakute, Aichi 480-1198, Japan

Shingo Mabu
Graduate School of Science and Engineering, Yamaguchi University, Tokiwadai 2-16-1, Ube, Yamaguchi 755-8611, Japan

Masanao Obayashi
Graduate School of Science and Engineering, Yamaguchi University, Tokiwadai 2-16-1, Ube, Yamaguchi 755-8611, Japan

Mohd Azuwan Mat Dzahir
Shibaura Institute of Technology, Department of Bio-Science Engineering, 307 Fukasaku, Minuma-ku, Saitama City, Saitama 337-8570, Japan
Universiti Teknologi Malaysia, Faculty of Mechanical Engineering, UTM Skudai, Johor Bahru 81310, Malaysia

Shin-ichiroh Yamamoto
Shibaura Institute of Technology, Department of Bio-Science Engineering, 307 Fukasaku, Minuma-ku, Saitama City, Saitama 337-8570, Japan

Hicham Chaoui
School of Electrical Engineering and Computer Science, University of Ottawa, 800 King Edward Avenue, Ottawa, ON K1N 6N5, Canada

Wail Gueaieb
School of Electrical Engineering and Computer Science, University of Ottawa, 800 King Edward Avenue, Ottawa, ON K1N 6N5, Canada

Mohammad Biglarbegian
School of Engineering, University of Guelph, Guelph, ON N1G 2W1, Canada

Mustapha C. E. Yagoub
School of Electrical Engineering and Computer Science, University of Ottawa, 800 King Edward Avenue, Ottawa, ON K1N 6N5, Canada

Abhilash Pandya
Department of Electrical and Computer Engineering, Wayne State University, Detroit, MI 48202, USA

Luke A. Reisner
Department of Electrical and Computer Engineering, Wayne State University, Detroit, MI 48202, USA

Brady King
Department of Pediatric Surgery, Children's Hospital of Michigan, Detroit, MI 48201, USA

Nathan Lucas
Department of Electrical and Computer Engineering, Wayne State University, Detroit, MI 48202, USA

Anthony Composto
Department of Electrical and Computer Engineering, Wayne State University, Detroit, MI 48202, USA

Michael Klein
Department of Pediatric Surgery, Children's Hospital of Michigan, Detroit, MI 48201, USA

Richard Darin Ellis
Department of Industrial and Systems Engineering, Wayne State University, Detroit, MI 48202, USA

Jacky C. K. Chow
Department of Geomatics Engineering, Schulich School of Engineering, University of Calgary, 2500 University Drive N.W., Calgary, AB T2N 1N4, Canada

Derek D. Lichti
Department of Geomatics Engineering, Schulich School of Engineering, University of Calgary, 2500 University Drive N.W., Calgary, AB T2N 1N4, Canada

Jeroen D. Hol
Xsens Technologies B.V., Pantheon 6a, Enschede 7521 PR, The Netherlands

Giovanni Bellusci
Xsens Technologies B.V., Pantheon 6a, Enschede 7521 PR, The Netherlands

Henk Luinge
Xsens Technologies B.V., Pantheon 6a, Enschede 7521 PR, The Netherlands

Ken Saito
Department of Precision Machinery Engineering, College of Science and Technology, Nihon University, 7-24-1 Narashinodai, Funabashi-shi, Chiba 274-8501, Japan

Kazuaki Maezumi
Precision Machinery Engineering, Graduate School of Science and Technology, Nihon University, 1-8-14 Kanda Surugadai, Chiyoda-ku, Tokyo 101-8308, Japan

Yuka Naito
Precision Machinery Engineering, Graduate School of Science and Technology, Nihon University, 1-8-14 Kanda Surugadai, Chiyoda-ku, Tokyo 101-8308, Japan

Tomohiro Hidaka
Precision Machinery Engineering, Graduate School of Science and Technology, Nihon University, 1-8-14 Kanda Surugadai, Chiyoda-ku, Tokyo 101-8308, Japan

Kei Iwata
Precision Machinery Engineering, Graduate School of Science and Technology, Nihon University, 1-8-14 Kanda Surugadai, Chiyoda-ku, Tokyo 101-8308, Japan

Yuki Okane
Precision Machinery Engineering, Graduate School of Science and Technology, Nihon University, 1-8-14 Kanda Surugadai, Chiyoda-ku, Tokyo 101-8308, Japan

Hirozumi Oku
Precision Machinery Engineering, Graduate School of Science and Technology, Nihon University, 1-8-14 Kanda Surugadai, Chiyoda-ku, Tokyo 101-8308, Japan

Minami Takato
Precision Machinery Engineering, Graduate School of Science and Technology, Nihon University, 1-8-14 Kanda Surugadai, Chiyoda-ku, Tokyo 101-8308, Japan

Fumio Uchikoba
Department of Precision Machinery Engineering, College of Science and Technology, Nihon University, 7-24-1 Narashinodai, Funabashi-shi, Chiba 274-8501, Japan

Franco Tedeschi
Department DICEM, University of Cassino and Southern Lazio, Via di Biasio 43, Cassino 03043 (Fr), Italy

Giuseppe Carbone
Department DICEM, University of Cassino and Southern Lazio, Via di Biasio 43, Cassino 03043 (Fr), Italy

Auday Al-Mayyahi
Department of Engineering and Design, University of Sussex, Famer, Brighton, BN1 9QJ, UK

William Wang and Phil Birch
Department of Engineering and Design, University of Sussex, Famer, Brighton, BN1 9QJ, UK

Muhammad Ilhamdi Rusydi
Department of Mechanical Engineering, Gifu University, 1-1 Yanagido, Gifu City 501-1193, Japan
Department of Electrical Engineering, Andalas University, Limau Manis, Padang City 25163, Indonesia

Takeo Okamoto
Department of Mechanical Engineering, Gifu University, 1-1 Yanagido, Gifu City 501-1193, Japan

Satoshi Ito
Department of Mechanical Engineering, Gifu University, 1-1 Yanagido, Gifu City 501-1193, Japan

Minoru Sasaki
Department of Mechanical Engineering, Gifu University, 1-1 Yanagido, Gifu City 501-1193, Japan

Wuming Jing
School of Mechanical Engineering, Purdue University, 585 Pudue Mall, West Lafayette, IN 47907, USA

David Cappelleri
School of Mechanical Engineering, Purdue University, 585 Pudue Mall, West Lafayette, IN 47907, USA

XiaoqinWang
Department of Electrical and Computer Systems Engineering, Monash University, Melbourne 3800, Australia

Y. Ahmet Şekercioğlu
Department of Electrical and Computer Systems Engineering, Monash University, Melbourne 3800, Australia

Tom Drummond
Department of Electrical and Computer Systems Engineering, Monash University, Melbourne 3800, Australia

Marsel Mano
Graduate School of Science and Engineering for Education, University of Toyama, Gofuku Campus, 3190 Gofuku, Toyama 930-8555, Japan

Genci Capi
Faculty of Engineering, University of Toyama, Gofuku Campus, 3190 Gofuku, Toyama 930-8555, Japan

Norifumi Tanaka
Department of Computational Intelligence and Systems Science, Tokyo Institute of Technology, 4259 G5-17, Yokohama-shi, Kanagawa 226-8503, Japan

Shigenori Kawahara
Department of Life Sciences and Bioengineering, University of Toyama, Gofuku Campus, 3190 Gofuku, Toyama 930-8555, Japan